Design and analysis of fatigue resistant welded structures

Design and analysis of fatigue resistant welded structures

DIETER RADAJ, Dr-Ing habil

Professor of Mechanical Engineering
Braunschweig Technical University

Senior Staff Manager
Daimler-Benz Corp, Stuttgart

HALSTED PRESS
a division of JOHN WILEY & SONS, Inc.
605 Third Avenue, New York, N.Y. 10158

New York • Toronto

Published by Abington Publishing,
Woodhead Publishing Ltd, Abington Hall,
Abington, Cambridge CB1 6AH, England

First published 1990

Published in the Western Hemisphere by Halsted Press; a division of
JOHN WILEY & SONS, INC. NEW YORK.

ISBN 0-470-21695-6

Designed by Geoff Green (text) and Chris Feely (cover),
typeset and printed by Crampton & Sons Ltd, Sawston, Cambridge CB2 4BQ, England

Contents

Preface

Jointing components by welding is the most widespread manufacturing method for metal structures. It provides an almost infinite number of design variants. This advantage can be used effectively for design optimisation, but of course the variety of options available can equally open the way to poor design. To avoid this a rational design based on theory and testing is necessary. Additionally, the close interrelationship between design, material and manufacturing aspects of welded structures has to be taken into account.

This book considers the fatigue strength of welded structures. Fatigue strength is very important in those structures which are subjected to time-dependent loads. It also has some relation to statically loaded structures, the quality of which is increased by fatigue relevant design measures. At the same time, fatigue strength analysis and assessment for welded structures require great expertise with respect to methods and use of data.

Design which takes fatigue into account is necessary for all moving structures such as vehicles, aeroplanes, cranes and engines and to some degree also for structures which support movement such as bridges, craneways or engine housings. A lightweight design is especially important for moving structures to make operation more efficient. Variable loading, in relation to static loading, generally increases with decreasing weight. Guidelines and methods for fatigue resistant design, because of its practical importance, are not only a matter for a few experts but are obligatory knowledge for any welding engineer and are included in relevant design codes.

With respect to method, traditional global approaches to strength assessment are more and more being supplemented or even substituted by local approaches. Calculations for determining dimensions on the basis of simple engineering formulae are now extended to numerical simulation of structural behaviour using finite element and boundary element methods. The aim of simulation is to support design decisions and optimise design. The numerical simulation methods are part of the computer aided design (CAD) process. Today simulation methods are being included in expert systems which incorporate knowledge bases and inference methods. High demands are put on systematically applied, precise knowledge in connection with these developments. Such demands are quite new for welding engineers.

The book presents the various knowledge requirements, components and development methods for fatigue resistant welded structures in a systematic and comprehensive manner. It aims to provide an integration of theoretical and practical knowledge. The peculiarities and limitations of the different methods are emphasised. As the subject matter is still at the development stage, particularly with respect to local approaches, a final and complete presentation is not possible. However, the fundamental structure is provided to which further detail can be added when available.

The book has been written for designers and structural analysts, inspection and manufacturing engineers, researchers, teachers and students. It should be of value not only in the high technology countries but also in countries which are at the threshold of higher technology.

'Design and analysis of fatigue resistant welded structures' is a revised and extended English edition of the German standard works 'Festigkeitsnachweise' (1974) and 'Gestaltung und Berechnung von Schweißkonstruktionen, Ermüdungsfestigkeit' (1985). It has been extended to include the local approaches for spot welded joints in which field substantial progress has been made in recent years. In addition, the notch stress assessment for seam welds could be prepared for inclusion into design codes. A list of symbols and a subject index has been added to this edition. In its extended form therefore this book has much of value even for German readers with access to the original works.

The book is rooted in the pioneering works of two early German schools on fatigue resistant design. A Neumann and his team supplied a solid basis for the design of fatigue resistant welded structures in terms of design recommendations and nominal stress considerations. The 'shape strength philosophy' developed by A Thum, and the 'service fatigue strength philosophy' developed by E Gaßner have greatly influenced the content of the book. My motivation to concentrate on the strength of welded structures was reinforced by J Ruge who asked me early in my professional career (1968) to lecture on this at the Technical University of Braunschweig. My own scientific contributions to the strength of welded structures are mainly related to the theoretical aspects *i.e.* mainly to structural analysis on the basis of notch stress theory, fracture mechanics, finite element and boundary element methods. On the other hand my daily occupation as a senior research manager at Daimler-Benz Corp in Stuttgart (including the automobile manufacturer Mercedes-Benz) has given me useful insight into practical problems and design situations.

I want to express my gratitude to W Asshauer, DVS-Verlag for the contractual arrangements for the English edition and to R Smith, The Welding Institute, for organising the translation. I am particularly grateful for the co-operation extended by Abington Publishing.

DIETER RADAJ
Stuttgart, January 1990

List of symbols

The list of symbols used in equations, figures and text is arranged first according to the Latin and then according to the Greek alphabet. The upper case letter is always quoted first, the lower case second. Within each letter group, quantities with the same dimensions are arranged consecutively.

A uniform and unique notation, which is additionally comprehensible to the English reader, could not always be introduced because of different habits in the different fields of engineering science compiled here in one book. Some symbols are multiply assigned if mistaken usage can be excluded. The symbols with multiple assignments are listed separately if the meaning is quite different and in the same line if the meaning is similar.

The former notations σ_Y, $\sigma_{0.1}$, σ_U are retained instead of using the newer standard notations R_e, $R_{p0.1}$, R_m because they are consistent with the stress symbols and are therefore more descriptive.

It is not possible to list the symbols with every combination of subscripts occurring in the text.

$A_{11}, A_{12}, A_{22}, A_{33}$	(-)	Material constants in equation (182)
A, A_0, A_U	(mm^2)	Area of cross section, initially and at rupture
A^*	(mm^2)	Area circumscribed by hollow section
a	(mm)	Thickness of fillet weld
a, a_i, a_0	(mm)	Crack length or depth, slit length, initial crack length or depth
a	(mm)	Distance from weld toe (Fig.157)
b, b_f	(mm)	Specimen width, width between webs, flange width
b	(-)	Fatigue strength exponent in equation (108)
b_1, b_2, b_3	(mm)	Distance from weld toe (Fig.157)
C, C', C_0	(N, mm)	Material constants for crack propagation
C	(-)	Neuber constant (Fig.215)
c	(mm)	Crack width
c	(mm)	Leg width of fillet weld

c	(-)	Cyclic ductility exponent in equation (108)
$c, c_\perp, c_\parallel$	(-)	Notch loading coefficients in equations (77), (78), (79)
D	(mm)	Diameter of chord tube
D	(mm)	Diameter of circular plate
d	(mm)	Diameter of weld spot
d	(mm)	Diameter of brace tube, of pipe, hole, vessel
d_1	(mm)	Diameter of nozzle pipe
d_A, d_E, d_N, d_{HAZ}	(mm)	Diameter of adhesive face, electrode, nugget, heat affected zone
Δd	(mm)	Local diameter deviation
E	(N/mm^2)	Elastic modulus
E_p, E_c	(N/mm^2)	Elastic modulus of plate and of core
e	(mm)	Eccentricity of centre line
e	(-)	Euler number, 2.7183
F	(N)	Force (tensile or compressive)
F'	(N)	Running-through tension force
$F_\parallel, F_\perp$	(N)	Tensile shear force, cross tension force
$\Delta F, \Delta F_F$	(N)	Shear force range, the same as endurance limit
F_f, F_{cr}	(N)	Spring force, buckling force
F_A, F_{AF}	(N)	Endurable shear force amplitude for finite and infinite life
f	(-)	Function according to equation (25)
f	(-)	Force ratio F/F_{cr}
f_k, f_{k0}	(-)	Notch factor, surface factor
f_{k1}, f_{k2}, f_{k3}	(-)	Weld correction factors in code
f_{kl}	(-)	Function in equation (159)
f_T, f_m	(-)	Temperature factor, mean stress factor
g	(mm)	Gap between brace tubes (Fig.105)
H_i	(-)	Frequency on stress level, i
h	(mm)	Height of girder, double bottom, vessel end, tie
I	(-)	Irregularity factor
i	(-)	Number of load or stress level
$\Delta J, \Delta J_{en}$	(N/mm$^{3/2}$)	J-integral, endurable J-integral
j	(-)	Number of load or stress level
K	(-)	Stress concentration factor
K_k, K_{nk}	(-)	Notch stress concentration factor referring to σ_n and σ_{ns}
K_s, K_{ns}	(-)	Structural stress concentration factor referring to σ_n and σ_{ns}
$K_f, \overline{K}_f$	(-)	Fatigue notch factor referring to σ_n and $\bar{\sigma}_n$
K_{keq}	(-)	Equivalent notch stress concentration factor
K_{s0}, K_{cw}, K_{ch}	(-)	Structural stress concentration factors (equations (32), (33))

Symbol	Unit	Description
K_{sl}	(-)	Structural stress concentration factor at lower notch
$K_{sa}, K_{sb}, K_{s\tau}$	(-)	Structural stress concentration factors (equation (45))
$K_{0.2}^{*}$	(N/mm^2)	Shape dependent strain limit
K'	(N/mm^2)	Cyclic hardening modulus
K_k^{*}	(-)	Corrected notch stress concentration factor
K_{k1}, K_{k2}, K_{keq}	(-)	Notch stress concentration factors for σ_{k1}, σ_{k2}, σ_{keq}
$K_{k\perp}, K_{k\parallel}$	(-)	Notch stress concentration factors referring to $\sigma_\perp$ or $\tau_\parallel$
K_σ, K_ε	(-)	Stress and strain concentration in elastic-plastic notch root
K_{s1}, K_{s2}	(-)	Structural stress concentration factors for σ_{s1}, σ_{s2}
K_U	(-)	Ultimate load stress concentration factor
K_I, K_{II}, K_{III}	(N/mm$^{3/2}$)	Stress intensity factors, mode I, II, III
K_{eq}	(N/mm$^{3/2}$)	Equivalent stress intensity factor
$\Delta K, \Delta K_J$	(N/mm$^{3/2}$)	Stress intensity range, the same derived from ΔJ
$\Delta K_{th}, \Delta K_{th0}$	(N/mm$^{3/2}$)	Threshold stress intensity range, the same for $R = 0$
$\Delta K_{I\,en}$	(N/mm$^{3/2}$)	Endurable stress intensity factor
K_u, K_l	(N/mm$^{3/2}$)	Upper and lower stress intensity
K_{Ic}, K_c	(N/mm$^{3/2}$)	Critical stress intensity factors, fracture toughness
k	(-)	Gradient exponent of S-N curve
k	(-)	Gradient exponent of life curve
k	(-)	Factor in equation (31)
k	(N/mm)	Spring constant
L	(mm)	Length of chord
l_0	(mm)	Distance to zero crossing point of residual stress
l, l_0, l_U	(mm)	Specimen length, initially and at rupture
Δl	(mm)	Increment of specimen length
l_f	(mm)	Support spacing of flange
l_s	(mm)	Weld spot pitch
M_b	(Nmm)	Bending moment
M_b'	(Nmm)	Running-through bending moment
M_t	(Nmm)	Torsional moment
M_{me}	(N)	Meridional bending moment per length unit
M	(-)	Constraint factor
m	(-)	Number of half waves
m	(-)	Exponent of crack propagation rate
N	(-)	Number of cycles

Symbol	Unit	Description
N_f	(-)	Number of cycles to failure
N_i	(-)	Number of cycles to crack initiation
N_i	(-)	Number of level i crossings
$\overline{N}, \overline{N}^*$	(-)	Number of mean load crossings
ΔN_j	(-)	Number of level j cycles
N_p	(-)	Number of peaks
N_F	(-)	Number of cycles at endurance limit
N_{F_j}	(-)	Number of cycles endured on level j
N_{ci}	(N/mm)	Circumferential normal force per length unit
N_{me}	(N/mm)	Meridional normal force per length unit
n'	(-)	Cyclic strain hardening exponent
n_z	(-)	Load multiple
$n_1 \ldots n_6$	(-)	Exponents in equation (31)
P_f, P	(-)	Failure probability
P_s	(-)	Survival probability
P_o	(-)	Occurrence probability
P, P_{SWT}, P_{HL}	(-)	Damage parameter, according to Smith, Watson and Topper, or Haibach and Lehrke
$P, \Delta P$	(N)	Load, load range
P_u, P_l	(N)	Upper load, lower load
$P_i, \Delta P_i$	(N)	Load on level i, class width on level i
$\overline{P}_m$	(N)	Mean load
$\overline{P}_a, \Delta P_{ai}$	(N)	Maximum load amplitude, load amplitude on level i
P_u, P_l	(N)	Upper load, lower load
P_{Up}	(N)	Upper load fatigue strength
$\overline{P}_u, \overline{P}_l$	(N)	Maximum upper load, maximum lower load
p	(N/mm^2)	Internal pressure
p	(-)	Load spectrum coefficient
p	(mm)	Dimension for brace overlap in Fig.105
q	(-)	Load spectrum coefficient
q	(mm)	Dimension for brace overlap in Fig.105
R	(-)	Limit stress ratio, limit stress intensity ratio
R_c	(-)	Limit compressive stress ratio
R	(mm)	Radius of tube
R, R_1	(mm)	Radius of weld spot
R_2	(mm)	Radius of bonded face
r	(mm)	Radial distance from centre
r	(mm)	Radial distance from crack or slit tip
r_c, r'_c	(mm)	Corner radius of cover plate or core
S	(-)	Sample standard deviation
S, S^*	(-)	Safety factor on stress and on number of cycles
S, S^*	(-)	Total fatigue damage, total creep damage
S_j	(-)	Partial damage on level j

s	(mm)	Slit length
s	(-)	Microstructural support factor
T	(N)	Transverse force or load
ΔT_F	(N)	Cross tension force range (endurance limit)
T	(°C)	Temperature
T	(mm)	Thickness of chord tube
t	(mm)	Depth of shallow notch
t'	(mm)	Thickness of strip core, of covered plate
t	(s)	Time
Δt_j	(s)	Stress duration on level j
t_{Fj}	(s)	Time for creep failure on level j
t	(mm)	Thickness of plate, shell, pipe, (brace) tube
t_{sh}, t_{pl}, t_1	(mm)	Thickness of shell, pipe, nozzle pipe
t_0	(mm)	Initial deflection
x, y, z	(mm)	Co-ordinates
$\overline{\alpha}$	(-)	Non-dimensional crack length (equation (129))
α	(-)	Geometrical parameter (equation (26))
β	(-)	Geometrical parameter (equation (27))
β	(-)	Safety index
γ	(-)	Geometrical parameter (equation (28))
$\gamma, \overline{\gamma}$	(-)	Fatigue strength reduction factors referring to σ_n and $\overline{\sigma}_n$
γ_A, γ_p	(-)	Reduction factors for alternating and pulsating load
$\gamma_f, \gamma_t, \gamma_r$	(-)	Reduction factors for flange, toe, root
γ, γ_{max}	(-)	Slanting angle of weld spot
δ	(mm)	Crack opening displacement
δ	(-)	Size ratio, thickness ratio
δ	(-)	Initial deflection t_0/t
$\varepsilon_a, \varepsilon_A$	(-)	Strain amplitude, strain amplitude strength
$\Delta\varepsilon, \Delta\varepsilon_r$	(-)	Strain range, radial strain range
$\Delta\varepsilon_{en}, \Delta\varepsilon_{eff}$	(-)	Endurable and effective strain range
$\varepsilon_k, \varepsilon_{kel}, \varepsilon_{kpl}$	(-)	Notch strain, the same elastic and plastic
ε_U	(-)	Strain at rupture l_U/l_0
ε_U'	(-)	Cyclic ductility coefficient
ε	(-)	Geometrical parameter in equation (30)
θ	(-)	Slope angle at weld toe, chord-brace angle
$\kappa, \overline{\kappa}$	(-)	Adjustment factors in equations (122), (123), (130)
κ^*	(-)	Surface roughness factor on fatigue strength
λ	(-)	Adjustment factor in equation (7)
λ	(-)	Biaxiality factor σ_2/σ_1
ν	(-)	Poisson's ratio
ν	(-)	Adjustment factor in equation (7)

ρ, ρ_g, ρ_r	(mm)	Notch radius, the same at groove and root
ρ^*	(mm)	Microstructural support length
ρ_f	(mm)	Fictitious notch radius
ρ^*	(-)	Radius-thickness ratio ρ_f/t
σ_{A0}, σ_{AW}	(N/mm^2)	Alternating fatigue strength, the same with weld
$\bar{\sigma}_p$	(N/mm^2)	Pulsating local fatigue strength
σ_p, σ_{PW}	(N/mm^2)	Pulsating fatigue strength, the same with weld
σ_{Pc}	(N/mm^2)	Pulsating compressive fatigue strength
σ_F, σ_{F0}	(N/mm^2)	Fatigue strength, the same for R = 0
σ_B, σ_W	(N/mm^2)	Fatigue strength of base material, the same with weld
σ_A, σ_{AF}	(N/mm^2)	Amplitude fatigue strength, the same as endurance limit
σ_A^*	(N/mm^2)	Amplitude fatigue strength with rough surface
$\bar{\sigma}_A, \bar{\sigma}_A^*$	(N/mm^2)	Endurable maximum stress amplitude, the same for $\overline{N}^*$
σ_{Up}, σ_L	(N/mm^2)	Upper and lower stress fatigue strength
σ_U, σ_{Uc}	(N/mm^2)	Ultimate tensile and compressive strength
$\sigma_Y, \sigma_{0.1}, \sigma_{0.2}$	(N/mm^2)	Yield strength, 0.1% and 0.2% offset yield stress
$\Delta\sigma_F, \Delta\sigma_{F0}$	(N/mm^2)	Fatigue strength range for infinite life, the same for R = 0
σ_R	(N/mm^2)	Rupture strength
σ_U'	(N/mm^2)	Fatigue strength coefficient
σ_0	(N/mm^2)	Non-singular stress components in equation (159)
σ_0	(N/mm^2)	Maximum residual stress
$\sigma, \Delta\sigma, \sigma_a$	(N/mm^2)	Stress, stress range, stress amplitude
σ_n, σ_{ns}	(N/mm^2)	Nominal stress, nominal structural stress
σ_s, σ_{sb}	(N/mm^2)	Structural stress, structural bending stress
$\sigma_k, \bar{\sigma}_k$	(N/mm^2)	Notch stress, fatigue effective notch stress
$\sigma_n, \bar{\sigma}_n$	(N/mm^2)	Nominal tensile stress, the same displacement-related
σ_{na}, σ_{nb}	(N/mm^2)	Nominal axial and bending stress
σ_1, σ_2	(N/mm^2)	First and second principal stress
σ_{1t}, σ_{1b}	(N/mm^2)	First principal stress at top, side and bottom side
$\sigma_b, \sigma_t, \sigma_\omega$	(N/mm^2)	Bending stress, tensile stress, warping stress
$\sigma_{tot}, \sigma_{res}$	(N/mm^2)	Total stress, residual stress
σ_i, σ_o	(N/mm^2)	Inside stress, outside stress
$\Delta\sigma_{eff}$	(N/mm^2)	Effective tensile stress range
$\sigma_{el}, \sigma_{elpl}$	(N/mm^2)	Elastic stress, elastic-plastic stress
$\sigma_{en}, \sigma_{per}$	(N/mm^2)	Endurable and permissible stress
σ_s, σ_k	(N/mm^2)	Structural stress, notch stress

$\sigma_m, \bar{\sigma}_m$	(N/mm^2)	Mean stress, sample mean stress
σ_u, σ_l	(N/mm^2)	Upper stress, lower stress
$\Delta\sigma$	(N/mm^2)	Stress increment
σ_{ai}	(N/mm^2)	Stress amplitude on level i
σ_{ai}, σ_{ac}	(N/mm^2)	Stress amplitude with initial and critical crack size
$\bar{\sigma}_a, \bar{\sigma}_a^*$	(N/mm^2)	Maximum stress amplitude
$\sigma_{eq}, \sigma_{eq\,en}$	(N/mm^2)	Equivalent stress, endurable equivalent stress
$\sigma_t, \sigma_r, \sigma_{ax}$	(N/mm^2)	Tangential stress, radial stress, axial stress
σ_{me}, σ_{ci}	(N/mm^2)	Meridional stress, circumferential stress
$\sigma_\perp, \sigma_\parallel$	(N/mm^2)	Stress in base material, transverse and longitudinal to weld
$\sigma_{\parallel W}$	(N/mm^2)	Longitudinal tensile stress in weld
$\sigma_{cr}, \sigma_{cr}^*$	(N/mm^2)	Critical buckling stress
$\sigma_{ei}, \sigma_{eo}, \sigma_{mi}, \sigma_{mo}$	(N/mm^2)	Structural stresses in flange points (equations (177)(178))
σ^*	$(-)$	Stress ratio σ_i/σ_o
τ_n	(N/mm^2)	Nominal shear stress
τ_q, τ_l	(N/mm^2)	Transverse and longitudinal shear stress
τ_P, τ_{PW}	(N/mm^2)	Pulsating shear fatigue strength, the same with weld
$\Delta\tau_F$	(N/mm^2)	Shear stress range (endurance limit)
τ_{per}	(N/mm^2)	Permissible shear stress
τ_{q1}	(N/mm^2)	Transverse shear stress in σ_1 plane
$\tau_\parallel$	(N/mm^2)	Shear stress longitudinal to weld
$\tau_{\parallel W}$	(N/mm^2)	Longitudinal shear stress in weld
τ	$(-)$	Geometrical parameter in equation (29)
ϕ	(mm)	Quadratic mean of spectral roughness
φ	$(-)$	Angle of misalignment
φ, φ_A	$(-)$	Fatigue strength reduction factor
φ_0	$(-)$	Crack propagation angle
ψ	$(-)$	Supplementary oblique angle of notch
ψ	$(-)$	Reduction in cross sectional area at rupture
Ω	$(1/mm)$	Road frequency

1

Introduction

1.1 The phenomenon of fatigue

Fatigue strength* is a particularly important application characteristic of welded structures, which can only be maintained at a safe level by broad knowledge and experience as well as by calculation and testing. In this book the phenomenon of fatigue is used as a starting point. The basis for the formal assessment of fatigue strength in the second half of the book (from Chapter 7 onwards) is produced via the large amount of technical and procedural data presented in the first half. This assessment is carried out in several different ways (on the basis of nominal stress, structural stress, notch stress, safety as determined by fracture mechanics, and limit load determined by testing) and for each of these ways only some of the necessary data are certain, otherwise they are estimated — material parameters, loading conditions, weld joint geometries and defect characteristics.

In the presence of time variable, frequently repeated (cyclic) loading, plastic deformation occurs at the microscopic and macroscopic level, which decreases future ability to withstand stress and initiates cracks. The cracks, initially microscopic, propagate, finally leading to a (monotonic) final fracture. Therefore, fatigue is the initiation and propagation of cracking, whereby, counted in cycles, the phase involving stable propagation of cracking can include a major part of the total life. Microscopic and macroscopic notches with their localised notch stress concentrations, and changes of section with their more extensive increases in structural stress are decisive here. Surface and environmental influences (surface roughness, corrosion, temperature) also have a particularly marked effect. A large number of further parameters such as design, material and manufacture also influences the fatigue process in a variety of combinations.

Fatigue viewed as a multiple parameter problem which cannot be seen in isolation not only results in the well known scatter of strength values within a

---

* Fatigue: the historically original and conceptually more appropriate designation; 'cyclic strength' is less comprehensive; see also section 1.3

series of tests and between various laboratories, but above all it prevents the quantitative forecast of the phenomenon which is the objective. Service fatigue strength predictions carried out in engineering on the basis of general published data and theory, without direct tests of fatigue strength under service conditions, are little more reliable than the medium term weather or economic forecasts, which are generally known to be problematical and error prone. However, unlike weather and economic forecasts, predictions of service performance can be improved in particular cases by detailed, well planned test methods, applied to the actual component with the precision usual in the mechanical sciences. Consequently, the generalised forecast with its quantitative uncertainty as mentioned above can only be used as an indicator for appropriate technical action as regards dimensioning, design, manufacture and operation. The interesting theoretical and experimental testing procedures involved in specific problems of service fatigue strength cannot be investigated in detail here.

From the point of view of the physics of metals, too, the phenomenon of fatigue appears to be extremely complicated. The effect of piled up groups of dislocations, which defines the static strength characteristics, does not explain fatigue fracture, which is known to occur well below the tensile strength. However, the formation and movement of dislocations are essential in fatigue fracture too as the first incipient microcracks are observed on the surface of slipbands. As a consequence, repeated etching to remove the slipbands increases the fatigue strength considerably. On the other hand, the microcracks combine to form macrocracks when subjected to further loading. Subsequent crack growth, which is stable at first, is described by slip processes at the tip of the crack, which occur in differing slip planes during loading and unloading.

1.2   Basic tests

The basic engineering fatigue strength test is the cyclic load test introduced by A Wöhler (1819-1914), in which a smooth or notched test specimen or the component itself is subjected to a periodically repeated (usually sinusoidal) load of constant amplitude (tension, pressure, bending or torsion). The number of cycles, N, which is endured before crack initiation or complete fracture occurs, is plotted against the load or stress amplitude (S-N curve), Fig.1. The stress amplitude which can be endured for any desired length of time without a fracture (the endurance limit) is called the fatigue strength for infinite life (for more than 2×10^6 to 10^9 cycles). Anything which fractures sooner, when subjected to a higher stress amplitude, exhibits the high or low cycle fatigue strength for finite life (the latter for more than 10^3 cycles) which in the limit tends to the 'static' strength at 0.5 cycles. Depending on the magnitude of the static mean load, differing S-N curves are produced. Most important are the curves for zero mean load (fatigue strength under alternating load) and for zero minimum load (fatigue strength under pulsating load). In the low cycle fatigue strength range, it is more appropriate to carry out cyclic tests within limits of

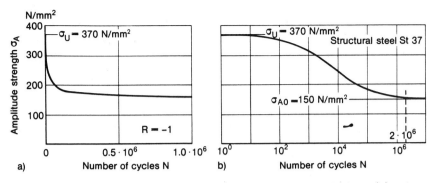

1 S-N curve for structural steel, St 37, number of cycles in linear (a) and logarithmic (b) scales, after Stüssi.[15]

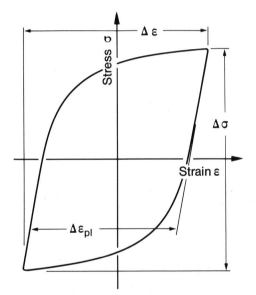

2 Hysteresis loop of the cyclic stress-strain curve.

displacement or strain than of load or stress. Instead of plotting the load or stress endured, the displacement or strain endured is plotted over the number of cycles (up to crack initiation).

The Wöhler test with a constant amplitude is not adequate for real conditions in which load sequences with variable amplitude are predominant and frequently there are also aperiodic processes. The random amplitude test has been introduced as an alternative to the Wöhler test under constant amplitude loading; in the former a strictly random load sequence is applied — with a fixed amplitude range and frequency content — until crack initiation or fracture occurs. However, the results of such tests are less suitable for generalisation, as can be seen from the aforementioned specification of amplitude range and frequency content and from the fact that the loads are indeed frequently aperiodic in practice, but are not strictly random. Fatigue tests with blocks of stepped amplitudes are of further assistance here (multilevel test, block program test), as are load history tests (follower tests).

The cyclic stress-strain curve of a smooth test specimen has proved to be an important basis for the macroscopic description of fatigue processes. The extreme points of the stationary hysteresis loops (Fig.2) which develop in a partly softening, partly hardening manner after a fairly large number of cycles (about 10% of the number of cycles to crack initiation) in a fatigue test between load or displacement limits delineate the cyclic stress-strain curve which may be used as the basis of a stress, strain and damage analysis at the root of notches subjected to cyclic loading.

1.3   Terms and definitions

The most important terms and definitions used in describing fatigue are explained below with reference to the German standard DIN 50100.[1]

Fatigue includes damage to materials, crack initiation and crack propagation under frequently repeated loading of variable amplitude. The loading sequence can be both determinate, periodic or aperiodic, and also stochastic (random). It can show a static mean stress σ_m. Fatigue strength is the ability to withstand load without crack initiation or propagation, expressed by the level and number of cycles of the loading endured. Terms and definitions, illustrated in Fig.3 and 4, are drawn from the fatigue test which is carried out using a periodic stress sequence with a constant amplitude. Applied stress values are indicated by lower case letter indices, strength values by upper case

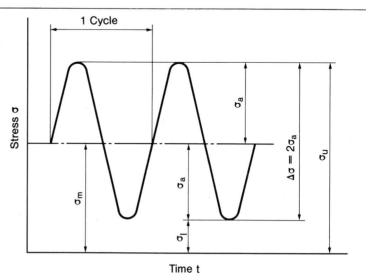

3 Loading parameters in the fatigue test:[1] upper stress, σ_u, lower stress, σ_l, mean stress, σ_m, stress amplitude, σ_a.

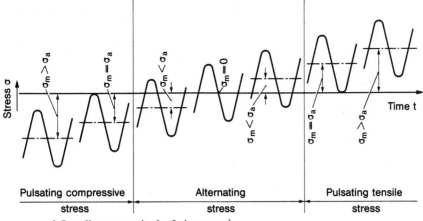

4 Loading ranges in the fatigue test.[1]

letter indices. The fatigue strength is the stress amplitude*, σ_A, endured without limitation on cycles at a preset mean stress or the corresponding upper stress[†], σ_{Up}. Fatigue strength for finite life (high or low cycle fatigue strength) is the (higher) stress amplitude endured with a limited number of cycles at a preset mean stress or the corresponding upper stress. The fatigue strength is dependent on the mean stress, the finite life fatigue strength is additionally dependent on the number of cycles, N, (see Fig.1 and 5). The fatigue strength**, σ_{A0}, under alternating load is characterised by zero mean stress, the fatigue strength**, σ_P, under pulsating load is characterised by the lower stress equal to zero. The stress amplitude is constant in a one level fatigue test; in a multilevel test it changes stepwise in a predetermined manner. In a simulated service fatigue test it follows a sequence which is similar to an operational one, *i.e.* completely or partially random. The service fatigue strength is determined in the latter case. The statistical analysis of scattering fatigue strength and fatigue life values uses the complementary terms 'failure probability', P_f, and 'survival probability', P_s, the numerical values of which complement each other giving the value 1.0.

1.4 Fatigue strength of unwelded material

The most important facts about fatigue strength, which are relevant to engineering applications, will be illustrated on unwelded material as an introduction. Here we will be talking about cyclic strength, as only the fatigue strength of specimens subjected to constant amplitude loading (Wöhler test) will be discussed.

---

*   Stress amplitude, σ_a, endured stress amplitude, σ_A, identical with amplitude strength σ_A (not a standard term), double stress amplitude $2\sigma_a$ identical with stress range $\Delta\sigma$.

†   Unlike fatigue strength for finite life, fatigue strength for infinite life is occasionally characterised by the additional index $_F$.

** Fatigue strength, σ_{A0}, under alternating load and fatigue strength, σ_P, under pulsating load, reserved for fatigue strength for infinite life according to Ref.1, are here also used for fatigue strength for finite life.

The fatigue strength for infinite life (number of cycles to fracture $N_f \approx 2 \times 10^6$) of steels and light metal alloys, determined from unnotched polished specimens in alternating axial or bending fatigue tests is dependent on their static tensile strength[††], σ_U (Fig.5). As a rough estimate $\sigma_{A0} \approx 0.4-0.6\sigma_U$, $\sigma_P \approx 0.6-0.8\sigma_U$ is valid for steels, $\sigma_{A0} \approx 0.3-0.5\sigma_U$ for aluminium alloys. The ratio of fatigue strength under alternating load, σ_{A0}, to yield strength[††], σ_Y, is correspondingly higher. In the case of steels, the representation of fatigue strength under alternating load as proportional to hardness is also possible. The precise values are dependent on the composition (alloy content, purity) and on the manufacturing process (melting, casting, hot or cold working, heat treatment). In the case of bending stress, somewhat higher values are determined than in tensile loading, which can be explained by the higher 'support effect' in specimens which are subjected to inhomogeneous stress (*i.e.* stress gradients).

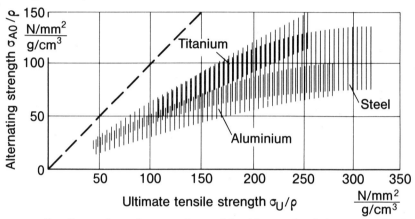

5   Comparison of groups of materials with regard to fatigue strength, after Hempel.[21]

The condition of the surface has a considerable influence on fatigue strength because fatigue damage usually starts on the surface. Fatigue strength is reduced by surface roughness, surface factor κ^* (Fig.6). The depth of roughness determines the reduction. An as-rolled finish decreases the fatigue strength more than a machined coarse finish, caused by a greater degree of roughness and detrimental decarburisation of the surfaces. Corrosive environments considerably reduce fatigue strength. In the case of unalloyed steels a reduction of about one third occurs in moist air and a reduction of about two thirds in salt water. On the other hand, surface hardening plus residual compressive stress in the surface layer has a strong strength-increasing effect.

Fatigue strengths of small and large specimens are somewhat different, caused by the dimensions alone (geometrical influence of size), or also caused by changed grain size, yield strength and surface properties (metallurgical influence of size).

[††]   For ultimate tensile strength σ_U is used instead of R_m in accordance with ISO, and for yield strength σ_Y instead of R_e or $R_{p0.2}$ in accordance with ISO, so as to be consistent with the conventional designation for stresses.

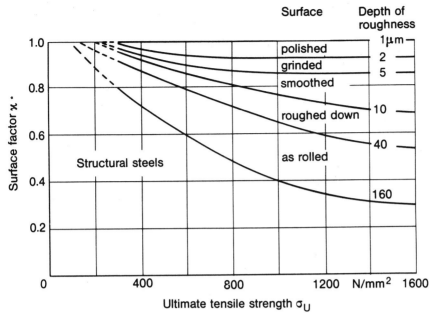

6   Influence of the surface condition on fatigue strength, after Wellinger and Dietmann.[27]

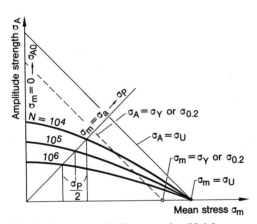

7   Fatigue strength diagram, after Haigh.

Fatigue strength is reduced as a result of static (tensile) pre-stressing (*i.e.* mean stress), σ_m, graphically illustrated in fatigue strength diagrams of various kinds (Fig.7-10). The fatigue strength in the compressive range can, however, be increased considerably compared with the tensile range (Fig.11). Figure 12 gives the fatigue strength of structural steels for calculation purposes. The linearised curves show the mean values of a statistical evaluation of the available test results. Another method for determining generalised (synthetic) S-N curves via formulae based on hitherto published test data is presented in Ref.217.

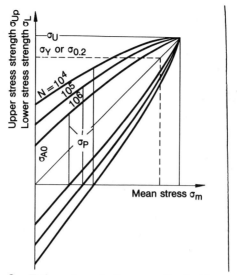

8   Fatigue strength diagram, after Smith.

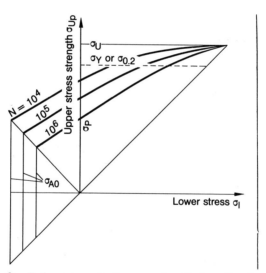

9   Fatigue strength diagram, after Gerber, Goodman, Kommerell, Roṣ, Launhard-Weyrauch.

The fatigue strength curves (Fig.7-10) may be seen to be dependent on the number of cycles corresponding to the rise in the S-N curve from the high cycle fatigue strength ($N_f \lesssim 2 \times 10^6$) to the low cycle fatigue strength ($N_f \lesssim 10^3$) and finally to the static tensile strength σ_U (Fig.1). Wöhler strain curves which are independent of mean stress for pressure vessel structural steels, with $\sigma_U = 336\text{-}846 \, N/mm^2$ are illustrated in Fig.13. In the case of load spectra, increased fatigue strengths are produced (*i.e.* service fatigue strengths) expressed by the endurable highest stress amplitude of the spectrum with a preset total number

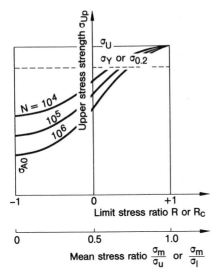

10   Fatigue strength diagram, after Moore, Kommers, Jasper, Pohl.

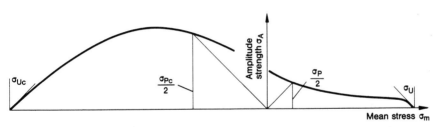

11   Fatigue strength (endurance limit) in the tensile and compressive
strength range for cast iron.

of cycles (*i.e.* the service life curve). The extent of the increase corresponds to
the quantity of lower stress amplitudes in the spectrum. The influence of
surface roughness and corrosion is reduced in the finite life and service fatigue
strength ranges.

In the frequency range 0.1-200cps, which is technically particularly
important, fatigue strength of steels is only slightly dependent on the frequency
of the load cycles. A rise in strength with frequency is observed on cooled
specimens, which is, however, changed to a fall in the absence of cooling. In
aluminium alloys, the influence of frequency is more marked.

At low temperatures the fatigue strength of materials increases in a similar
manner as the static strength and it decreases at high temperatures. Structural
steels show high fatigue strength up to approximately 400°C, aluminium alloys
up to approximately 150°C. Particular attention should be paid to creep and the
corrosion and oxidation processes in the presence of elevated temperatures.

Under multi-axial stresses, the von Mises distortion energy hypothesis or
the Tresca maximum shear stress hypothesis, which approximates to it,

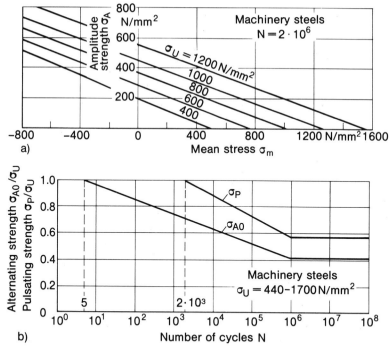

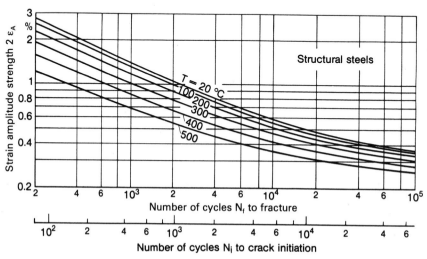

12   Fatigue strength (endurance limit) for structural steels, curves linearised after statistical evaluation, after Lang:[216] fatigue strength diagram (a) and S-N curves (b).

13   Strain S-N curves for structural steels at varying temperatures, after Schwaigerer.[9]

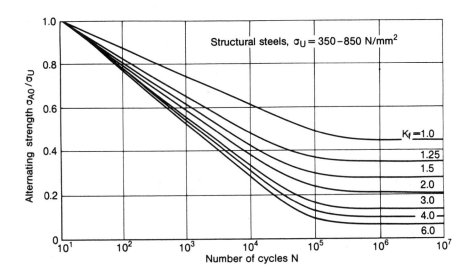

14 S-N curves for structural steels with varying notch effect,
conservative approximation, after Schwaigerer.[9]

determines the fatigue strength. In the case of a phase lag between the cycles of
the applied stress components, an additional reduction in strength can arise.
The fatigue strength is then given by the mean shear stress of all stress plane
orientations ('shear stress intensity') or, simplified (conservatively) by the
greatest differences in the principal stresses independent of their phase
positions.

The notch condition of a specimen or a component has an overriding
(negative) influence on its fatigue strength. The stress concentration at the root
of the notch is almost completely effective during fatigue loading. In the case of
sharp notches and cracks, an elastic support effect (microstructural support
effect) arises in the fatigue strength for infinite life range limiting the effect of the
notch stress concentration, which is particularly high. In the finite life fatigue
range, an additional support effect arises as a result of plastic deformation at the
root of the notch (macrostructural support effect). Then it is no longer stress
alone which determines fatigue fracture, but also the plastic strain at the root of
the notch. Simplified crack initiation S-N curves for notched structural steels
with various fatigue notch factors, K_f, are illustrated in Fig.14 (fracture S-N
curves are steeper).

As the condition of the notch has such a strong influence on fatigue strength
and as the fatigue crack initiates on the surface of the root of the notch, surface
hardening, generation of residual compressive stress in the surface and surface
coating, can increase the notch fatigue strength considerably.

1.5   Methodological aspects and stress assessments

The fatigue strength of welded joints is determined predominantly by:

— crack initiation processes, which are dependent on local notch stresses at the weld toe and weld root;
— crack propagation processes, which are dependent on the local stress intensity factor of the crack including notch effects.

The notch effect of the weld toe and weld root coincides locally with extremely inhomogeneous material. Here mixed material consisting of the melted electrode and parent material meets with parent material which has not been melted. As a result of the particularly rapid cooling of this zone after welding, tiny gas pores, slag inclusions, lack of fusion defects and undercuts arise, which do not represent welding defects as yet, but which can accelerate the crack initiation and propagation process in the presence of the relatively high notch stress (Fig.15).

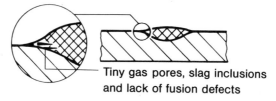

Tiny gas pores, slag inclusions
and lack of fusion defects

15   Micronotch effect at the weld toe, after Neumann.[3]

The maximum notch stress values and crack stress intensity factors, which determine the fatigue strength of the component, are considered for practical purposes, in so far as they are proportional to the basic stress, to be dependent on the nominal stress in the adjacent parent material and less dependent on the nominal stress in the weld seam or weld spot itself, which is controlling under static loading. In general, the fatigue strength is not proportional to the thickness of the weld seam or the diameter of the spot, and the measures taken in design and dimensioning assume that fatigue fracture will appear in the parent material rather than in the cross section of the weld seam or spot. The majority of fatigue fractures which appear at seams or spots do indeed pass through the parent material or the heat affected zone. The nominal stress in the parent material is known, even for joints which are not directly load bearing, which is not the case for the nominal stress in the seam or the spot.

Starting from the fact of the dominating influence of the localised maximum stress, fatigue strength of welded joints and structures can be dealt with on four different levels: assessment of nominal stress; of structural stress; of notch stress and of safety based on fracture mechanics. These assessments should partly complement each other and should lead to the same final result independently of each other.

In the nominal stress approach (see section 7.2), it is shown by simple calculations that the nominal stress (axial force divided by area of cross section, bending or torsional moment divided by the relevant section modulus) in the load bearing (parent material) cross section, is less than the (endurable) nominal stresses for fatigue fracture. The endurable or permissible nominal stresses are stated (in so far as they are known) as a tabular or graphic summary in the codes (see Chapter 7) dependent on geometry of the joint, type of weld, type of loading and manufacturing influences (*e.g.* surface dressing, defect content). The nominal stress approach has been introduced and recognised universally.

In the structural stress approach (see section 7.7) the inhomogeneous stress distribution in the welded structure subjected to external loads without the notch effect of the welds is determined additionally to the nominal stresses either by calculation in accordance with the engineering theories of structures (*e.g.* theories of beams, girders, frames, plates and shells, mostly involving a finite element approximation), or by testing involving measurement of stress, in the latter case with the occasional problem of being certain of excluding the weld notch stress. Marked concentrations of structural stress arise, for example, at nozzles, tube and beam joints. Deviations of the structural stresses from the nominal stresses in a less marked form appear in almost all practical cases. Notch stress analysis, which goes further, can be carried out starting from the structural stresses; endurable maximum values of structural stress can be stated for defined classes of components, or as an approximation the fatigue strength of the simple butt or fillet joint, when subjected to alternating axial loading, can be established as local structural strength. Predecessors of the structural stress approach for welded structures are the systematic structural stress measurements by W Kloth,[83] which were particularly advantageous to the automotive industry. E Haibach[53] laid a safe, methodical, and quantitative foundation. The structural stress approach was used at an early date in pressure vessel construction. The tubular intersections of drilling platforms are designed in accordance with structural stresses (here they are called 'hot spot stresses'). In shipbuilding, determination of structural stress is desirable for more realistic strength predictions. In general, the development of the finite element method helped the structural stress approach to be accepted and practical proof of its value is only now receiving wider acceptance.

In the notch stress approach (section 8.1), the concentration of the notch stress is determined for the weld toe and (as far as possible) for the weld root, additionally to the nominal stresses and structural stresses. The concentration is determined either by measurement, by means of strain gauges on the toe or via plane photoelastic models, or by calculation, by the finite element or boundary element method. The calculations are primarily limited to plane cross sectional models for economy, but are approximately transferable to the actual three dimensional conditions. Starting from the boundary stresses in the weld seam lines of the global structure (*i.e.* structural stresses) the notch stresses of the local structure, the 'joint', are determined. The endurable notch stresses are set equal to fatigue strength values determined on smooth specimens made from homogeneous (parent) material. However, the maximum notch stresses are to be reduced according to the notch support effect. The endurable notch strain amplitudes were also determined directly for structural steels and aluminium alloys on the (inhomogeneous) weld toe. In the case of service load sequences with variable amplitudes, which are closer to reality, a more sophisticated stress, strain and damage analysis should be carried out for the root of the notch (see section 8.3). This approach, which has been successful for unwelded structural components, has only been carried out in exceptional cases up to now on welded structures.

In the fracture mechanics approach the unavoidable cracks and defects, especially at the weld toe and the weld root, are evaluated based on the stress intensity factor at the crack tip. The gap in a double fillet joint can also be assessed directly as a macrocrack alternatively to notch stress analysis. The stable crack growth with number of cycles and the critical size of the crack, at which unstable final fracture occurs can be determined from the amplitude of the stress intensity factor at the crack tip. This is a safety assessment regarding unavoidable defects or cracks, but it should not be used as a design and dimensioning tool. It is applied mainly on high quality welded structures but still needs further empirical verification.

Nominal stress assessment is the basis of Chapters 2-4, 6 and 7 (apart from section 7.7). Chapter 5 (in an indirect manner) and section 7.7 are devoted to structural stress assessment. Finally, notch stress assessment is the basis for Chapters 8 and 10, while the fracture mechanics approach is the basis for Chapters 9 and 10. The further methodological points of view below are introduced with regard to nominal stress assessment, which is mainly used at present. Reference is made only to joints with seam welds.

1.6   Nominal stress assessment for fatigue resistant joints

The conventional procedure used is to work with differing endurable or permissible nominal stresses in the parent material for differing 'notch classes' of the joints. The joints are allocated to the notch classes depending on the type of joint, weld type, loading type, and quality of manufacture and/or absence of defects. Here the results of fatigue tests (not stress analyses) are used as a basis. The name 'notch class' is therefore only a partially accurate choice, the notch effect actually remains hidden and is not quantified. More precisely expressed, these are classes of endurable or permissible nominal stresses in the parent material. The joint in question is allocated to one of several standardised S-N curves of endurable or permissible stresses. The disadvantage of this method is that only a few joints of practical importance, which can be uniformly manufactured, can be allocated to the S-N curves with sufficient certainty on the basis of the fatigue tests carried out (mainly butt and cruciform joints, transverse and longitudinal stiffeners). Moreover, it is disadvantageous that for more complex joints (*e.g.* tubular joints) there are frequently several possibilities (of equal value) for defining the nominal stress, which then leads to differing endurable or permissible nominal stresses. Nonetheless, this allocation procedure is the one primarily used in practice if there is sufficient empirical support.

If endurable or permissible nominal stresses in the parent material are established in this way, it is an unstated precondition that the nominal weld seam stresses can be disregarded, at least for welds which are adequately dimensioned statically. The endurable or permissible nominal stresses for the parent material, especially for butt and fillet welds, as well as for transverse and longitudinal stiffeners, will of course turn out somewhat differently depending on the type of joint on which they were established (*e.g.* endurable or permissible transverse nominal fillet weld stress in the cruciform joint compared with the lap joint); the most important influence of shape, namely of the butt joint, which is mainly concentric and has a weak force flow diversion, compared with the fillet weld, which is extremely eccentric and has a strong force flow diversion, is, however, clearly expressed. The successful combination in the presence of static loading of the butt and fillet weld characteristics via the nominal stress in the middle plane of the welds is not possible with respect to fatigue, as nominal weld seam stresses are expressly discarded because of the dominant influence of shape. In the same way all the other considerations concerning the nominal weld seam stresses must be dropped at least at the first attempt, and can only be drawn on as an addition in special cases, namely, for explaining the varying endurable nominal stresses in the parent material in the presence of the less frequent weld seam fractures (they can be avoided by ample weld seam dimensioning). It is to be noted that, nonetheless, in the technical literature endurable weld seam nominal stresses are also occasionally used for general characterisation of fatigue strength, especially with fillet welds. As fillet welds generally reveal a greater weld cross section than parent metal cross section, the endurable nominal stresses for fillet welds come out correspondingly lower than the relevant nominal stresses for the parent material.

The emphasis on the type of weld rather than the type of joint (the weld is always part of the joint) despite using the nominal stresses of the parent material, is based on the fact that the number of even elementary types of joint and their load cases is so large and therefore requires a simplified description precisely by means of the type of weld. In addition, the strength of the type of weld can be transferred to the structure as a rough approximation if the local basic stress (*i.e.* structural stress) is known there.

All the information on the fatigue strength of joints has been gained from fatigue tests on typical welded specimens. Thus the specification of the above-mentioned weld type strength must be based on these specimens. Quantitative data on the fatigue strength of joints are difficult to interpret because results within a fatigue test series are scattered to such a large extent that statistical evaluation is indispensable and is the norm nowadays. However, the statistical mean values determined by various research laboratories are fairly widespread too when faced with the same fatigue problem (order of magnitude: factor 2 for endurable stresses). From this it follows that, as far as representation in a textbook is concerned, only a few key values can be given, and more emphasis is to be put on the trend of the different design and manufacturing measures which can be taken. Here, considerations of notch stress have the greatest value in explanations. The design engineer cannot avoid either staying well away from the smallest possible strength values or undertaking a few fatigue tests which are close to reality (service fatigue strength tests) for the problem in hand.

Further to this, as a part of the considerations concerning the dominating influence of the local (notch and crack) stresses, the following points are important regarding representation of the endurable or permissible nominal stresses in the parent material with weld:

— butt weld with weak force flow diversion versus fillet weld with strong force flow diversion (corner and edge welds are treated as fillet welds);
— load bearing versus non-load bearing welds, the first includes the higher notch effect with the same geometry;
— continuously welded (either with ends, *e.g.* welded plate specimens or without ends, *e.g.* circumferential welds on cylindrical components) compared with intermittent welded joints, the latter with high notch stresses at the ends of the weld in all types of loading;
— loading in the structural plane, transverse ($\sigma_\perp$), diagonal ($\tau_\parallel$), or longitudinal ($\sigma_\parallel$) to the weld (simply loaded joints), transverse shear loading perpendicular to the structural plane which can be disregarded because it does not have a very large notch effect, bending ($\sigma_{b\perp}$, $\sigma_{b\parallel}$, $\tau_{b\parallel}$) in the structural plane, which is comparable with the corresponding tensile state via the maximum bending stress (conservatively);
— loading transverse, diagonal and longitudinal to the weld simultaneously (multiply loaded joints);
— degree of freedom from defects (weld quality), important in welds which are otherwise notch-free (ground flush) or welds with only a slight notch (butt welds with root sealing run under transverse loading, and butt and fillet welds under longitudinal loading).

In the statements below it is always an unstated precondition that the welded structures and joints have been adequately designed and dimensioned for static loading.

1.7   General references to relevant literature

The following publications, on which short comments and summaries are given below, deal (exclusively or in part) with fatigue strength in general, with the fatigue strength of welded structures in particular, as well as with the relevant questions concerning calculations, design, and testing.

The strength of the book by Gurney[2] on the fatigue strength of welded structures, is, apart from its successful didactic element, the many sided and meticulously accurate representation and comparative commentary on test results from different sources. On the other hand, the principles and procedural guidelines for the strength assessment of welded structures are somewhat too brief. The book concentrates more on the strength of joints than on the strength of structures.

The welding engineering manual by Neumann *et al*,[3] in several volumes, is aimed at the designer of welded structures without neglecting the scientific basis. In addition to the principles of strength, the areas of application: structural steel engineering, machinery construction, tank construction, and vehicle engineering are dealt with using a wealth of design examples and manufacturing information. Those elements of knowledge which are typical and capable of generalisation are shown clearly. However, a result of the wealth of information presented is that there are overlaps and repetitions, particularly in the older editions because different authors produced individual contributions. The new editions have been revised and tightened up. The value and importance of this work for design practice is considerable. A shortened version is available.[4]

The welding manual by Ruge[5] concentrates on the primary information requirements of the design and field engineer. It is a modern reference work for the practitioner and the engineer.

Three complementary titles from the DVS textbook series should be mentioned. The book 'Assessments of strength' by the author,[6] published in 1974, represents a forerunner of the current work. It is still important because of its second volume, 'Special approaches', which is not continued in the current work and because of the older literature which is quoted in detail. The book 'Design and calculation of welded structures' by Erker *et al*[7] may still make some valuable suggestions today since it is a book which deals in detail with practical matters and machinery construction. Finally, the book by Sahmel/Veit[8] on the design of welded structures represents a widely read, brief introduction, particularly from the point of view of steel engineering.

Amongst those publications which are more technically specific, the works by Schwaigerer[9] (boiler, vessel, and pipeline design), Stüssi/Dubas[10] (steel engineering), and Hertel[11] (aircraft design) contain valuable information on the technical and scientific principles of fatigue strength and on its realisation in design (including welded structures).

The publications Ref. 12-36 also deal with fatigue strength in general and the fatigue strength of welded structures in particular. They are recommended for introductory, educational and reviewing purposes.

The technical literature mentioned in the rest of this book is limited to an incomplete list of useful and stimulating sources. In a period when lists of technical literature are stored electronically and information from it can be obtained via appropriate keywords, the classical bibliographical outline is no longer required. On the contrary, the limitation is a matter of not being drowned in the flood of information. The bibliographic databases available to the national institutes of welding are recommended to those readers who need a more complete set of references on special items.

2

Fatigue strength for infinite life of welded joints in structural steel

2.1 Simply loaded joints

2.1.1 *Defining the reduction factor*

For easier survey and as a first estimate, the fatigue strength for infinite life (endurance limit) of the welded joint* is characterised by the reduction factor, γ, using the pulsating fatigue strength as a basis:

$$\gamma = \frac{\sigma_{PW}}{\sigma_P} \text{ (for } \sigma_\perp \text{ and } \sigma_\parallel)\tag{1}$$

$$\gamma = \frac{\tau_{PW}}{\tau_P} \text{ (for } \tau_\parallel)\tag{2}$$

Here σ_P and τ_P are the tensile and shear pulsating fatigue strengths of the mill finished sheet of base material without a weld and σ_{PW} and τ_{PW} the tensile and shear pulsating fatigue strengths of the mill finished sheet of base material with a weld.

The relation between σ_P and τ_P is derived from the von Mises distortion energy hypothesis, which is valid for the base material in the presence of multi-axial loading:

$$\sigma_P = \sqrt{3}\tau_P\tag{3}$$

The reference value, σ_P, for the mill finished structural steel St 37 is selected according to Neumann.[3]

$$\sigma_P = 240\,\text{N/mm}^2\tag{4}$$

The mill finish has a strength-reducing effect as a result of the roughness and decarburisation of the surface. The pulsating fatigue strength of a polished specimen of the base material without the mill finished surface is higher, at

---

* Welded joint: The joint connected by a seam weld, excluding spot welded, friction welded, flash welded and stud welded joints, see Chapter 6.

270 N/mm^2, according to Ref.3. As a polished rod, the deposited weld metal has a pulsating fatigue strength greater than this value, but this is unimportant in practice as the fatigue crack generally runs in the diluted parent material close to the edge of the fusion zone.

The above reference value, $\sigma_P = 240 \text{ N/mm}^2$, is determined for failure probability, $P_f = 0.1\text{-}0.01$. A reference value about 1.5 times as large is valid for $P_f = 0.5$.

The reference value of 240 N/mm^2, derived from St 37, is also retained for higher tensile steels, although their σ_P value is higher. This is, of course, incompatible with equations (1) and (2) but offers the advantage of giving approximately the same numerical values for γ for higher tensile steel (joints made from higher tensile steels have a pulsating fatigue strength which is only slightly higher). In the case of aluminium alloys, however, the procedure in accordance with equations (1) and (2) is followed correctly, whereby the strength-increasing oxide skin has to be considered instead of the strength-reducing mill finish.

The closeness of the pulsating fatigue strengths to the global or local yield stress (non-linear processes) and the global or local mean stress, which varies with the magnitude of the pulsating fatigue strength, suggests that the reduction factor should not be established by using the pulsating fatigue strength as a basis. The values for alternating fatigue strength appear to be more suitable to be used as a basis. However, the documented testing results mainly confirm the selected basis.

The reduction factors which are quoted have been checked against the fatigue strength surveys produced by Neumann,[3,16] Gurney,[2] and Olivier/ Ritter[37] amongst others (failure probability $P_f \approx 0.1$), in which process, however, contradictory data also had to be reconciled. The reduction factors are determined for an environment consisting of air (without corrosion) and for room temperature. They are quoted for plate thicknesses of about 10mm (a drop of 25% in fatigue strength occurs with increasing thickness to 50mm whilst preserving geometrical similarity).[38]

A distinction is made below between load-bearing and non-load bearing welds. In the former a further distinction is made between butt welds, including single bevel and double bevel joints, which are of similar strength, and fillet welds, including T, corner and edge joints, which are also of similar strength.* The governing criteria are the amount of force flow diversion and the resulting notch effect. The successful simplification used in static load design, which is based on weld fracture area, is not very helpful for fatigue because of the greater importance of the local notch effect.

The concentration below on welded joints of simple shape and (basic) loading, made of structural steel, has to take place because there is no information available capable of generalisation for differing materials (apart from aluminium alloys), shapes and loadings.

---

* *Weld* (conventional linguistic usage) is frequently used in the considerations on strength below in the sense of *welded joint* (more correct term); the *butt weld* is actually a *butt welded joint*, the corner weld actually a *corner welded joint*; the definitions given in DIN 1912 are inadequate here.

2.1.2   *Butt weld*

The butt weld is a connecting weld between plates which is largely free from force flow diversion and hence has a low stress concentration factor. It makes a joint in the same plane or only at a slight angle. It has the following variants: square butt weld, single V butt weld, single U butt weld with or without root face, U and V butt welds also arise as double welds (double V weld identical to X weld), and the raised edge butt weld.

The primary loading on the base material is tension or compression in the direction of the weld, $\sigma_\parallel$, perpendicular to it, $\sigma_\perp$, and shear, $\tau_\parallel$, in the direction of the weld. Bending and transverse shear are dealt with as secondary effects.

The usual test specimen is a transverse weld specimen for $\sigma_\perp$, a longitudinal weld specimen for $\sigma_\parallel$ (continuous weld) or a gusset plate specimen for $\sigma_\parallel$ (discontinuous weld), see Fig.16.

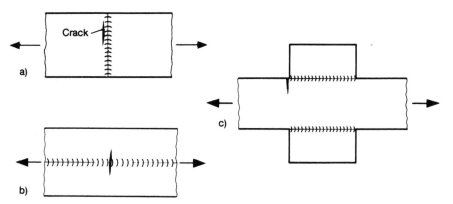

16   Butt weld specimens: specimens with transverse and longitudinal welds (a, b), gusset plate specimen (c).

Critical crack initiation points are undercut and root notch for $\sigma_\perp$, electrode stop/start points within the length of the weld or noticeable weld ripples for $\sigma_\parallel$ and continuous welds, and ends of the weld for $\sigma_\parallel$ and discontinuous welds, see Fig.17 (and 16c). The leading edge of the fracture runs perpendicular to the principal loading direction.

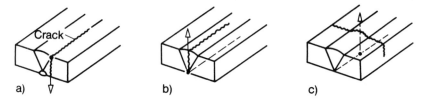

17   Failure initiation points in butt welds: weld toe (a), weld root (b), electrode stop/start point or weld ripples (c), after Gassner and Haibach.[251]

The reduction factors for the continuous butt weld for low strength structural steel are between 0.5 and 0.9 for transverse tensile stress, $\sigma_\perp$, between 0.7 and 0.9 for longitudinal tensile stress, $\sigma_{||}$, without a weld end, and between 0.6 and 0.9 for shear, $\tau_{||}$. Thus the butt welded joint can be subjected to higher levels of loading than the riveted joint, which is allocated the factor, $\gamma = 0.5$-0.6.

Measures for improving the weld strength in the presence of transverse loading, $\sigma_\perp$, meet with particular interest because of the considerable reduction in this loading case and will be dealt with first of all. They also improve weld strength in the presence of shear loading, $\tau_{||}$.

The single V or square butt weld, welded on one side only, with or without a root face (Fig.18a), gives a reduction factor of 0.5 for $\sigma_\perp$. Tensile bending stresses at the root of the weld are particularly critical.

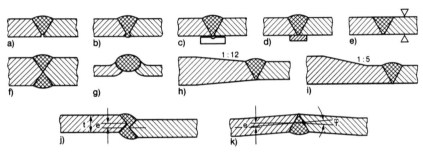

18 Types of butt weld in manufacture: with root face (a), with sealing run (b), with ceramic backing strip (c), with backing plate (d), machined after welding (e), welded on both sides (f), raised edge joint (g), tapered joint at change of section (h), tapered joint with weld removed from change of section (i), joint with axial misalignment (j), joint with angular misalignment (k).

Grooving out the weld root and depositing a sealing run (Fig.18b), welding with the root being shaped on a ceramic backing strip (Fig.18c), or welding through on to a backing plate (Fig.18d), improve the reduction factor to 0.7 assuming that there is a low level of weld reinforcement and that there are no weld defects. There is still disagreement about the extent of the improvement when welding through is carried out.

Machining after welding (machined flush) (Fig.18e) can produce a further improvement to 0.9-1.0 (grinding flush is less advantageous because of the risk of grinding marks). After all the rest of the notch effect has been removed, only microdefects in the diluted parent material in the fusion zone can initiate failure.

The double V and double U weld for thick plates (Fig.18f) has the same fatigue strength reduction value as the V weld with a root sealing run.

The raised edge butt weld (Fig.18g) in the gas welding of thin plates shows a marked notch effect and eccentricity, with a reduction factor of 0.5.

Craters at the end of a weld pass have a strength-reducing effect. They can be avoided by use of run-on and run-off plates, which are subsequently removed (Fig.19).

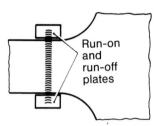

19   Preventing craters at the ends of a weld pass by run-on and run-off plates.

Butt joints between chamfered plates of different thicknesses do not produce any additional reduction if the slope is slight (1:5-1:20 is usual) or if the weld is positioned outside the change in thickness area (Fig.18i). Single bevel or double bevel butt joints can be used without chamfering the plates where the thickness changes, if the thickness ratio is limited.

Butt joints between plates which meet at slight angles give a reduced strength as a result of secondary bending and notch stresses.

Butt joints between plates with different widths produce a significant strength reduction which can be diminished provided that there is an adequate radius and chamfer, *i.e.* a smooth profile and if the weld is removed sufficiently far from the transition area (Fig.19).

Inaccuracy of fit, axial misalignment (Fig.18j) and angular misalignment (Fig.18k) reduce the fatigue strength as a result of secondary bending and notch stresses, which increase with the ratio of eccentricity, e, to plate thickness, t. In longitudinal seam welding of pipes, which are manufactured from sheet metal strips, 'ridge formation' as a result of insufficient bend forming of the sheet metal edges which butt together is a well known phenomenon.

To improve fatigue strength, a weld profile free from sharp notches should be the objective. In addition to the measures which have already been mentioned, careful smoothing of the toe area, a small toe angle and small reinforcement of the weld also play a part. Well smoothed weld toes can be obtained by using electrodes giving fine droplet melting and a viscous slag (titania electrodes, thick flux coating, and manual welding in a flat position are beneficial, fully mechanised submerged-arc and CO_2 welding are not). Better smoothing of the weld toe can be achieved by grinding or TIG dressing. Deep penetration electrodes, which increase the fatigue strength of the weld root substantially, do not have any identifiable advantage as regards the weld toe.

The welded lap joints and 'reinforcing cover plates' adopted initially from riveting techniques reduce strength markedly because of their strong notch effect and are considered out of date.

Weld toes can be treated locally to increase strength (see section 3.4). Heat treatment by fusing the transition area between the base metal and the weld bead using the TIG or plasma beam causes local hardening and removes local inhomogeneities of shape or material. Use of shot or hammer peening also increases strength.

Welding defects, particularly poor fusion, undercuts, root notches, slag inclusions and weld cracks considerably reduce strength, the more so in joints free of other discontinuities and notches. Thus, weld defects are to be avoided, especially in high quality welds and it must be shown that they are not present by non-destructive testing (radiographic and ultrasonic testing, surface inspection). In the presence of relatively large undercuts or slag lines, $\gamma < 0.4$.

Joint edge preparation (differing geometries, machining or flame cutting) influences strength only indirectly because of possible weld defects. A flame cut edge is more prone to defects.

Preliminary tack welding encourages cracking and welding defects. It should therefore be avoided on high quality fatigue resistant welds.

The size of the component or specimen influences fatigue strength both via the usual size effect on notch strength (expressed by stress gradient, equivalent radius of curvature, and fatigue notch factor) and by the transverse residual stresses in the weld, caused by welding, which are mainly present in large components and not so much in specimens. Compressive transverse residual stresses have a strength-increasing effect in the presence of a pulsating tensile load. Tensile transverse residual stresses have a strength-reducing effect.

Fatigue strength in the presence of shear loading, $\tau_{\parallel}$, is reduced less than in the presence of transverse loading of the weld, $\sigma_{\perp}$, with the same weld profile because of the weaker force flow diversion and notch effect. Therefore, a butt joint specimen subjected to axial load achieves somewhat higher fatigue strength with an inclined butt weld than with a transverse butt weld, provided that the ends of the weld are removed.

The fatigue strength in the presence of longitudinal loading, $\sigma_{\parallel}$, which encounters no marked weld shape notch effect, is determined by weld defects, by rippling in the weld surface and (dominating if present) by the discontinuity of the surface and the weld root at the electrode stop/start points. Fine ripples are promoted by fine droplet melting and viscous slag. As a result of this, thickly coated titania electrodes (rutile electrodes) produce a higher longitudinal weld strength than basic lime electrodes.

In the case of longitudinal weld loading the (tensile) longitudinal residual stresses in the weld, caused by welding, have a strength-reducing effect.

The end of a butt joint on a gusset plate (discontinuous butt welds are otherwise to be avoided) has an extremely high notch effect because of its position in the re-entrant corner, which leads to a reduction factor, $\gamma < 0.3$. Design and manufacturing measures in accordance with section 5.1.1 are appropriate.

Most triple plate T joints (Fig.20) should be considered as butt welds, because they are mainly loaded horizontally by $\sigma_{\perp}$, $\tau_{\parallel}$ and $\sigma_{\parallel}$ in the plates

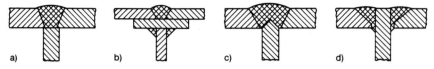

20   Triple plate T joints for loading which is principally horizontal.

which butt against each other, whilst the third vertically positioned plate remains largely free of loading. The triple plate T joints which are illustrated can be classified as unmachined welded joints with backing plate or ceramic backing (single bevel butt welds).

2.1.3   *Single bevel and double bevel butt welds*

The single bevel and double bevel butt weld with its variants, with and without a root gap, with and without a fillet weld cover pass, V or double V groove, used on a cruciform joint (Fig.21), on a (double plate) T joint and on a butt joint with a change in thickness, are to be considered butt welds in the broadest sense. The notch effect caused by force flow diversion in the simple fillet weld is avoided so that strength characteristics similar to those of the butt weld are produced. In the above mentioned cruciform and T joints it is presumed that there is a tensile loading from two opposite sides, and tensile loading from one side with bending of the transverse member is excluded. The latter case results in increased notch effect and is therefore to be allocated to the fillet welds.

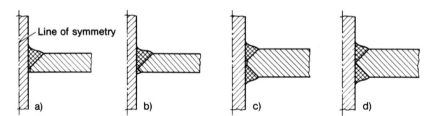

21   Single bevel and double bevel butt welds: with root face (a, c), without root face (b, d).

However, single bevel and double bevel butt welds cannot generally be produced as well as high quality butt welds. Grooving out the weld root and depositing a run on the back are more difficult so that these are mostly carried out with a remaining root gap. Ceramic backing strips or metal backing plates can be used, but the weld is more prone to defects. Non-destructive testing for defects (radiographic and ultrasonic testing, surface inspection) is also more difficult and liable to error. Finally, machining is only possible locally on the weld toe.

As regards failure initiation points, the same applies as for the butt joint. The reduction factor in the presence of transverse tension, $\sigma_\perp$, is between 0.4 (with root gap) and 0.7 (without root gap and machined). In the presence of longitudinal tension, $\sigma_\|$, it is between 0.6 and 0.8.

The individual data given for the butt joint regarding the factors influencing fatigue strength can also be used by analogy for single bevel and double bevel butt welds, the margin for improving strength being lower, according to the statements above. It is most important to achieve a good weld profile. Smaller toe angles and better smoothing of the toe are advantageous (*i.e.* concave fillet weld as cover pass, weld toe machined or TIG dressed and small or no root gap).

2.1.4. *Fillet weld*

The fillet weld is a connecting weld with force flow diversion and therefore has a strong notch effect in the re-entrant corner of the perpendicularly butting plates.

The re-entrant corner can also be formed in a different way, namely as an overlapping plate edge or as a step between plates of different thickness which butt against each other. As a result of the strong notch effect in fatigue, fillet welds are less advantageous as regards fatigue strength than butt welds. Their economic advantage lies in the fact that no joint preparation is required and less fit-up work is needed.

A fillet weld can be made using concave, flat, or convex fillets (Fig.22). It can be made on both sides of the plate or on one side (Fig.23), continuous or intermittent, in the latter case likewise positioned or staggered. Contrary to common thinking, a single sided fillet weld can also be used successfully where it is subjected to fatigue loading (as can a double sided fillet weld) provided that bending around the weld root which opens the root gap can be avoided. In

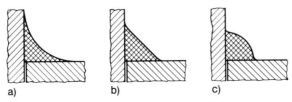

22   Fillet weld profiles: concave (a), flat (b), convex (c).

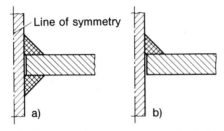

23   Fillet weld arrangements: on both sides (a), on one side (b).

current design practice this is dealt with by means of transversely positioned plates or ribs, which increase the bending resistance of the joint considerably. This design measure is also necessary for bending which closes the root gap, as poor fit-up must always be allowed for in practice. Intermittent welds are to be avoided if possible when there is fatigue loading present because the weld ends have a marked strength-reducing effect. The thinner continuous weld is much superior to the thicker intermittent weld compared on the basis of identical amounts of weld material. A convex fillet weld is to be avoided if possible because of its particularly severe notch effect.

The primary loadings on the base material are, as for the butt weld, $\sigma_\perp$, $\tau_\parallel$ and $\sigma_\parallel$. Bending and transverse shear in the base material are dealt with as secondary phenomena. Bending around the centreline of the one sided weld is unacceptable.

Normally, test specimens with fillet welds, which are directly load bearing (Fig.24), are made as a cruciform joint, or alternatively (less usual nowadays) as

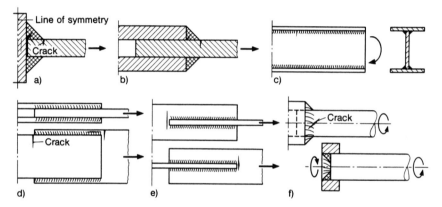

24   Fillet weld specimens: cruciform joint (a), lap joint with transverse fillet weld (b), double fillet weld on I section girder (c), lap joint with longitudinal fillet weld (d), member splicing with longitudinal fillet welds (e), circumferential fillet weld (f).

a double face, occasionally also single face lap joint with edge fillet weld for $\sigma_\perp$, as a double fillet weld on an I section girder for $\sigma_\parallel$ and as a lap joint with side fillet welds, or less usually as a spliced plate joint or as a circumferential weld on a cylindrical specimen for weld shear stress, $\tau_{\parallel w}$. The specimens are subjected to tensile, bending or torsional loading. The lap joint shows considerably less notch effect at the weld root than the cruciform joint as the unfused root gap is longitudinal rather than transverse. In the case of tensile loading the ends of the unfused gap are relieved of some stress by friction contact of the plates. This increases the tensile pulsating fatigue strength. In contrast, in the presence of compressive loading, opening occurs because of plate bending so that the compressive pulsating fatigue strength and the alternating fatigue strength are lowered. A T joint specimen with bending load from transverse force, which is occasionally used (Fig.25), produces a less favourable force flow, a higher

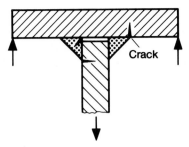

25 T joint specimen subjected to bending load from transverse force.

notch effect and greater strength reduction, so that the toe on the transverse member forms an additional critical point. On the other hand, pure bending load is resisted relatively well in joints with fillet welds as a result of the external position of the fillet welds (increased bending resistance).

Critical crack initiation points are the toe and root notches for loading by $\sigma_\perp$ and $\tau_\parallel$, and the electrode stop/starts or a noticeable rippling for loading by $\sigma_\parallel$ (Fig.26). A discontinuous weld fails at the weld ends. The leading edge of the fracture runs perpendicular to the principal loading direction.

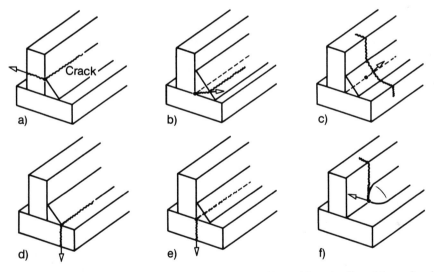

26 Fracture initiation points in fillet welds: weld toe (a, d), weld root (b, e), electrode stop/start point or weld ripples (c), weld end (f), after Gaßner and Haibach.[251]

Unlike a butt weld a fillet weld offers the weld throat thickness as a free parameter for the designer, which can be favourably selected as regards fatigue strength. In the first place, the question of the critical fillet weld thickness arises, that is, the thickness above which the fracture no longer appears in the weld first but rather in the toe cross section. The objection is often made, incorrectly, that the weld thickness has no effect on the failure (and if it does it is a negative one). This is actually not true as regards failure in the weld. The stress intensity factor at the weld root and the crack propagation time to reach the weld surface are primarily dependent on the weld throat thickness. It is only in the case of weld toe fractures that the weld thickness has no effect. The critical weld thickness is dependent on joint shape, joint dimensions, weld arrangement and local notch conditions. It is not simply identical to the values obtained with static loading. Care is therefore advised as regards a critical weld thickness which is stated to be generally applicable.

The critical (nominal) weld thickness for the cruciform joint, analysed from a fracture mechanics point of view, and verified by testing, is dependent on the plate thickness and the amount of penetration (see section 9.4). For zero penetration the critical ratio of weld size to plate thickness is between 0.6 and 0.8 (for plate thicknesses of 6-50mm). It is therefore not very different from the critical ratio 0.7-1.0 for static loading.

The critical (nominal) weld thickness for a (double face) lap joint with transverse fillet welds is dependent on the ratio of the cover plate thickness to the base plate thickness and on the type of loading (pulsating or alternating). The latter is because of the favourable frictional contact forces in the presence of tensile pulsating loading. The critical thickness also depends on the depth of penetration and on the notch severity of the weld toe (toe angles less than 45° and grinding the weld toe impede toe failure and therefore encourage weld failure). The critical ratio of weld to plate thickness is somewhat lower than for a cruciform joint.

The critical weld thickness for a lap joint with longitudinal fillet weld has only been investigated occasionally because of the large number of shape parameters, amongst other factors. Weld failure is avoided above a ratio of weld fillet section to cover plate cross section of 0.6-1.6. It is therefore recommended that the weld fillet section selected is twice the size of the cover plate cross section.

The value of the reduction factor for a continuous fillet weld on low strength structural steel is between 0.3 and 0.5 for $\sigma_\perp$, between 0.4 and 0.6 for $\tau_\parallel$, between 0.5 and 0.7 for $\sigma_\parallel$. The factor for a discontinuous longitudinal fillet weld which is directly load bearing has a value between 0.2 and 0.4; the intermittent double fillet weld which is not directly load bearing has a value between 0.3 and 0.5.

The fatigue strength of a (continuous) fillet weld subjected to transverse loading, $\sigma_\perp$, is determined above all by the weld profile, including the toe and root notches, because of the strong notch effect, which is not a matter of the dimensions themselves but rather of the ratios of the dimensions, *e.g.* the ratio of weld to plate thickness. The actual weld profile must also be checked by measurement to ensure the correctness of the strength assessment.

A flat and rounded weld toe is advantageous, *i.e.* a concave fillet (produced by welding in a flat position), which can additionally be machined by grinding. Convex fillet welds are unsuitable in the presence of fatigue loading. The slope angle of a flat fillet weld should not exceed 45°. A weld with a flatter slope and unequal legs is particularly good. Deep penetration at the weld root has a particularly advantageous effect (increase of the actual weld thickness up to a factor of 1.3 is possible). Such penetration can be obtained by manual welding with deep penetration electrodes and by fully mechanised welding methods (submerged-arc and CO_2 processes). Too wide a root gap impairs strength because the actual weld throat thickness decreases and because gap closure in compression is prevented.

The weld toe can be dressed to increase strength in the same way as the butt weld, by toe remelting, shot or hammer peening, or grinding.

Weld defects play a considerably smaller part in fillet welds than in butt welds because the notch effect of the joint itself is so large that additional discontinuities are not significant. Particular care need only be exercised in the avoidance of larger lack of penetration and undercut defects. A large undercut can reduce the fatigue strength to about half. As the absence of root defects in fillet welds cannot be established by non-destructive testing, production measures are recommended for avoiding them, such as pre-welding a small root pass. Poor fit-up has a smaller effect on the fatigue strength of fillet welds than of butt welds.

Precisely because a fillet weld has a particularly strong notch effect, its removal at the design stage to areas of low basic stress or its replacement by a butt weld, has an especially beneficial strength-increasing effect. Single or double face lap joints are out of date.

As has already been mentioned for the single and double bevel butt joint, the notch effect as a result of stronger force flow diversion is also increased in a T joint with fillet welds subjected to transverse loading. The risk of failure at the weld toe of the transverse member must be investigated separately, using the bending stress in the transverse member as a basis.

The fatigue strength of a (continuous) fillet weld subjected to shear loading, $\tau_\parallel$, is less strongly reduced than when subjected to transverse loading, $\sigma_\perp$; because of the milder degree of force flow diversion with an identical weld and joint profile, $\gamma = 0.4\text{-}0.6$. However, these values are not applicable to the curved fillet weld of the common bar specimen with circular cross section (fillet welded joint between bar and base plate). Smaller values appear here for reasons which have not been clarified as yet.

The fatigue strength of a continuous fillet weld subjected to longitudinal loading, $\sigma_\parallel$ (double fillet weld on an I section girder, $\sigma_\parallel$ determined for the distance of the double fillet weld from the neutral bending plane) is reduced by a relatively large amount, $\gamma = 0.5\text{-}0.7$, in manual welding because a surface crater and a root slag inclusion occur at the electrode stop/start point. Butt joints, single bevel and double bevel butt joints are helpful here, enabling the root defect to be machined out and a weaker surface notch to be produced, or fully mechanised welding without electrode stop/start and with a particularly smooth weld surface (submerged-arc and CO_2 welding processes). A crater is nonetheless a problem in the latter case if there is an unplanned interruption of welding; the problem may be alleviated by tapered finishing of the weld end and subsequent welding over, in accordance with Ref.2. As a result of these measures the reduction can be improved to at least $\gamma = 0.7$.

When the specimen is subjected to longitudinal loading of the weld, the high residual (tensile) welding stresses in this direction have a negative effect. They reduce the tensile pulsating strength.

In the case of intermittent double fillet welds (staggered or non-staggered) on I section girders (found occasionally), $\gamma = 0.3$-0.5 is to be expected as a result of the weld end notch effect when subjected to pure bending, tensile or compressive loading. Intermittent welds are prohibitive with higher shear loading of the girder. A thinner continuous weld is considerably more advantageous than a thicker intermittent weld compared on the basis of identical amounts of weld material.

The fatigue strength of a discontinuous fillet weld, subjected to weld shear stress, $\tau_{\parallel W}$, is also very much reduced as a result of the weld end notch effect. The specimen of this type most frequently investigated is the lap joint with longitudinal fillet welds. It is known that in longitudinal fillet welds of the type under consideration, the weld shear force is distributed in such a way that the highest values appear at the ends of the welds (increase factor about 3). However, with current (over) dimensioning practice, fracture does not occur in the weld at all but in the base material, in the cover or base plate transverse to the weld end. The surface profile in the direction of the weld at the end crater with its uncontrolled shape is the decisive factor. These crater areas can be ground to round them for greater safety and strength. For practical purposes the design stress is specified as the endurable nominal stress in the weaker cover plate, as fracture of the cover plate is the more frequent and more probable event and as, from practical experience, the endurable nominal stresses in the base plate are dependent on the width and thickness of the cover plates. The reduction factor, $\gamma = 0.2$-0.4, is established with reference to the nominal stress of the cover plates, the higher values are reached only if the weld ends are dressed and the connection is relatively slender (ratio weld length to cover plate width large).

The dependence of the fatigue strength on the shape parameters already includes six independent dimensions with regard to the global shape even for this relatively simple component, which is assumed to be symmetrical in several ways: base and cover plate thickness, base and cover plate width, weld thickness and weld length. If the local shape of the weld is added, as is necessary for a more precise analysis of strength, the number mentioned above increases, quite apart from the combinations with transverse edge welds and rounded or bevelled cover plates which are possible in practice. Reliable general data for the fatigue strength concerned cannot possibly be given in these circumstances.

In the case of lap joints of section bars on flange plates it is usual to choose the weld lengths in reverse proportion to the distance from the bar centreline, so that they are adjusted to match the force flow. An increase in fatigue strength of about 10% could be proved for an angle section bar joint with unequal weld lengths compared with equal weld lengths.

With a single sided cover plate in the lap joint considerable superimposed bending occurs which causes a reduction in strength to about half.

A longitudinal fillet weld specimen with spliced plate joint reduces the number of geometrical parameters and achieves a further increase in symmetry where there is splicing on both sides, elimination of friction coupling and removal of the remaining transverse gap to the central cross section, which is subjected to evenly distributed low stresses. After failure in the weld has been overcome by increasing the thickness of the weld, only fracture of the base plate remains probable. Then the strength of the specimen is still dependent on its having a smooth profile and bevelling of the weld and rod ends.

The double plate longitudinal fillet weld specimen adopted for static loading tests should not automatically be considered suitable for fatigue strength tests because the notch effect of this joint is extremely high.

2.1.5   *Corner weld and edge weld*

Welds between two plates which meet perpendicularly or at an angle are considered as corner welds (Fig.27). Versions which are similar to butt welds (square, single V, double V, single bevel and double bevel groove) can be distinguished by the groove shape from versions with a single or double sided fillet weld. Both types of design lead to high levels of bending and notch effect, caused by the strong force flow diversion. A corner joint can be considered as a T joint with transverse force support to one side only. With regard to fatigue strength it behaves in a similar manner as a fillet weld (with even greater

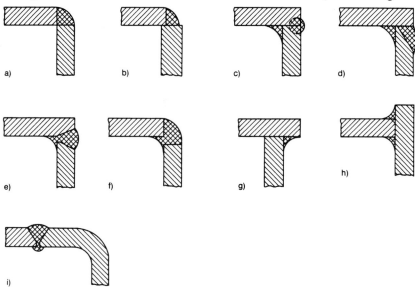

27   Corner joints: V butt weld (a, b), square butt weld made with deep penetration electrode from outside (c), single bevel and double bevel weld with fillet weld (d, e, f), single and double sided fillet welds (g, h), curved corner plus butt joint (i).

reduction). In the case of components subjected to fatigue loading, a single sided fillet weld can be used only if there is compressive support over the open root gap, with root stress relieving transverse plates, or with low level loading. The critical factor affecting the fatigue strength of the welds with root pass is the quality of the rounding at the internal fillet; in addition, a small root gap, or none at all (version similar to a butt weld) is an advantage. The most advantageous is forming the plate by bending and positioning of the butt weld outside the rounded corner. The quantitative strength analysis to be carried out in individual cases for the corner weld must include the bending and notch effect. There are no test results available on fatigue strength which can be generalised.

Single V edge welds and edge joint welds (Fig.28) are to be considered particularly notch effective when subjected to loading by $\sigma_\perp$ and $\tau_\perp$ perpendicular to the weld. There are no test results available on fatigue strength.

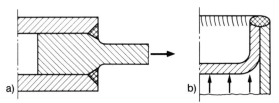

28   Fillet edge weld and flanged edge weld.

2.1.6   *Transverse, longitudinal and cover plate stiffeners*

Plates attached to a base plate by a fillet weld, but not integrated into the direct force flow of this base plate are designated 'stiffeners' because one of their purposes is stiffening of member cross sections. They cause a relatively strong local notch effect so that particular considerations are necessary here with regard to fatigue strength (unlike static strength).

A transverse stiffener can be attached on both sides, giving a more marked notch effect, or on one side, with a less marked effect (Fig.29). It can be the same width as the base plate (edges machined) or slightly narrower with the weld continued all around. Alternatively, a rectangular cover plate is also used with fillet welds on all sides on a wider base plate, allocated here to the transverse stiffener group because failure occurs at the toe of the transverse weld. The test results relate to tensile loading, $\sigma_\perp$, in the base material. The fatigue fracture occurs at the weld toe (at the corners for cover plates) and propagates into the base plate; if there is a relatively thin weld or a single sided weld it can also occur at the weld root.

The reduction factor for the transverse stiffener (or cover plate stiffener) is $\gamma = 0.4$-0.8. The high values apply only to the more advantageous single sided stiffener with additional shape improvement and toe dressing.

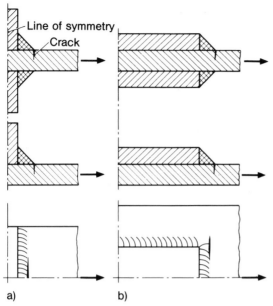

a)   b)

29 Transverse stiffeners: on both sides and one side (a), rectangular cover plate (b).

With regard to the influence of shape, the following measures increase strength: favourable weld profile with small toe slope angle and large radius of curvature (welding in flat position, grinding, TIG dressing), small weld and stiffener thickness in relation to the base plate thickness, and for a cover plate stiffener, small thickness, width and length of the cover plate and rounded cover plate corners.

A longitudinal stiffener can also be attached on both sides giving rise to a more severe notch effect, or on one side giving a less severe notch effect (Fig.30). It can be attached to the centre of the base plate or, more unusually, to its edges, in the same plane or at right angles to it. The longitudinal fillet welds

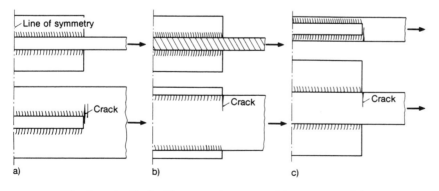

a)   b)   c)

30 Longitudinal stiffener: on the centre of the base plate (a), on the edge of the base plate, at right angles or in the same plane (b).

are frequently not closed at the end of the stiffener in test specimens because this reduces scatter in the results without making a significant change in the mean value. In the welded structure the welds are closed to avoid crevice corrosion. Test results relate to tensile loading, $\sigma_\parallel$, in the base plate, which is connected with the weld stresses, $\tau_{\parallel W}$ and $\sigma_{\parallel W}$ in the weld. In contrast to specimens with transverse stiffeners, specimens with longitudinal stiffeners include the influence of the high longitudinal residual stresses in the weld. The tensile pulsating fatigue strength of specimens in the as-welded condition is noticeably smaller than that of stress relieved specimens. Fatigue fracture in longitudinal fillet welds occurs in the base plate at the end of the stiffener either from the undercut near the end of the longitudinal weld or from the toe of the short transverse weld.

The reduction factor for the longitudinal stiffener, $\gamma = 0.3\text{-}0.6$, is below that for the transverse stiffener. Once again, the high values only apply to the more advantageous single sided stiffener with additional shape improvement and dressing.

With regard to the influence of shape, the following measures have a strength-increasing effect: reducing the thickness, length and height of the stiffener; bevelling and smoothing the weld toe at the stiffener end. Longitudinal stiffeners should be designed in such a way that they begin and end in areas with low basic stress.

2.1.7   *Bead on plate weld, arc strike, flame cut surface*

A bead on plate weld parallel or transverse to the basic loading, $\sigma_\parallel$ or $\sigma_\perp$, and arc strikes (Fig.31), can cause a reduction in strength to $\gamma = 0.6\text{-}0.7$ despite a relatively weak notch effect, which is a result of weld defects being generated. Welding of this type should therefore be avoided on components which are subjected to fatigue loading. Values of $\gamma = 0.9\text{-}1.0$ can be achieved by post-machining (grinding flush) as for a butt weld.

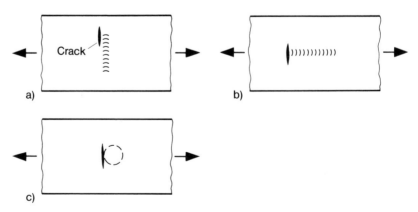

31   Bead on plate weld: transverse (a), longitudinal (b), arc strike (c).

Surface drag lines on flame cut surfaces have a strength-reducing effect, whilst compressive residual stresses and hardening of the surface have a strength-increasing effect. The resulting reduction factor is about $\gamma = 0.65\text{-}1.0$ depending on the quality of the flame cutting. The high values imply that a flame cut surface is not more damaging than a mill finish. The low values occur if an edge is too strongly fused (hot cracks) or if defective burning through takes place with slag residues and subsequent cracking of the edges. It is then worth machining the edges, but it is not worth removing the drag lines over the whole surface when the flame cutting is of high quality. Relieving of compressive residual stresses in the flame cut surface by machining, cold working or annealing has a strength reducing effect. In the case of flame cut holes, there is an additional strength reducing effect introduced by the notch stress concentration at the hole.

2.1.8   Survey of reduction factors

The reduction factors given in sections 2.1.2-2.1.7 are compiled in Table 1. A few missing values (*i.e.* not documented by fatigue tests) for shear loading are estimated and supplemented on the basis of notch stress considerations.

When making a comparison with the numerical values given by other authors, a check must be made on the extent to which the reference stress is identical as regards type and magnitude, for what materials, mean stresses, and numbers of cycles the values are available, whether and how scatter is treated,

Table 1. Survey of reduction factors, γ, low strength structural steels, tensile pulsating fatigue strength, $\sigma_P = 240$ N/mm^2, failure probability, $P_f \approx 0.1$

Type of weld	Loading on base material		
	$\sigma_\perp$	$\tau_\parallel$	$\sigma_\parallel$
Load-bearing welds			
Butt weld	0.5-0.9	0.6-0.9	0.7-0.9
Single and double bevel welds	0.4-0.7	0.5-0.7***	0.6-0.8
Fillet weld	0.3-0.5*	0.4-0.6	0.5-0.7
			0.3-0.5**
			0.2-0.4[†]
Corner weld	0.3-0.5[††]	0.4-0.6[††],***	0.5-0.7
Non-load bearing welds			
Butt welded gusset plate	0.2-0.3		
Fillet welded transverse, longitudinal			
and cover plate stiffener	0.4-0.8	0.4-0.7***	0.3-0.6
Bead on plate weld and arc strike	0.6-0.9	0.6-0.9	0.6-0.9

*	Tensile loaded cruciform joint and lap joint with transverse fillet weld
[†]	Longitudinal fillet weld in lap joint
**	Interrupted double fillet weld on I section girder
[††]	Value also considerably smaller depending on support condition
***	Estimate on the basis of notch stress analysis.

whether endurable or permissible nominal stresses are related to each other, and whether the nominal stresses refer to the base material or the weld.

When using the reduction factors in the course of calculation and dimensioning, the problem of scatter and the further influence of parameters mentioned above, should be taken into account, as they can change the reduction factor. In addition, possible welding defects should be taken into consideration.

2.1.9   *Influence of welding defects*

Welding defects are primarily geometrical imperfections in the welded joint caused by manufacture: cracks, pores, solid inclusions, incomplete fusion and unsatisfactory penetration, undercuts, misaligned edges as well as further shape imperfections (classification, designation and explanation as per DIN 8524[39] or IIW Doc 340-64[42]). The minimum requirements for the quality of manufacture are subdivided into four groups of requirements for butt welds and three groups of requirements for fillet welds (definition corresponding to DIN 8563, see section 7.5). This definition is supplemented by manufacturing standards which are specific to the technical fields involved. However, the groups of requirements cannot be correlated with the notch classes of fatigue strength because defects take effect in very different ways. The degree of fulfilment of the requirements ('manufacturing quality') is established by inspection methods. In addition to checking the external shape with the naked eye, supplemented by positional measurements, the less obvious smaller defects inside and on the surface of welded joints are found by special non-destructive testing methods: dye penetrant tests, magnetic particle tests, radiographic and ultrasonic tests, and acoustic emission tests, see Ref.6.

Numerous investigations have been carried out on the fatigue strength of welded joints with defects. In investigations of this type it is difficult to produce defects in a reproducible form as regards their type, magnitude and position. There is also disagreement as to whether effects of defects should be investigated in unmachined welds or welds which have been ground flush. The results which are capable of generalisation are summarised below and attention is drawn to Ref.2, 40 and 41 with regard to individual investigations and their interpretation. The reduction of fatigue strength by small welding defects produced by normal welding practice is already included in the reduction factors stated in the section above determining their lower limit values.

Cracks

Cracks are planar separations in the material with a correspondingly sharp crack tip. They tend to occur transverse or parallel to the weld, but are occasionally star shaped without a preferred direction (Fig.32). The end crater is particularly at risk of cracking as a result of the localised accumulation of

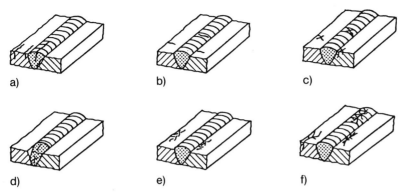

32 Welding defects, hot and cold cracks: longitudinal and transverse (a, b), radiating (c), crater (d), group of disconnected cracks (e), branching cracks (f).[39]

microdefects combined with high residual stress. Macrocracks are recognisable with the naked eye, microcracks only with magnifying equipment.

Cracks occur after welding and thermal or mechanical post-treatment for various reasons, which are expressed by their designations. Hot cracks, including solidification and liquation cracks, are initiated at high temperatures where solid and liquid phases occur together. Cold cracks, including hydrogen-induced, hardening, ageing and precipitation cracks, occur at low temperatures after the material's ductility is exhausted. Lamellar tears occur in rolled out layers of segregation when there is loading present in the direction of the plate thickness. Cracks markedly reduce fatigue strength, particularly under transverse loading. There are no systematic investigations available of specimens with reproducible hot and cold cracks. However, a quantitative estimate of strength on the basis of fracture mechanics is possible (see section 9.5). The reduction in strength is dependent above all on the size of crack so that quality standards for manufacturing mainly limit this parameter. Hot and cold cracks can be avoided by appropriate manufacturing measures, among others, by reducing the degree of restraint, by checking solidification and cooling speeds and by excluding hydrogen.

Pores and shrinkage cavities

Pores are cavities (frequently spherical) with gas residues, which cannot escape when solidification takes place rapidly. Depending on their size (diameter 0.01-1.00mm), shape, and number, these are described as single pores, linear porosity, localised porosity, uniformly distributed porosity and wormhole pores (Fig.33). Cavities as a result of shrinkage during solidification are designated shrinkage cavities. Pores occur in welding because of too high a sulphur content in the base material and the filler metal, because of too high a moisture content in the electrode coating and because of nitrogen entering through the arc shield. Welding through primer paints can also cause pores, to a dangerous extent in fillet welds. Systematic investigations have been carried out on the effect of porosity on fatigue strength in butt and fillet welds. Some of

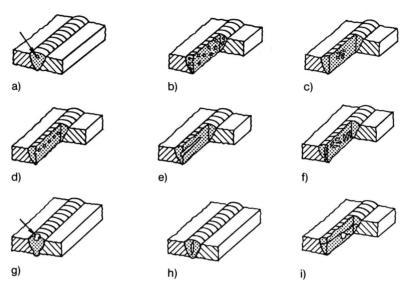

33 Welding defects, cavities: gas pore (a), uniformly distributed porosity (b), localised porosity (c), linear porosity (d), elongated cavity (e), wormhole pore (f), surface pore (g), shrinkage cavity (h), crater pipe (i).[39]

these determined the pore proportion at the failure cross section from the darkening of the X-ray film used in radiography. For transverse welds, the investigations show a steep decrease in fatigue strength at first, followed by a flatter one with an increasing proportion of pores in the failure cross section;[43] in contrast, in a longitudinal weld the influence of porosity is only minor.[44]

Solid inclusions

Solid inclusions are foreign materials which are embedded in the weld, in the case of steels they are usually slags, in light alloys usually foreign metals (*e.g.* tungsten, copper), which can appear singly, in lines or in local clusters. Slag inclusions occur when slag on a previous weld pass was not removed adequately. Slag inclusions can be produced in a reproducible manner for fatigue testing. The continuous slag layer of a weld pass is only removed (by grinding or milling) where no slag inclusions are to occur under the cover pass. In a different method, slag pockets are produced at the groove face as a result of moving the electrode in a particular way. Slag inclusions reduce fatigue strength. The residual compressive stresses inside multilayer welds have a favourable effect on fatigue strength with inclusions, and stress relieving heat treatment has a negative effect.

Lack of fusion defects and inadequate penetration

Unfused interfaces between filler metal and base metal (groove face) or between different layers of the filler material are characterised as fusion defects

(Fig.34). The most frequent cause of fusion defects is foreign matter on the surface to be welded, slag or mill scale for steel, oxide film for light alloys. In the case of steel an incorrect welding current can be the cause, in the case of light alloys it can be too large a molten pool. Fusion defects reduce fatigue strength in a similar way to cracks.

a)   b)   c)   d)

34   Welding defects, lack of fusion and penetration: lack of fusion at the weld root (a), lack of inter-run fusion (b), lack of sidewall fusion (c), lack of penetration (d).[39]

Inadequate penetration means that the weld pool does not reach the weld root, therefore a root gap is left, intentionally or accidentally. Poor fit-up and use of an incorrect welding method for the type of joint are the cause of unintentional lack of penetration.

Lack of fusion and penetration defects can be generated in a reproducible manner for fatigue tests, *e.g.* as in Fig.35. The edges of two plates with double sided, quarter circle grooves are abutted against each other at the root face. First and second weld passes are laid over the root face on both sides. One of these layers is sectionally removed together with the unwelded root face. Later, the remaining root face has the effect of a lack of fusion or penetration defect. The grooves are subsequently completely filled by further layers. Surface defects can also be reproduced in this way. Using a different method, a discontinuous root face is used with the initial groove. It has been shown using

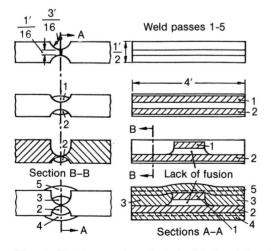

35   Artificial generation of a lack of fusion defect in a butt weld specimen, after Gurney.[2]

steel and light alloy specimens with appropriate defects that fatigue strength decreases steeply both with a reduction of the cross sectional area and with an increase in size of the defect in the direction of plate thickness. Additionally there was a noticeable influence from residual stress. Lack of fusion and penetration defects behave in a similar manner to cracks.

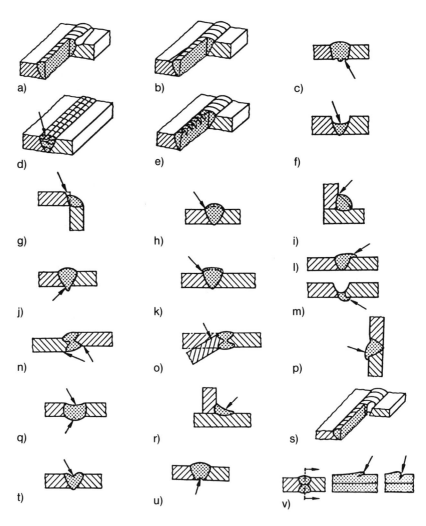

36 Welding defects, shape imperfections: continuous undercut (a, b, c), inter-run and transverse undercut (d, e), underfilling of groove (f), melting of the edge (g), excessive convexity and penetration (h, i, j), incorrect weld profile (k), overlap (l, m), linear and angular misalignment (n, o), sagging (p, q, r), burnthrough (s), top bead depression (t), root concavity (u), poor restart (v).[39]

Undercuts, misalignment and other shape imperfections

Undercuts are groove shaped depressions at the weld toe, at the root of the weld (on both sides) or between weld beads. The cause of undercuts is incorrect welding (*e.g.* use of wrong process parameters). Undercuts reduce the cross sectional area and increase the notch effect. The relative reduction is about double the magnitude of the ratio of the notch depth to plate thickness.[46]

Axial and angular misalignment arises as a result of inaccuracy in assembly and of welding distortion. The effect of a defect of this type on fatigue strength can be estimated on the basis of the changes in stress distribution caused as a result of the deviation in shape. Thus misalignment produces superimposed bending stresses and additional notch stresses. Fatigue strength is reduced considerably.

A large number of other shape imperfections are listed in Fig.36 together with those mentioned above. Their effect on fatigue strength can be estimated on the basis of their notch effects.

2.2   Multiply loaded joints

Loading by more than one of the basic stress components, $\sigma_\perp$, $\sigma_\parallel$, $\tau_\parallel$, is designated 'multiple loading'. In pulsating cyclic loading, the relevant stress components may be applied in or out of phase. Test results are only available for loading in phase. It is known from investigations on base materials that an additional reduction can occur in the presence of out of phase loading.

The strength behaviour of a butt or fillet weld is dependent on direction (anisotropic) as can be seen from the differing reduction factors for different loading directions. Orthogonal anisotropy of strength is the consequence of an orthogonally aligned weld geometry. The distortion energy hypothesis for orthogonal strength characteristics (it does not matter whether this is caused by orthotropy of the material or the geometry) can be transferred as follows to the behaviour of a welded joint (the reduction factors in the denominator refer to the basic stress components above them in the numerator for simple loading). σ_{eq} is the endurable equivalent stress in the base material:

$$\sigma_{eq} = \left[\left(\frac{\sigma_\perp}{\gamma_\perp} \right)^2 + \left(\frac{\sigma_\parallel}{\gamma_\parallel} \right)^2 - \frac{\sigma_\perp \sigma_\parallel}{\gamma_\perp \gamma_\parallel} + 3 \left(\frac{\tau_\parallel}{\gamma_\tau} \right)^2 \right]^{1/2} \tag{5}$$

Using the endurable stresses for simple loading, equation (5) can also be expressed as follows:

$$1 = \left(\frac{\sigma_\perp}{\sigma_{\perp en}} \right)^2 + \left(\frac{\sigma_\parallel}{\sigma_{\parallel en}} \right)^2 - \frac{\sigma_\perp \sigma_\parallel}{\sigma_{\perp en} \sigma_{\parallel en}} + \left(\frac{\tau_\parallel}{\tau_{\parallel en}} \right)^2 \tag{6}$$

These relationships are represented by a diagonal ellipsoid in the $\sigma_\perp$-$\sigma_\parallel$-$\tau_\parallel$ space, whose intersection with the $\sigma_\perp$-$\sigma_\parallel$ plane ($\tau_\parallel = 0$) is illustrated in Fig.37. These relationships are confirmed by earlier fatigue tests[47,48] on butt and fillet welds in pressure vessels (adopted at that time by the Swiss Design Code) and

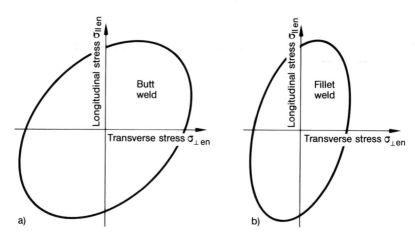

37   Endurable fatigue loading under combined action of $\sigma_\perp$ and $\sigma_\parallel$: butt
welded joint with $\gamma_\perp = \gamma_\parallel$ (a), fillet welded joint with $\gamma_\perp = 0.6\gamma_\parallel$ (b).

by more recent tests on specimens with a welded stiffener.[3] Tests on double
fillet welds of bending girders subjected to $\sigma_\parallel$ and $\tau_\parallel$ and on circumferential butt
and fillet welds of tubular specimens subjected to $\sigma_\perp$ and $\tau_\parallel$ suggest slight
modifications to the formulae above in order to make a more precise
adjustment to specific test results, *e.g.* according to Ref.3:

$$1 = \left(\frac{\sigma_\perp}{\sigma_{\perp en}}\right)^v + \left(\frac{\sigma_\parallel}{\sigma_{\parallel en}}\right)^2 - \lambda\frac{\sigma_\perp\sigma_\parallel}{\sigma_{\perp en}\sigma_{\parallel en}} + \left(\frac{\tau_\parallel}{\tau_{\parallel en}}\right)^2 \qquad (7)$$

The adjustment factors have values $v = 1.0\text{-}2.0$ and $\lambda = 0.85\text{-}1.0$.

The phenomenological basis of the selected formulation for the distortion
energy hypothesis is the fact that the initial fatigue cracks form on slipbands, the
occurrence of which is controlled by the local distortion energy, which can be
proved. The transference of the formulae from local to global conditions is well
grounded but deviations are to be expected.

2.3   Influence of mean stress

2.3.1   *The influence of mean stress from loading*

Fatigue strength is dependent on mean stress. The previous data described
above were for tensile pulsating fatigue strength, σ_P, where the tensile mean
stress is $\sigma_m = \sigma_P/2$. Retaining the previous definition of fatigue strength,
namely the endurance limit at 2×10^6 cycles, what is the fatigue strength under
loadings other than pulsating tension?

Fatigue strength at different mean stresses is represented in different ways
(see section 1.4). In machinery construction the diagram according to Smith is
used, plotting the upper and lower stresses against the mean stress. In

structural steel engineering the diagram after Moore/Kommers/Jasper/Pohl is preferred, plotting the endurable upper stress against the limit stress ratio[*], R: (linguistically more precise 'extreme value stress ratio'), *i.e.* the ratio of the lower stress, σ_l, to the upper stress, σ_u:

$$R = \frac{\sigma_l}{\sigma_u}, (\sigma_m \geqslant 0) \tag{8}$$

The advantage of this manner of representation is that alternating fatigue strength $(R = -1)$, pulsating tensile strength $(R = 0)$ and static tensile strength $(R = +1)$ appear vertically one above the other, each independent of their numerical values, which makes comparison easier. The disadvantage is that a separate plot is necessary for compressive mean stress. Another manner of representation in structural steel engineering is to plot the endurable upper stress against the lower stress ('stress house'); this is less commonly found nowadays. The manner of representation generally preferred for fundamental investigations is Haigh's method, which is widely used in the aircraft industry, in which the endurable stress amplitude, σ_A, is plotted against the mean stress. The endurable upper stress, σ_{Up}, results from the endurable stress amplitude, σ_A, according to:

$$\sigma_{Up} = \frac{2}{1-R}\sigma_A \tag{9}$$

Stüssi[15] has proposed plausible functions for the relationship $\sigma_{Up} = f(\sigma_m)$ or $\sigma_{Up} = f(R)$ for welded joints (including the further relationship $\sigma_{Up} = f(N)$ of the S-N curve for fatigue strength for finite life). The important equation detail for notched components including welded joints — the general theory cannot be explained here — states that the reduction ratio

$$\varphi = \frac{\sigma_W}{\sigma_B} \tag{10}$$

of the endurable nominal stress of the welded joint (index W) and the base material uninfluenced by welding (index B) is dependent on the mean stress σ_m in a linear manner, so that the minimum values, φ_A, appear at $\sigma_m = 0$ (alternating fatigue strength) and rise (in a straight line) to the maximum value $\varphi = 1.0$ at $\sigma_m = \sigma_U$ (ultimate tensile strength). The reduction ratio φ is identical to the reciprocal of the fatigue notch factor, K_f, (see section 8.1.2) in so far as the strength reduction in the welded joint can be interpreted as a notch effect:

$$\varphi = \frac{1}{K_f} \tag{11}$$

For demonstration purposes, a most suitable representation appears to be the one which illustrates a few results from older fatigue tests by the German Railways on typical welded joints in structural steel engineering in an evaluation.[49] The results refer to stresses endured for infinite life in welded joints made in St 37 (static strength values more like St 44). The joint types, investigated under pulsating tensile loading, are shown in Fig.38 together with

[*] Limit stress ratio, R instead of S, in accordance with Ref.1 has been introduced internationally.

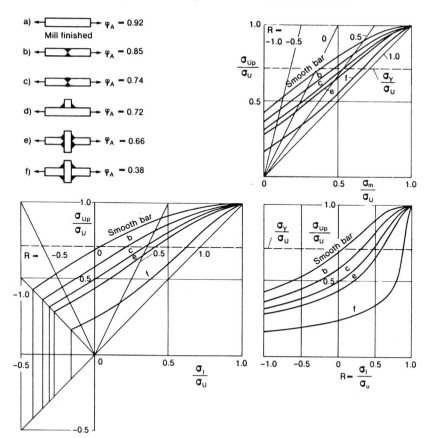

38 Fatigue strength diagrams for base material and welded joints, St 37, after Stüssi.[49] Smooth bar identical to polished specimen: curves a and d, omitted for a better general view, can be constructed graphically by interpolation in accordance with above values φ_A.

their reduction ratio, φ_A, under alternating loading and with all the fatigue strength values plotted using the alternative diagrams described above. As stresses in excess of yield stress, σ_Y, are not generally used in practice, the strength curves are frequently cut off at the yield stress, indicated by a broken line.

The question arises as to whether and to what extent the reduction factors, γ, quoted for pulsating loading, change dependent on the mean stress, where the ratio of the fatigue strength of the welded joint to the fatigue strength of the mill finished base material for the same limit stress ratio, R, is denoted by the reduction factor, γ. Obviously γ and φ are largely identical for R = 0. The difference lies only in the reference to the base material with polished surface for φ, which produces the factor 270/240 = 1.13 for St 37.

$$\gamma = 1.13\varphi, (R = 0) \tag{12}$$

The answer to the question above is found most quickly by evaluation of the curve values in Fig.38, whilst an algebraic representation proves to be rather complicated. The difference in the value of the reduction factor, γ, between alternating and pulsating loading is 0.04 at the most. The difference in the reduction factor for R = 0.5 is only slightly larger. Therefore the approximation to $\gamma = 1.0$ only takes place close to the tensile strength. In contrast, a more accurate rough calculation for $0.2 \leqslant \gamma_P \leqslant 0.8$ results in:

$$\gamma_A \approx \gamma_P - 0.1 \tag{13}$$

The γ values are thus somewhat dependent on the mean stress.

The reference stress of γ in the presence of alternating loading for mill finished base material, St 37, is

$$\sigma_A = 145\,\text{N/mm}^2 \tag{14}$$

from which $\sigma_A/\sigma_P = 0.6$ is deduced as a guide rule.

The question of the influence of compressive mean stress has not yet been considered; it is omitted when plotting takes place against the (tensile) limit stress ratio and is not shown in the other diagrams in Fig.38 either. The fatigue strength of welded joints for compressive mean stress $\sigma_m < 0$ has only been determined on very rare occasions because it is higher than for the corresponding tensile mean stress (therefore on the conservative side) and because in practice high (global) compressive mean stresses are not acceptable for reasons of stability.

The interrelationships one should principally be aware of both in welded joints and in the base material can best be explained using Haigh's diagram (Fig.39). Starting from the ultimate compressive strength, σ_{Uc}, which is higher than the ultimate tensile strength, σ_U, both under static loading, tensile and compressive limit curves of the same type are drawn for the loading amplitudes. The latter intersect well within the compressive range; in consequence the compressive pulsating strength is generally determined from the tensile curve and thus is significantly higher than the tensile pulsating strength. The

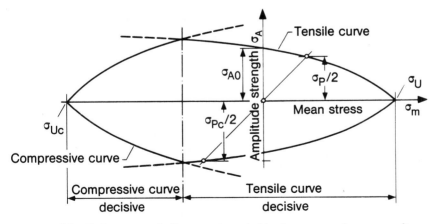

39 Fatigue strength diagram, expanded in the compression range, after Stüssi.[15]

difference is particularly marked for severely notched components and for cast grain materials, and thus especially for welded joints too.

Plotting over R in the compressive mean stress range requires a special definition. When defining R according to equation (8) it was assumed that $\sigma_m \geqslant 0$. For $\sigma_m < 0$, limit stress ratios are produced with the identical numerical value as regards magnitude and sign if the loading cycle is reflected around zero into the compressive stress range and the ratio of the upper stress to the lower stress is now understood as the compressive limit stress ratio, R_c:

$$R_c = \frac{\sigma_u}{\sigma_l}, (\sigma_m < 0) \tag{15}$$

The curves of the endurable lower stress, σ_L over R_c in the compressive range, are generally higher according to the above statements than the curves of the endurable upper stress, σ_{Up} over R in the tensile range.

2.3.2   Influence of residual stress

Finally, in welded structures, there remains the question of the influence of residual stresses, in particular the influence of residual welding stresses on fatigue strength. A distinction should be made here between three groups of components according to the condition of their manufacture:

— Welded components with high tensile residual stresses (the stresses parallel to the weld are particularly high for unalloyed steels, namely at the yield stress level);

— Welded components which are stress relieved, e.g. by heat treatment;

— Components with deliberately introduced high compressive residual stresses, particularly at the failure initiation point (by means of the methods described in section 3.4, supplemented by a favourable welding sequence).

It should be noted that the residual stresses are altered after the first load or overload cycle by local yielding with stress redistribution. That is also the reason why, in the range of lower cycle fatigue (for $N \leqslant 10^5$), no difference can be established between the differently manufactured welded components above. On the other hand, in the high cycle fatigue range the residual stress state can be altered because of cyclic relaxation.

Components with high tensile residual stresses and superimposition of the stresses from the applied load in the same direction largely show endurable stress amplitudes of constant low magnitude in the whole applied mean stress range. The fatigue strength can be further reduced by a corrosive environment. Stress relieving by heat treatment only increases strength significantly under compressive applied mean stresses. In contrast, deliberately introduced compressive residual stresses increase the stress amplitude in the whole applied mean stress range. The data above apply to the high cycle fatigue strength. An explanation of the increases in strength based on crack closure under compressive total stress is given in section 9.4.

The distortion resulting from welding is more important than welding residual stresses with respect to the fatigue strength of the components. Distortion produces local secondary stresses in the case of external loading which have a fatigue strength reducing effect.

3

Fatigue strength for finite life and service fatigue strength of welded joints

3.1 Fatigue strength dependent on number of cycles

The influence of the number of cycles N on the fatigue strength for a fixed mean stress, lower stress or limit stress ratio is represented by the S-N curve. The endurable upper stress, σ_{Up}, or the endurable stress amplitude, σ_A, is plotted dependent on the number of cycles to fracture, N, (best based on a logarithmic scale). The fatigue strength decreases from the tensile strength, σ_U, (or compressive strength σ_{Uc}) at N = 0.5 (static loading and unloading) to the endurance limit, $\sigma_F = \sigma_{A0}$, Fig.1, passing through the ranges of low cycle fatigue strength (0.5 < N ⩽ 10^3), medium cycle fatigue strength (10^3 < N ⩽ 10^5) and high cycle fatigue strength (10^5 < N ⩽ 2 × 10^6). The endurance limit of welded joints made of structural steel is reached above N = 10^7 but is set at the standardised value of N = 2 × 10^6 for more economical testing. This substitute endurance limit is thus about 20% higher than the actual endurance limit.

3.1.1 *Medium cycle and high cycle fatigue strength*

Medium cycle and high cycle fatigue strength includes cycles to failure in the range 10^3 < N ⩽ 2 × 10^6, *i.e.* right up to the endurance limit. Plastic deformations are localised and remain small, so that, unlike the relation in the low cycle fatigue strength range, the global stresses and strains remain proportional to each other.

The results of high cycle fatigue strength tests have a large scatter, so that only statistical evaluation of a sufficiently large number of specimens provides results which can be generalised.[50-52] The scattering numbers of cycles to fracture for preset stress levels are generally assumed as Gaussian distributed in logarithmic scale. S-N curves are determined for different failure probabilities, P_f. With welded joints made of structural steel, in practice, the normalised representation[31,53,54,63] has proved successful, Fig.40, for high cycle and endurance limit curves (for N ⩾ 10^4), plotted in logarithmic scales. The curves of the endured stress amplitudes, σ_A, for probabilities P_f = 0.1, 0.5

and 0.9 (which correspond to survival probabilities P_s = 0.9, 0.5 and 0.1) appear as uniform straight lines for any desired welded joint and for any desired limit stress ratio, R, between − 1.0 and + 0.5. It is a precondition that the upper stress does not exceed the yield stress, σ_Y (or $\sigma_{0.2}$). Normalisation has been achieved by referring the endurable stress amplitude, σ_A, to the endurance limit, σ_{AF} (P_f = 0.5, N = 2 × 10^6), of the joint in question. This normalised S-N diagram has proved its worth for low and high tensile structural steels, including service fatigue strength tests[55,56] on specimens with different notch effects, Fig.41, for limit stress ratios close to R = 0, R = − 1 and R = +0.5. The ratio of the stress amplitudes of the pulsating and alternating fatigue strengths of welded joints made of structural steel is given as constant by $(\sigma_A)_{R\,=\,0}/(\sigma_A)_{R\,=\,-1}$ = 0.6.

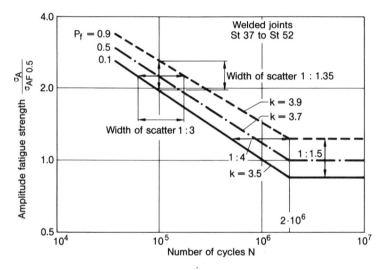

40 Normalised S-N curve with scatter bands for welded joints made of structural steels St 37 to St 52, any notch classes and limit stress ratios, after Haibach.[53]

The straight line for the fatigue strength for finite life is given by the formula below:

$$\sigma_A = \left(\frac{N_F}{N}\right)^{1/k} \sigma_{AF} \tag{16}$$

In this formula k = tan α, whereby α indicates the angle of the fatigue strength for finite life line with the vertical in equalised double logarithmic scales. The fatigue strength line with P_f = 0.1 which is frequently considered to be suitable in practice (a safety factor is additionally introduced) is characterised by k = 3.5. P_f = 0.9 is more suitable for some scientific investigations because interference effects are suppressed to a higher degree.

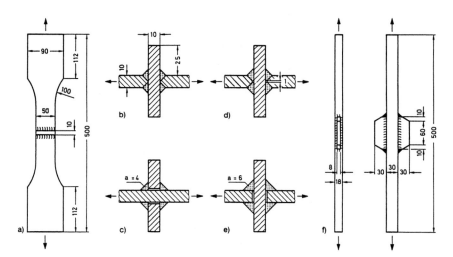

41   Welded joints of varying notch severity used for Wöhler and service
fatigue tests, after Bierett:[56] Cruciform joint with double bevel butt weld,
with double bevel butt weld with root gap and with fillet weld (a, b, d, e),
transverse stiffener (a, c) and longitudinal stiffener (f).

The equation above implies, with the k value stated, that a small change in stress
amplitude results in a large change in the number of cycles endurable (more
precisely: half the stress means approximately 10 times the number of cycles).
The equation allows estimates of the stress reduction required for a given
increase in life or reduction in failure rate for a mass produced component.

The normalisation of the S-N curve for welded joints which has been
illustrated includes the fact that both the type of joint and the limit stress ratio
change the endured stress amplitudes, σ_A and σ_{AF}, in the same ratio. It follows
from this that the reduction factors, γ, introduced for the purpose of survey on
endurance limits can also be transferred to the high cycle fatigue strengths
independently of R.

A peculiarity, which should be taken note of in practice, occurs in welded
joints where high (tensile) residual welding stresses and (tensile) applied
stresses in the same direction are superimposed, for instance, in a longitudinal
stiffener. In the endurance limit and high cycle fatigue strength ranges, the
residual stresses are not at all, or only slightly, relieved because the sum of the
load stress and the residual stress does not reach the yield stress or only reaches
it at a few points. The fatigue strength is reduced corresponding to the
superimposed (tensile) mean stress. A strong relief of residual stress takes place
in the area of the medium cycle fatigue strength range as a result of local
yielding. The fatigue strength is increased corresponding to the reduced
(tensile) mean stress. As a result of the different mean stresses, an S-N curve is
determined for the welded joint, which consists of a lower part affected by
residual stress and an upper part independent of residual stress with a
transition area, Fig.42.

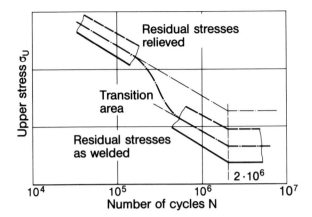

42 S-N curve for superimposed applied stresses and residual stresses acting in the same direction, welded joint as longitudinal stiffener, curved transition as a result of residual stress relief in the medium cycle fatigue strength range, after Olivier and Ritter.[37]

3.1.2   Low cycle fatigue strength

The low cycle fatigue strength includes the range of $N \leqslant 10^3$ endured cycles, in which the S-N curve runs largely horizontally, close to the ultimate tensile strength, σ_U. Extensive plastic deformation occurs mainly in the notch area which, however, remains constrained in an elastic manner, both by the less deformed adjacent areas and by the elastic strain component in the strongly plastically deformed area itself. The key to a quantitative description of the low cycle fatigue of notched components, also called 'plastic fatigue' lies in the total strain at the notch root (elastic plus plastic component) in combination with the cyclic stress-strain curve of smooth reference specimens. The relevant equations are set out together in section 8.3.2 as part of the chapter on the notch effect of welded joints. A continually decreasing strain S-N curve (*e.g.* Fig.13) is used instead of the stress S-N curve which begins horizontally and then falls. The fact that with $\Delta\varepsilon = 0.1, N > 10^3$ cycles can be achieved without cracking, may be used as a reference value for structural steel.

Test results on welded joints in the range of the low cycle fatigue strength, in some cases at elevated temperature, are only available in small numbers. They refer to butt joints subjected to longitudinal and transverse loading, transverse and longitudinal stiffeners and specimens with simulated weld seam microstructure, in each case using high tensile or heat resistant structural steels.[2,57] The welded joints do not quite achieve the lifetime of the parent materials (factor of 0.8 to 0.2) in the as-welded condition but this can largely be reversed by subsequent heat treatment. The low cycle fatigue strength is greatly reduced by environmental hydrogen. The influence of weld position (transverse or longitudinal), post-weld heat treatment and weld defects on the low cycle fatigue strength is reviewed in Ref.58 including the influence of corrosion and higher temperatures.

3.1.3 Staircase and Locati method, arcsin $\sqrt{P}$ transformation

In the current Wöhler method, series of specimens are tested to failure using stepped levels of stress and the resulting numbers of cycles are evaluated statistically.[50-52] In tests on the fatigue strength for finite life and endurance limit of welded joints, the staircase method, the Locati method, and the arcsin $\sqrt{P}$ transformation are also used occasionally.

In the staircase method the (0.5 survival probability) fatigue strength (or another damage occurrence e.g. crack initiation) is determined for a given number of cycles; thus, this is a reversal of the Wöhler method, where the number of cycles endured is determined for a given level of stress. First a level of stress is estimated for the desired number of cycles. The first specimen is tested at this stress level. If it fails before the desired number of cycles is reached, the stress is lowered for the next specimen. If it fails after the desired number of cycles, the level of stress is raised. Thus the stress on the next specimen is always determined by the stress and test result of the previous one. To make this a systematic procedure, equidistant stress levels are preset for the levels of loading on the specimens, starting from the first stress estimate. If the increment between stress levels is sufficiently small, the results are scattered on either side of a mean value which can be determined by statistical methods.

Mean value and scatter are determined by

$$\bar{\sigma} = \sigma_{i=0} + \Delta\sigma \left(\frac{\Sigma iH_i}{\Sigma H_i} + \frac{1}{2} \right) \tag{17}$$

$$S = 1.62\Delta\sigma \left[\frac{\Sigma H_i \Sigma i^2 H_i - (\Sigma iH_i)^2}{(\Sigma H_i)^2} + 0.029 \right] \tag{18}$$

with sample mean stress, $\bar{\sigma}$; sample standard deviation, S; stress increment, $\Delta\sigma$; number of stress level, i; frequency of stress level, H_i.

An example of an evaluation from Gurney[2] is represented in Table 2, from which, after insertion into equations (17) and (18) the values $\bar{\sigma} = 138$ N/mm^2 and $S = 12.6$ N/mm^2 are calculated.

The advantage of the staircase method is that the test concentrates automatically on the sample mean value, which increases the accuracy of the result. The disadvantage is the relatively large number of specimens to

Table 2. Example of evaluation by the staircase method, after Gurney[2]

Stress level, i	Stress, σ, N/mm^2	Sequence and result of tests: × = broken before 2 × 10⁶ cycles ○ = unbroken at 2 × 10⁶ cyles	Summation broken	Summation unbroken	Frequency of unbroken specimens H_i	Frequency of unbroken specimens iH_i	Frequency of unbroken specimens i^2H_i
3	165	×	1	0	0	0	0
2	150	○ × × × ×	4	1	1	2	4
1	135	× ○ ○ × ○ ×	3	3	3	3	3
0	120	○ ○ ○	0	3	3	0	0
Total			8	7	$\Sigma H_i = 7$	$\Sigma iH_i = 5$	$\Sigma i^2H_i = 7$

determine a single result point with scatter (according to the method's inventor 40 to 50 specimens). In particular, the potential for obtaining further information from the specimens which did not fail is not utilised as they are not taken to the point where they fracture. There is thus little to recommend the staircase method in preference to the current Wöhler method.

In accordance with the Locati method, the endurance limit in particular, but the whole S-N curve also, should be able to be determined by a single specimen, Fig.43. An S-N curve is estimated together with two parallel lines corresponding to the range of scatter. Commencing with a stress level somewhat below the endurance limit being sought, loading blocks which have a constant number of cycles and a stress level which is increased step by step are imposed on a single specimen until fatigue fracture occurs. Linear damage accumulation calculations are carried out for the loading blocks on the basis of the three estimated S-N curves, so that a damage total can be allocated to each. The S-N curve which produces the damage total 1.0 is considered to be apt. For this end the damage totals and endurance limits of the three S-N curves are interpolated directly.

The Locati method assumes three preconditions which are largely unfulfillable. The gradient of the S-N curve must be known in advance, the linear damage accumulation hypothesis must be applicable and it must be possible to ignore scatter in fatigue strength. As these preconditions cannot be fulfilled, the Locati method is unsuitable for qualified investigations.

In the arcsin$\sqrt{P}$ method,[59] the Gaussian distribution is substituted by an arcsin square root function, where P is the failure probability. This transformation has proven to be a robust procedure especially for determining

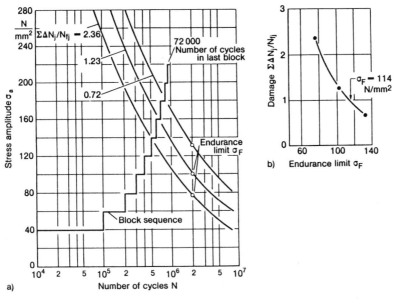

43 Locati method for determining the endurance limit, after Gurney:[2] block program and hypothetical S-N curves (a), interpolation of the damage totals and endurance limits (b).

the fatigue strength for infinite life. The failure probability $P = r/n$ with r number of fractured specimens and n total number of specimens on the stress level considered (in the endurance limit range) is transformed according to $\arcsin\sqrt{P}$ resulting in a linear dependency of the stress levels on the transformed failure probabilities. Contrary to the Gaussian distribution, the stress levels for $P = 0$ and $P = 1.0$ have realistic finite values.

3.1.4   *Maennig's boundary line method*

In contrast, Maennig's more recent boundary line method[60,61] can be considered a more efficient method of determining the endurance limit and the fatigue strength for finite life. It takes its name from the two boundary lines of the scatter band of the failures (or the equivalent damage occurrences) in the S-N diagram compared with the lower area of perfect strength and the higher area of complete absence of strength. The representation, Fig.44, is applicable to metal alloys with volume centred microstructure (*e.g.* structural steel and

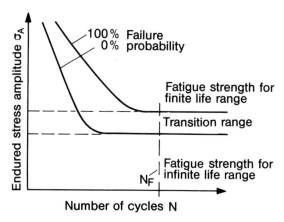

44   S-N diagram with distinct endurance limit for metal alloys with volume centred microstructure, after Maennig.[61]

titanium alloys). It shows a horizontal scatter band with parallel boundary lines following a relatively wide transition area (limiting number of cycles $N_F = 5 \times 10^6$ for establishing the endurance limit). Figure 45 is applicable to metal alloys with area centred microstructure (*e.g.* austenitic steels and copper alloys). It shows a flattened slope scatter band with parallel boundary lines following a uniformly narrow transition area (substitute limit number of cycles, $N_F = 2 \times 10^6$-10^9, to establish the substitute endurance limit). In the practical realisation of the boundary line method the scatter band boundary lines correspond to the failure probability, $P_f = 0.1$ or 0.01 and $P_f = 0.9$ or 0.99, whereby the distribution in the scatter band follows a Gaussian cumulative frequency curve.

The specific point about the boundary line method is that the width of the scatter band is determined with a minimum number of specimens for a given

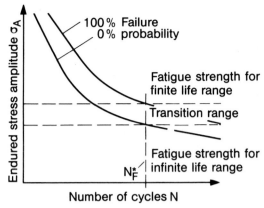

45 S-N diagram without distinct endurance limit for metal alloys with area centred microstructure, after Maennig.[61]

number of cycles (thus following a vertical line in the S-N diagram). The test is carried out for a given number of cycles on two load levels with 10 specimens each, which are to lie close to the upper or lower boundary of the band, hence further specimens are required to establish the initial load level. The method is illustrated in Fig.46 for the two possible cases, that the specimen survives in the first trial (test series 1) or fails in the first trial (test series 2), which results in a step by step raising or lowering of the load level in the subsequent trial until failure occurs for the first time or until one of the specimens runs out without failure. The lower load level is estimated from experience on the basis of the upper load level and its test results. Finally, plotting the fracture probability for the two load levels produces the mean value and the scatter. The evaluation close to the scatter band boundary is characteristic of the method and reduces the residual error with a given number of specimens.

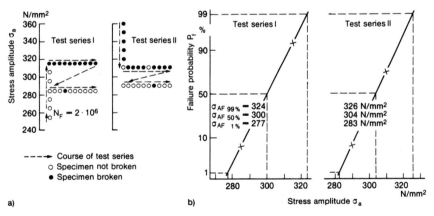

46 Boundary line method, after Maennig,[61] for fatigue strength at a given number of cycles, two test series with different starting values, σ_a, $P_f = (3r - 1)/(3n + 1)$ on the four stress levels with n = 10 specimens and number of failures, r.

The boundary line method has been used successfully on homogeneous and inhomogeneous (surface treated) specimens in practice both in the fatigue strength for finite life and in the transition area of the scatter band. Use on welded joints can therefore be recommended.

3.2   Load spectra

Welded joints subjected to fatigue loading are only rarely subjected to a constant load amplitude in practice, *i.e.* that which forms the basis of the testing and calculation procedure (S-N curve, fatigue strength for finite life and endurance limit). In practical use, mainly variable amplitudes occur, which are represented as a load spectrum for the purposes of extended testing and calculation procedures (service life line, service fatigue strength). The basis of the load spectra is a method of consideration and representation deduced from statistics which will be explained in greater detail below, as the basis for statements on service fatigue strength.

Load spectra were measured for the first time in 1935 by Kloth/Stroppel on agricultural machinery and from 1940 onwards were introduced on a broader basis into aircraft and vehicle engineering by Gaßner. Until this time it was usual to use the yield limit alone (for statically loaded structures) or the endurance limit alone (for structures subjected to fatigue loading) as the basis for the dimensioning of a component. From the above date onwards an attempt has been made to cover the range between the two strength limits too to reduce the weight of the structure for economic or functional reasons (lightweight design).

3.2.1   *Load-time function*

The load, P, on the structure or the component is generally variable with the time, t, in a more or less irregular manner. The mean value over time can vary as well as the momentary amplitude. The relationship $P = P(t)$ is designated the load-time function. Typical load-time functions from actual operation are represented in Fig.47. A distinction is made between deterministic and stochastic load-time functions. Periodic or non-periodic processes, which take place in a demonstrably determinate manner are designated deterministic. Processes, the determinacy of which cannot be proved, are designated stochastic (also called random process). Working processes (controlling, manoeuvring, manufacture, transport) are examples of deterministic processes, environmental influences (gusts, sea waves, uneven ground, noise) are examples of stochastic processes.

Random processes are dealt with by statistical characteristic values. In the stationary random process the characteristic values are not dependent on time, in the non-stationary random process they are time dependent. The load-time functions of technical components generally represent a superimposing of deterministic and stochastic processes. Like the purely stochastic load-time

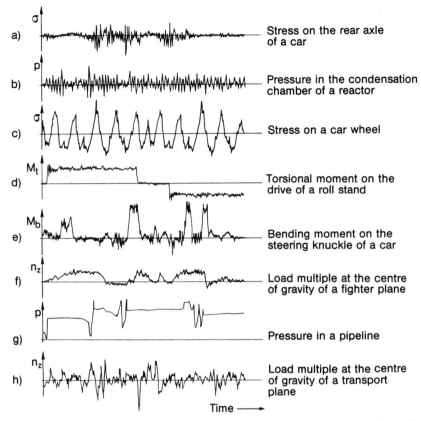

a) Stress on the rear axle of a car

b) Pressure in the condensation chamber of a reactor

c) Stress on a car wheel

d) Torsional moment on the drive of a roll stand

e) Bending moment on the steering knuckle of a car

f) Load multiple at the centre of gravity of a fighter plane

g) Pressure in a pipeline

h) Load multiple at the centre of gravity of a transport plane

Time ⟶

47 Typical load-time functions from actual operation, after Buxbaum.[28]

functions they are evaluated statistically and then give correspondingly limited statistical information.

A separation between measured load-time functions depending on their cause is often thought desirable as a basis for calculations and tests. Separation criteria can be frequencies, curve shape or amplitude ranges.

3.2.2   Amplitude spectra and spectral composition

By far the most important characteristic parameter of the load-time function with regard to fatigue strength is the amplitude content. It is illustrated as an amplitude spectrum called the 'load spectrum'. This is the totality of the service load amplitudes recorded statistically according to magnitude and frequency. In the simplest case, Fig.48, it is determined from the number N_i by which the load-time function is crossing above or below predefined load levels or load amplitude classes, selected to be equidistant, of the level, P_i, and the width, ΔP_i = ΔP (with i = ... − 2, − 1, 0, 1, 2 ...). N_i is plotted over P_i using logarithmic scale, however, in an unusual horizontal representation, Fig.49. One and the same curve is produced for the numbers of crossings above or below the load

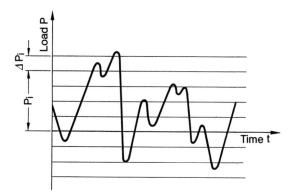

48   Load classes and load-time function: Class level, P_i, and class width, ΔP_i.

levels. Furthermore this curve, the graphic representation of the load spectrum, is considered as the function $P_i = f(N_i)$. The mean load P_m must be estimated before fixing the load levels or load amplitude classes.

The design load spectrum is defined by the following characteristic values, which have a bar to distinguish them from the corresponding values in the Wöhler test:

- mean load, $\overline{P}_m$;
- number of mean load crossings (or spectrum size), $\overline{N}$;
- maximum load amplitude, $\overline{P}_a = \overline{P}_u - \overline{P}_m$, for $N_i = 1$ with $\overline{N} = 10^6$;
- spectrum shape, $P_{ai} = f(N_i)$ or $P_{ai}/\overline{P}_a = f(N_i/\overline{N})$ with $\overline{N} = 10^6$;
  The spectrum referenced on $\overline{N} = 10^6$ is called the standard spectrum.

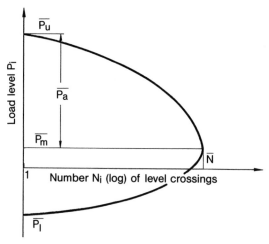

49   Load spectrum, number N_i of crossings above or below predefined load class levels, P_i.

It is assumed with this type of short form definition that the spectrum of crossings above or below is symmetrical to the mean load. Unsymmetrical spectra which do occur, for example with mean loads which vary unsymmetrically with time, are represented for dimensioning purposes by the fuller, less favourable half of the spectrum as regards fatigue strength.

The method of counting load level crossings is only one of several possible methods of recording load amplitudes with respect to fatigue strength.[62] Further methods are:

— counting the load peaks or troughs per class ('peak counting');
— counting the rising and falling load amplitudes per class independent of their respective mean values ('range counting' or 'range pair counting');
— counting the momentary load values per class in equidistant small time steps ('level distribution counting');
— counting only the highest load peaks or deepest load troughs per class between two mean load crossings.

The principles of these one parameter counting methods are illustrated in Fig.50. Further details are reviewed in the FVA Merkblatt.[62] The German

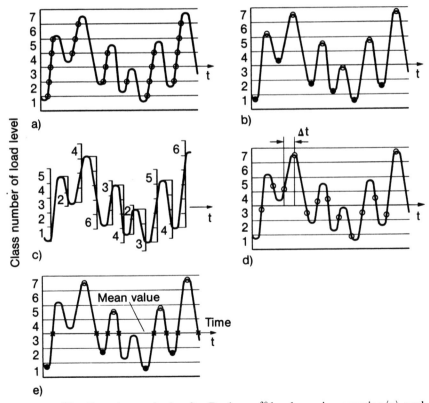

50 Counting methods, after Buxbaum:[28] level crossing counting (a), peak counting (b), range pair counting (c), level distribution counting (d), counting only one peak or trough between two mean level crossings (e).

standard DIN 45 667[63] on counting methods for random vibrations is not well
suited to service fatigue strength problems.

The two parameter counting methods can be considered as an expansion of
the above mentioned one parameter counting methods. In the simplest form
not used any more, the load mean value is counted in addition to the load
amplitude. The result of the counting is allocated to a matrix of amplitude
classes over mean value classes. Preferred today is the 'rainflow counting
method' which takes its name from rainflow over a pagoda roof, Fig.51. The
results of counting are hysteresis loops to be stored in special matrices which
serve as a basis for service fatigue strength evaluations.

The ratio of the crossing numbers, N_i/N, is identical to the ratio $H_i/\overline{H}$ of the
crossing frequencies in the limit of an infinitely small class width and an
infinitely large spectrum size. For the stationary Gaussian random process the
crossing frequency on the other hand is identical to the cumulative frequency of
the load peaks or load troughs and also to the cumulative frequency of the
amplitudes between peak and trough or vice versa. The relevant frequency
distributions characterise the sum of the peak or trough values either above or
below the considered load level. The identity of the crossing frequency with the

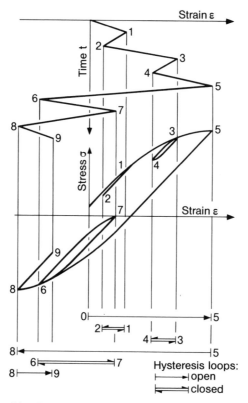

51   Interrelation between stress-strain function, cyclic stress-strain
      behaviour and counted hysteresis loops ('rainflow counting'), after
      Steinhilber and Schütz.[286]

cumulative frequency of the peak or trough values disappears to a greater or lesser extent with the only partially random or stationary technical load-time functions. Then the different counting methods lead to results which vary to a greater or lesser extent.

The load spectrum is determined, starting from the load-time function, which is generally recorded by in-service measurement. In special cases it can also be derived without measurement from operating plans, loading plans and timetables (*e.g.* for pressure vessels, crane bridges, railway bridges). From a technical point of view the procedure is to measure local strains with strain gauges, store the results on tape and finally digitise and classify them. In the elastic range the local strain is proportional to the local stress. The latter must then be converted to the external load, taking into account a possible vibration component. The strain gauge is also frequently calibrated directly to external loads. When converting or calibrating, care must be taken to ensure that the measuring signal only comes from the load component under investigation.

When the load-time function is replaced by the load spectrum, the information on the sequence of the load amplitudes and on the speed and/or frequency content of the loading process is lost. This information has to be carried separately in more precise strength investigations. In the simplest case, the irregularity factor

$$I = \frac{\overline{N}}{N_p} \tag{19}$$

can be used to characterise the relative spectral composition of the stochastic load-time function of the standard spectrum, where $\overline{N}$ is the number of mean load crossings and N_p is the number of peaks or troughs. The value $I \approx 0.99$ characterises a narrow banded excitation of the structural component close to an isolated natural frequency as often occurs in fatigue testing. The value $I \approx 0.30$ characterises a broad banded excitation of the structural component with the combination of several natural frequencies, which is usually the case in practice. The two processes are represented in Fig.52. In the first case the load curve crosses the mean load once between a peak and a trough, in the second case only occasionally.

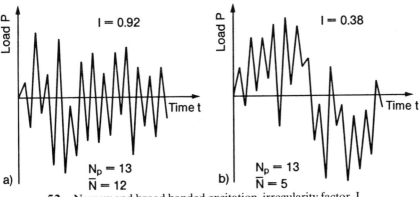

52   Narrow and broad banded excitation, irregularity factor, I.

Another method of representation with frequency information consists of plotting the amplitudes over the frequency according to spectral analysis, which gives the spectral density distribution of the amplitudes. In Fig.53 the roughness of road surfaces is represented in this way, from which the spectral density distribution of the loads can be determined using a vehicle vibration model, as can the load spectra on the structural components.

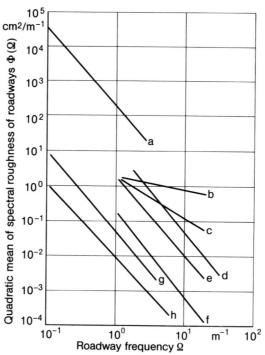

53 Spectral density distribution of the roughness of road surfaces: tank testing ground (a), ploughed field (b), furrow track (c), Belgian cobbles (d), country road (e), main road (f), aircraft landing strips (g, h).

3.2.3   Shape of the load spectrum

Typical load spectrum shapes, as they occur with actual load-time functions, are shown in Fig.54. Curve (a) arises by counting the crossing numbers for the constant amplitude process. The curves (b) characterise mixed spectra, which can be interpreted as the superimposing of constant amplitude and random processes. The curve (c) characterises the stationary random process and corresponds as a cumulative curve to the Gaussian normal distribution of the amplitudes. The straight line distribution, curve (d), and the logarithmic normal distribution, curve (e), represent mixed spectra, which can be interpreted as the superimposing of simple normal distributions. For example, the mixed spectrum can be composed from a spectrum with a small $\bar{P}_a$ and a large $\bar{N}$ and a spectrum with a large $\bar{P}_a$ and a small $\bar{N}$. Mixed spectra are characteristic of non-stationary random processes. The superimposed normal distributions occur one after the other or at the same time.

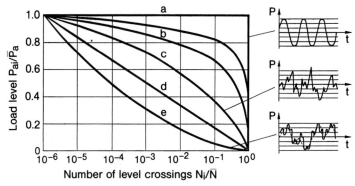

54 Load spectrum shapes: constant amplitude spectrum (a), mixed spectra (b), Gaussian normal spectrum (c), mixed spectra (d, e).

The mixed spectra, which lie above the normal distribution, are approximated by a normal spectrum in a given amplitude range (spectrum coefficient p) or in a given crossing number range (spectrum coefficient q) in formal practical use, the remaining part of the spectrum corresponding to a constant amplitude process. These mixed spectra are used standardised with $\overline{N}$ = 10^6, Fig.55. The coefficient p is the ratio of the smallest load amplitude which occurs $\overline{N}$ = 10^6 times to the largest load amplitude which occurs N_i = 1 times.

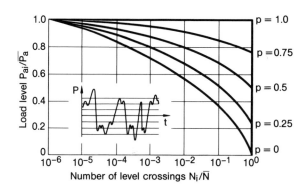

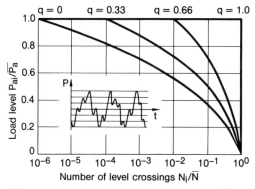

55 Standardised mixed spectra: p-value (a) and q-value (b).

The standardised mixed spectra serve the purpose of a uniform formal representation. In the first case, the mechanical background is a load-time function, which shows equal crossing numbers in the medium range for several load class levels, in the second case, a load-time function which never goes above or below the extreme values $\overline{P}_u$ or $\overline{P}_l$ however long an operating period is selected.

3.3   Fatigue strength with load spectra

3.3.1   Stress spectrum

In practice, constant load amplitudes only occur in special cases. Generally, the amplitudes are variable, partly determinate, periodic or aperiodic, and partly random. It is therefore necessary to examine the applicability of the fatigue strength values previously given for constant amplitude (single level tests) to those for stepwise variable amplitudes (multilevel test) or for randomly variable amplitudes (random test). The statistically evaluated load amplitudes, the load spectra, are the formal base for this transference, even when the loading pattern is not purely random.

In the same way as the load, which is frequently repeated and variable with time, is represented by a load spectrum using statistical methods, local stresses which correspond to this load are represented by a stress spectrum. The load spectrum is actually determined from a stress spectrum when using technical measuring procedures. Load and stress spectra are connected via the dynamic behaviour of the structure. By analogy with the load spectrum, the stress spectrum is characterised by the mean stress, $\tilde{\sigma}_m$, the stress amplitude, $\tilde{\sigma}_A$, for $N_i = 1$ (generally with $\overline{N} = 10^6$), the mean stress crossing number, $\overline{N}$, and the spectrum shape (e.g. spectrum coefficient p or q).

Where the mean stress is more or less constant, the stresses endured are characterised by the endured stress amplitude, $\tilde{\sigma}_A$, for differing values $\overline{N}$, $\tilde{\sigma}_m$ and p or q. For variable mean stress, the influence of the sequence of the variations in mean stress, if necessary their spectral composition too, should be taken into consideration.

3.3.2   Service fatigue test

In service fatigue tests, the endured (nominal) stresses are determined on smooth or notched specimens, on welded joints or on complete components, the loading amplitudes of which are variable, unlike the loading in the Wöhler test. A distinction is made between a multilevel or block program test, a load history test and a random test.

The historically earlier multilevel test[64] arose from the compelling need at that time (1940) to use fatigue testing machinery of a conventional design (resonance and hydraulic pulsators) with only slight modifications to the equipment. This type of testing is the cheaper alternative today too. The stress spectrum to be simulated in the test, which is steadily curved in general, is replaced by an equivalent spectrum with a stepped curve. Figure 56 only shows

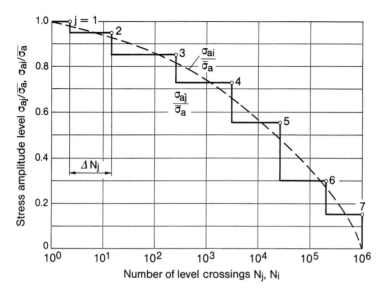

56 Stepped representation of the normal stress spectrum, steps j = 1-7, after Gaßner.[64]

the upper half of this. In the complete spectrum, the points of the upper stepped curve are connected with the opposite points of the lower stepped curve by stress amplitudes of a constant magnitude and a corresponding number, ΔN_j, per step. The stress amplitudes are applied to the specimen or the component in periodically repeated partial sequences. Each partial sequence consists of blocks with constant stress amplitudes with a fixed number of cycles. The block sequence is achieved in a program controlled manner (originally via punched tape), hence the term 'block program test'. As a rule, the test starts with a block with an average amplitude, subsequently rising to the highest amplitude, then falling to the lowest, Fig.57. Many variants on this block sequence have been used.

In reality a stress maximum from the upper spectrum curve can be followed by any stress minimum from the lower spectrum curve, so that the above mentioned procedure does not have a sound enough basis with regard to the amplitudes. In fact, however, the stress spectrum can also be interpreted in such a way that it reflects the amplitudes from maximum to minimum independent of the respective mean value. This interpretation is correct, strictly speaking, for a purely random process. Thus the justifiable approximation consists of disregarding the influence of the randomly scattered mean stress and of treating less random processes as random. The opinion that the connection of the stress extremes which lie opposite each other in the spectrum is an arbitrary method of procedure is, therefore, not correct.

The real loading processes can be reproduced more accurately using servo-hydraulically controlled testing machines (load history method). In practice, the procedure is often simplified so that the loading process, which has been

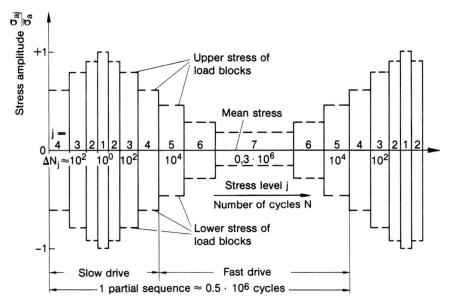

57   Stress amplitude sequence in the block program test, after Gaßner.[64]

measured over a relatively short period, is stored on magnetic tape and the ends of the tape are then joined together to form an endless loop. The load history test is relatively expensive as a result of high heat losses and heavy valve wear.

In the random test a random sequence of amplitudes corresponding to the given spectrum is generated servo-hydraulically using a noise generator. The spectral composition of the loading process produced in this way influences the results of the test. A distinction is made between narrow and broad banded loading processes. In testing practice, the narrow banded random test with defined spectral density distribution is dominant. The irregularity factor, I, as in equation (19) is also used as a simplification instead of the spectral density distribution. It is known from comparative investigations on welded joints tested under random loading and multilevel block program loading that the random process produces a lower lifetime by a factor of 2-5.

3.3.3   Service life curve

The dependency of $\overline{N}$, the number of mean stress crossings before failure, on $\bar{\sigma}_a$ for a given mean stress, spectrum shape and type of loading, constitutes the service life curve, Fig.58, when plotted in a similar way to the S-N curve. The endurable maximum stress amplitude, $\bar{\sigma}_A$ (connected with $N_i = 1$), for failure after $\overline{N}$ mean stress crossings is plotted versus $\overline{N}$ using logarithmic scales. The scatter in test results is expressed by the failure probability, P_f. The service life curve is also by analogy with the S-N curve, a falling straight line which approximates to the horizontal or slightly inclined end of the S-N curve in the endurance limit range.

The falling part of the service life curve can be expressed as follows:

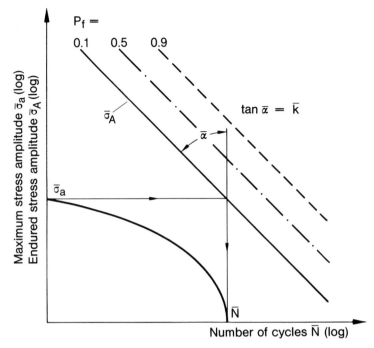

58 Service life curve for stress spectra, failure probabilities, P_f.

$$\bar{\sigma}_A = \left(\frac{\overline{N}^*}{N}\right)^{1/\bar{k}} \bar{\sigma}_A^*$$ (20)

starting from any known point $(\sigma_A^*, \overline{N}^*)$ on the line by analogy with equation (16).

The value of $\bar{k}$ in the exponent is given by $\bar{k} = 4.5\text{-}7.0$ for unwelded smooth and notched specimens and is therefore somewhat larger than the corresponding value of k in the equation for the S-N curve. In contrast $\bar{k} = 3.5$ is usual for welded specimens independent of type of joint and spectrum shape, *i.e.* the same value as for the S-N curve. This simplifies the normalisation of the life curves of welded joints for the purpose of design.

In the case of the normalisation of service life curves for welded joints made of structural steel, an expansion of the corresponding normalisation used for the S-N curve is required, whereby the spectrum shape coefficient p or q appears as an additional parameter. Service life curves are drawn in Fig.59 for p = 1.0, 0.5, 0.25 and 0, in each case for the failure probabilities P_f = 0.1, 0.5 and 0.9. Starting from the S-N curve (p = 1.0) the remaining lines with their scatter bands, to which the test results can be allocated consistently, are determined by the modified linear damage accumulation hypothesis (see section 3.3.4). The slope of the lines in the upper part of the diagram remains the same, with adequate accuracy for practical purposes, which is reflected in a constant $\bar{k}$ value.

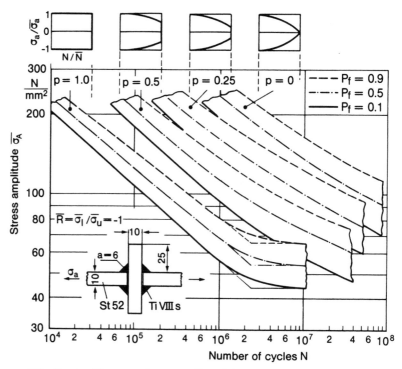

59   Service life curves for a cruciform joint, structural steel St 52, block
program test, after Gaßner.[65]

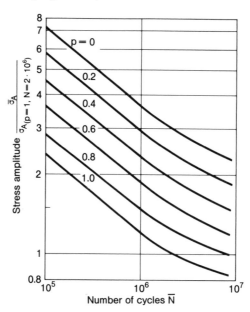

60   Standardised service life curves for welded joints of different notch
classes, structural steels St 37 and St 52, spectrum shape coefficient
p = 0-1, failure probability P_f = 0.1, after Bierett.[56]

A further simplification for design rules is represented by the introduction of equidistant lines for equidistant p values, Fig.60, drawn for $P_f = 0.1$. The diagram covers test results for the various welded joints, structural steels and mean stresses shown in Fig.40 and 41. The bottom line represents the S-N curve. In the German design rules the diagram is finally limited at the lower end by a horizontal endurance limit line which starts at $N = 2 \times 10^6$.

The publications[55,56,64-66] are highly recommended regarding the service fatigue strength of welded joints.

3.3.4   *Damage accumulation calculation*

Determination of service life curves (with scatter bands) using tests for different welded joints, materials, mean stresses and spectrum shapes, would lead to high expenditure. The detail given in the results would far exceed the capacity of generally valid design specifications. As a result of this a simple hypothesis is desired which permits the service life curve to be determined by calculation from the S-N curve and the (stepped) stress spectrum.

The damage accumulation hypotheses allocate a certain partial damage to each load cycle. The partial damage fractions add up to the total amount of damage, the maximum value of which is determined by the service life. Depending on the relationship of loading level, loading sequence, partial damage and total damage, a distinction is made between linear and non-linear hypotheses. Only the simple linear hypothesis by Palmgren (1924) and Miner (1945) has achieved practical importance, although it is inaccurate and has to be limited to a mean stress which is approximately constant.

The partial damage S_j from ΔN_j loading cycles of level j of the loading spectrum with amplitude σ_{aj} is determined from the number of cycles to failure, N_{Fj}, which correspond to σ_{aj} according to the S-N curve, Fig.61.

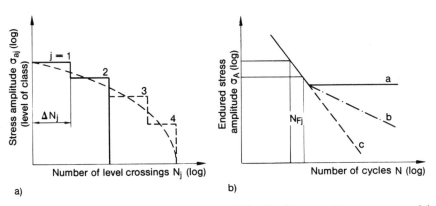

61   Calculation of damage accumulation for the stepped stress spectrum (a) and the S-N curve (b), conventional S-N curve (a), modified in accordance with Gatts/Haibach (b) and in accordance with NASA recommendation (c).

$$S_j = \frac{\Delta N_j}{N_{F_j}} \qquad (21)$$

The total damage S due to the stress cycles of the different levels $j = 1 \ldots n$ is given by

$$S = \sum_{j=1}^{n} S_j \qquad (22)$$

According to the damage accumulation hypothesis, fatigue failure should occur at $S = 1.0$. In practice the total damage can deviate considerably from this hypothetical value. Thus the values $S = 0.25\text{-}7.0$ have been obtained even under constant mean stress. Thus the hypothesis is frequently, but not always, on the safe side. The introduction of an empirically founded value $S \neq 1.0$ for defined problem groups is designated the 'relative Miner rule'. Besides the relative Miner rule, other numerical procedures[187] have been proposed on the basis of the available data from service fatigue tests. Further fatigue life prediction methods are reviewed in Ref.185.

According to the Palmgren/Miner hypothesis, the numerous cycles with stress amplitudes less than the endurance limit cause no damage and can be disregarded. It has turned out that this is not realistic. The damage caused by amplitudes below the endurance limit is perceptible if higher amplitudes follow them. In welding engineering practice the modified Palmgren/Miner hypothesis has proved itself, where the stress amplitudes below the endurance limit are taken into account in accordance with an extended S-N curve bisecting the slope angle to the horizontal, Fig.61. The modified hypothesis produces results which are sufficiently reliable. But the hypothesis is to be restricted to spectra with maximum amplitudes above the endurance limit. Spectra with amplitudes below the endurance limit can be sustained for infinite life.

Use of the Palmgren/Miner hypothesis on loading sequences with highly variable mean stress generally produces an unrealistic service life prediction so that the hypothesis has no practical value in this case. In the presence of variable mean stress, the 'notch stress approach' must be used to determine the service life (see sections 8.3.2 and 8.3.3).

In addition to damage caused by cyclic loading, creep damage occurs in the high temperature range. The amount of creep damage can be estimated starting from the creep strength for finite life curve (for life time to creep failure, t_{Fj}) and stress level, σ_j, plus the duration of the action of stress, Δt_j, in the same way as the fatigue damage fractions are determined and added up, starting from the S-N curve (for number of cycles to failure, N_{Fj}) and stress level, σ_{aj}, plus number of stress cycles, ΔN_j:

$$S_j^* = \frac{\Delta t_j}{t_{Fj}} \qquad (23)$$

$$S^* = \sum_{j=1}^{n^*} S_j^* \qquad\qquad (24)$$

Failure should occur at $(S + S^*) = 1.0$ for superimposed fatigue and creep damage.

Fatigue damage can also be expressed in a different way with reference to permissible stresses in design specifications by using the single level stress amplitude which produces the same amount of damage as the real stress spectrum. This requirement occurs when a design specification gives permissible stresses derived from the endurance limit without specifying permissible stresses with reference to the fatigue strength for finite life or service fatigue strength. The conversion of the considered stress spectrum to the single level stress amplitude on the basis of the Miner rule can be carried out in various ways:[185] equivalent number of constant amplitude cycles with the maximum stress of the spectrum; equivalent amplitude for constant amplitude loading with the total number of cycles of the spectrum; and finally, equivalent amplitude for constant amplitude loading with the standardised number of cycles for the endurance limit, $N_F = 2 \times 10^6$. Only the last method of conversion is consistent with the type of design specification and permissible stress mentioned above.

3.4   Manufacturing measures for increasing fatigue strength

It follows from the previous representation that the fatigue strength of welded joints in the high cycle range can be very low. Therefore the engineer often has the problem of raising the service fatigue strength at the design stage as a precaution, or when the design has been manufactured as a subsequent measure.

The most important measure for achieving sufficient service fatigue strength, namely adequate design, will be dealt with in Chapter 5. However, structures can only be improved in this way at the design stage. On the other hand, the most important strength increasing measure for static loading, the subsequent increasing of the load bearing cross section, is ineffective or even has a negative effect in the presence of fatigue loading because of the additional notch effects which arise around the edge of the 'reinforcement'. Finally, reduction of the operating loads causing failure is only an emergency solution for manufactured structures which are not adequate for service requirements.

Appropriate methods in manufacture for use on completed structures which increase the service fatigue strength, are of great importance. It is also possible to plan such measures at the design stage if a particularly high service life is to be achieved for the welded joint. These manufacturing measures, which have been discussed more exhaustively by Gurney[2] than they are below, are based on the following effects:

—   improvement of the weld shape resulting in a reduction of notch stress concentration;

— improvement of the material condition in the weld notch area, removal of micropores, microinclusions and microcracks;
— generation of favourable compressive stresses in the weld notch area;
— corrosion protection by coating the weld notch surface.

Various manufacturing processes are described below and their effects are explained in sequence. Quantitative data are not given until the end of the section.

3.4.1   Grinding

The notch effect is reduced by local grinding of the weld toes (or weld ends) using a grinding burr (grinding grooves in the direction of the profile) or grinding disc (grinding grooves transverse to the profile). At the same time, the micronotches in the weld metal to parent metal transition area are removed, which prolongs the crack initiation phase. Removal of residual notch stresses and surface hardening by grinding may also have a positive effect. It is important that toe material is removed by undercutting; a tangential profile is not sufficient, Fig.62. Transverse welds, in particular, can be improved in this way. Butt welds (and double bevel butt welds with fillet weld) can be improved to the strength of the unwelded mill-finished sheet metal, fillet welds to not quite such high values, which are given by failure initiation at the weld root. The improvement if the ends of longitudinal welds are ground is less marked. Grinding of critical weld notches is everyday practice.

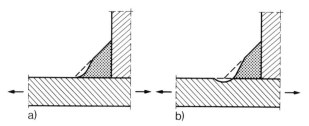

62   Grinding the weld toe: undercutting (b) better than a non-undercut profile (a) at the toe.

3.4.2   TIG dressing

Local melting of weld toes (or weld ends) using TIG (tungsten inert gas) welding torches or plasma welding torches improves the shape of the toe notch on butt and fillet welds (up to a radius of 5-10mm) and removes micronotches which may be present at the weld toe. This process is known as TIG dressing. The surface of the weld toe should be cleaned by brushing or grinding before dressing. Excessive local hardening can be a problem with this procedure. The ends of the dressing pass can be a further problem. Starting and stopping on the weld surface outside the toe notch is to be recommended. TIG dressing is primarily suitable for transverse welds. It is used successfully in practice, especially in connection with higher tensile steels.

3.4.3   Overloading

This method and those which follow it are based on the restraining effect of residual compressive stresses on fatigue processes. Whilst the fatigue strength of severely notched welded joints is hardly influenced by the magnitude of the tensile prestress (only a slight difference in the endured stress amplitudes between specimens with and without residual welding stresses, where the residual stress is tensile) it is considerably increased by compressive prestressing (see section 2.3.2). Stress amplitudes which are completely or partly in the compressive range do very little damage and may close cracks which are present. Therefore, considerable increases in strength can be achieved by residual compressive stresses (as high as possible) generated locally in the notch area. However, a single loading with excessively high

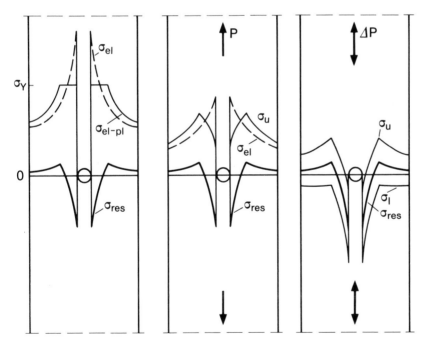

63   Residual compressive stresses at the root of the notch as a result of tensile overloading, after Gurney:[2] σ_{el} — elastic notch stress without residual stress; $\sigma_{el\text{-}pl}$ — elastic-plastic notch stress under overloading; σ_{res} — residual notch stress; σ_u — upper stress under cyclic loading with residual stress; σ_l — lower stress under cyclic loading with residual stress.

compressive stress can alter the residual notch stresses unfavourably as a result of local yielding. Loadings of this type must be excluded for this method to succeed in practice. On the other hand, (statically) higher strength steels are particularly suited to generating and maintaining high residual compressive stresses.

The desired compressive stress at the root of the notch can be generated by a single tensile loading. The sign of the residual notch stress after (elastic) unloading is always the opposite to the sign of the notch stresses under (elastic-plastic) loading, Fig.63. More precise quantitative investigations must take into account the variation of the yield stress in the area of weld notches. Bending overload is not suitable for increasing the fatigue strength because unfavourable residual tensile stresses are generated in the bending compression zone. On the other hand, multiple tensile loading is to be avoided because of the risk of low cycle fatigue. Overloading frequently takes place unintentionally if a proof test is carried out with an increased load.

3.4.4   *Surface compression*

In this method measures are combined which produce residual compressive stresses by means of locally concentrated surface pressure at the root of the notch resulting in yielding under restrained transverse deformation. Strain hardening has little or no effect on fatigue, but removal of the microseparations and improvement of the notch shape has a positive effect. Peening with hard metal particles, hammering the surface of the notch and rolling or pressing the root of the notch are possible. These methods are suitable for transverse welds but unsuitable if service load peaks occur which alter the residual notch stress state. Practical applications of the method to welds are rare.

3.4.5   *Spot compression*

The axially symmetrical residual stress state with continuous radial pressure, shown in Fig.64 for an infinite plate, is generated by squeezing a plate between two dies with a circular cross section ('spot compression'). The method is particularly suitable for the ends of longitudinal welds, Fig.65, but has been subjected to little practical testing. Longer transverse welds cannot be improved in this way because a part of the length of the weld would be subjected to residual tensile stresses. The method is successfully used in practice on spot welded joints ('post-weld compression'). An effective improvement of the method consists in allowing the dies to act between two clamping rings, which prevent thickening up close to the dies and enable a higher radial pressure to arise in the plate. Thus the endurance limit of spot welded joints can be increased by several times.

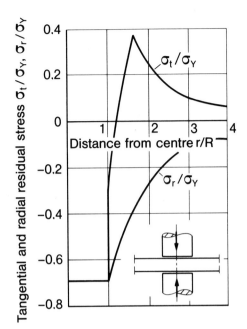

64 Residual stresses after spot compression: σ_t — tangential stress; σ_r — radial stress; σ_Y — yield stress; R — radius of compression spot.

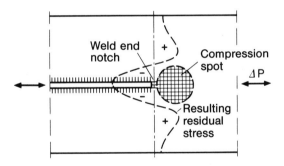

65 Compression spot at the end of the longitudinal weld, residual stresses in the tensile direction, after Gurney.[2]

3.4.6 *Spot heating*

Short term heating (a few seconds) of a circular area (spot) in the plate (generally using an oxyacetylene flame) until it passes the recrystallisation temperature, produces compressive yielding and, after cooling, radial tensile residual stresses (as a result of contraction of the heated spot) and tangential compressive residual stresses (representing the support of the radial tensile stresses), which are axially symmetrical for an infinite plate, Fig.66. The

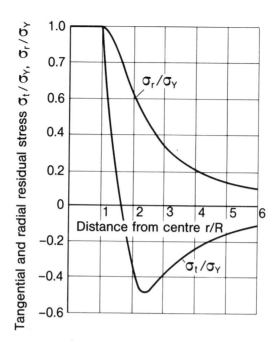

66   Residual stresses after spot heating, after Gurney:[2] σ_t — tangential stress; σ_r — radial stress; σ_Y — yield stress; R — radius of heat spot.

tangential compressive stresses can be used to increase the fatigue strength of the ends of longitudinal welds, if the distance and position of the heat spot are well selected. The most favourable distance can be adjusted on steel plates via the circle of red tempering colour (280°C). Good results were produced with a notch distance of 10-20mm from this circle. Unlike the compression spot, the heat spot must be positioned transverse to the weld end (relative to the tensile direction) to bring the tangential compressive stress favourably into line with the weld end, Fig.67. The radial tensile stresses, if brought into line, would have

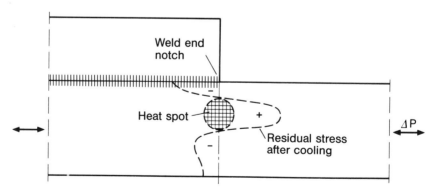

67   Heat spot at the end of longitudinal weld, residual stresses in the tensile direction, after Gurney.[2]

an unfavourable effect here. Tests for adjusting peak temperature, heating speed, heating area and distance between heat spot and weld end are indispensable for this method, which has hardly been applied in practice.

3.4.7   Notch quenching

Heating the notch area with subsequent local quenching of the surface by a water jet has a hardening effect and produces residual compressive stresses in the quenched surface both because of the subsequent contraction of deeper layers and because of phase transformation processes in the surface (e.g. formation of martensite). Notch quenching is less sensitive to adjustment than spot heating. Local concentration of quenching is important for the generation of high residual compressive stresses. Practical experience with the method is unknown.

3.4.8   Notch coating

The strength increasing effect of coating the surface of external weld notches with metal or plastics is mainly based on protection from corrosion (see section 11.2), to a lesser extent also on reduction of the notch stresses (the latter only if the coating has a high enough elastic modulus). Crack initiation is delayed.

3.4.9   Quantitative comparison

The Wöhler tests performed by various authors with regard to the different manufacturing processes used to increase the service fatigue strength of welded joints have been referenced and evaluated by Gurney.[2] The extent of the increase in service fatigue strength is not uniform in each case. The values determined at The Welding Institute in a research programme, which was carried out under uniform conditions, on the same type of specimens (transverse and longitudinal stiffeners) should be comparatively informative, Fig.68 and 69. The S-N curves apply to low tensile steels with the exception of the TIG dressing, which was carried out on high tensile steel. Notch quenching and coating are not included, neither are aluminium alloys on which spot compression and coating with plastics have been applied successfully. The effect of loading sequences with highly variable amplitudes is largely unknown.

Overloading is relatively ineffective, TIG dressing (at least for high tensile steel) is very effective. Surface compression, spot heating and spot compression lie in between. All the methods are most effective in the endurance limit range and are advantageous for $N \geq 10^5$ or at $\sigma_{Up} \lesssim \sigma_Y$ in the high cycle fatigue range.

Thus considerable improvements can be achieved on welded joints with relatively low service fatigue strength using the methods which have been described, particularly in the high cycle fatigue and endurance limit ranges. The methods based on residual compressive stresses are only to be recommended for very high numbers of cycles and where compressive overloading can be excluded whilst grinding and TIG dressing also remain effective for lower numbers of cycles. The methods are entirely or largely ineffective for failure initiation from internal defects or from the weld root.

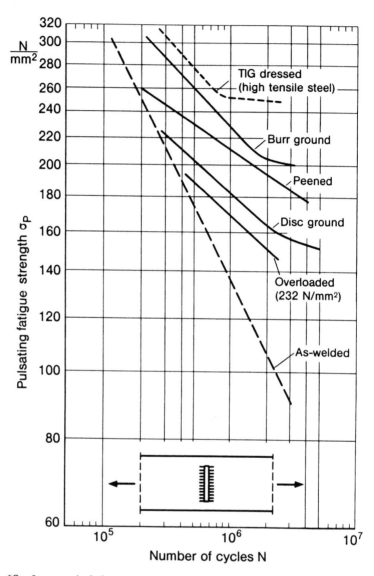

68   Increase in fatigue strength as a result of manufacturing measures, structural steel, transverse stiffener, pulsating loading, after Gurney.[2]

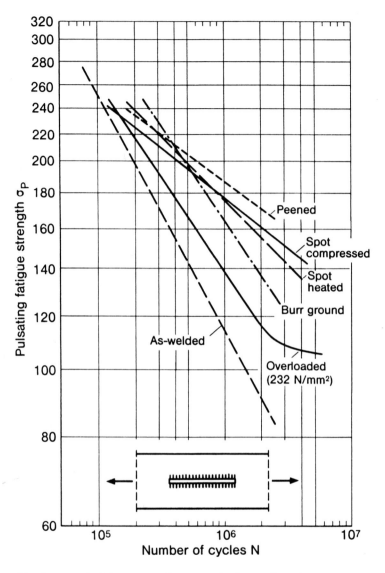

69 Increase in fatigue strength as a result of manufacturing measures, structural steel, longitudinal stiffener, pulsating loading, after Gurney.[2]

4

Fatigue strength of welded joints in high tensile steels and aluminium alloys

4.1   Welded joints in high tensile steels

High tensile steels (high tensile and ultra-high tensile structural steels, fine grained structural steels, case hardening and heat treatable steels, austenitic and martensitic steels) are used when the higher strength allows the quantity of materials or the component weight to be reduced to such an extent that the higher costs of these materials can be justified. They are suitable for components which are occasionally or permanently subjected to very high levels of stress. However, 'high tensile' only refers to the static strength of smooth specimens, *i.e.* to the yield point and tensile strength. A notched bar, even under static loading, may give a smaller increase in strength as a result of higher notch sensitivity. The problems caused by the latter as regards fatigue strength are even more marked. They increase in importance particularly in the course of weight reductions. High tensile steel does show a moderately increased endurance limit in smooth specimens. In contrast, the gain in fatigue strength in notched specimens decreases with the severity of the notch effect. Use of high tensile steel is only worthwhile if the notch effect is low. Finally, in the case of welded joints, this behaviour is so marked that the alternating or pulsating fatigue strength of a strongly notched welded joint (*e.g.* of a cruciform joint) is just as low for both low and high tensile steels.

Comparison of the endurance limits for the structural steels St 37, St 52 and St 60, versus limit stress ratio R, for smooth specimens with mill scale and for butt and cruciform joints (unmachined in each case), Fig.70 and 71, serves to illustrate this. When subjected to fatigue loading the high tensile steels lose the advantage of their high static strength all the more, the higher the notch effect and the smaller the static prestress.

The higher strength is effective, however, in the medium and low cycle fatigue strength range. The S-N curve for high tensile steel rises to a higher level, corresponding to the higher tensile strength, as does the yield stress, which serves as reference stress in the design codes, Fig.72. Normalisation of the S-N curve in the high cycle strength and endurance limit range according to section 3.1.1 is also valid for high tensile steels.[67-69]

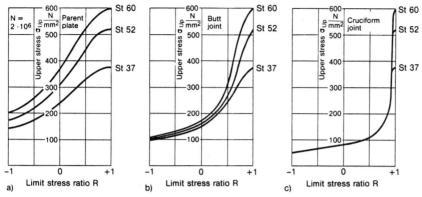

70 Endurance limit for welded joints made of low, medium and high tensile structural steel, dependent on limit stress ratio, after Neumann:[3] Mill finished sheet metal (a), medium quality butt joint (b) and cruciform joint (c), after Neumann.[3]

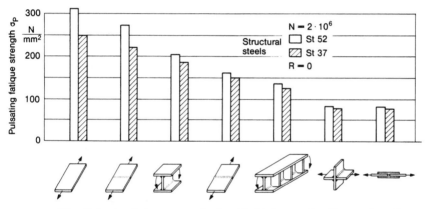

71 Pulsating fatigue strength of welded joints made of low and medium tensile structural steel, dependent on notch effect, after Neumann.[3]

Summarising, it can be stated[67] that use of high tensile steels in components with welded joints, which are subjected to fatigue, is only worthwhile if the static mean stress is high (*e.g.* bridges with large spans) and/or if only moderately large numbers of cycles arise (*e.g.* some high pressure vessels with piping systems, rotor discs, deep sea diving vessels). High tensile steel is also particularly suitable for components subjected to load spectra with load peaks which are extremely high (*e.g.* wheel mountings, plough frames). Moreover, the notch effect on components of this type must be reduced efficiently, for example by using butt or double bevel butt welds for preference. Shot peening or dressing the weld toe is particularly effective in high tensile steel.[68] However, it is to be noted that the rigidity and stability characteristics are not improved by higher tensile steels, but on the contrary are worse as a result of having smaller cross sections or thicknesses of the sheet metal while the elastic modulus remains the same.

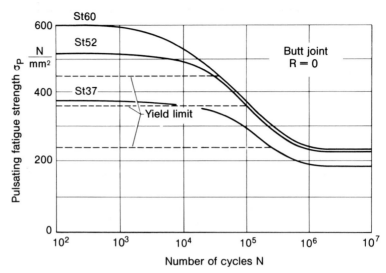

72   Low and medium cycle pulsating fatigue strength for butt joints made of
low, medium and high tensile structural steel, after Neumann.[3]

Furthermore, it should also be taken into account with high tensile steels,
that the residual welding stresses can be increased corresponding to the higher
yield stress so that the endurable load amplitude is reduced by this. However,
particularly in high tensile steel, there is also the possibility that the high
longitudinal residual stresses in the weld, caused by thermal contraction, are
lowered or have their sign reversed because of compressive transformation
stresses.

Use of high tensile steel outside the stated areas of application is not
worthwhile. The (low) reduction factors, γ, given earlier for low tensile steel St
37 with mill scale must be applied to high tensile steels also without increasing
the reference stress.

Finally, a question which has not yet been adequately clarified scientifically
remains to be discussed, namely why welded joints in high tensile steels give no
gain in alternating fatigue strength (see also Gurney[2]), although smooth
specimens, and to a lesser extent notched ones too, made from parent material
or weld material, do show moderately increased alternating fatigue strength. It
has already been stated that the local notch effect coincides with locally
inhomogeneous material in welded joints: molten material comprising a
mixture of weld and parent material interfacing with unfused parent material,
inclusions and grain boundary deposits of slag, may extend to a depth of up to
0.5mm below the surface, Fig.15. Microcracks may form at the inclusions and
grain boundary slag layers after relatively few cycles, so that the life then
consists mainly of crack propagation governed by fracture mechanics
parameters, which are found to be largely independent of tensile strength in
constant amplitude fatigue tests. The heat affected zone, which is itself rather
notch sensitive, has no influence on the failure process. Butt joints, which have

been ground flush, fail in the weld material (starting from internal defects) or in the base material but never in the heat affected zone. Molecular or atomic hydrogen, which plays such a major role in brittle fracture, has little effect on the fatigue phenomenon at least under inert environments. Hydrogen diffuses out of the surface layers, in which the cracks are initiated, relatively quickly.

4.2   Welded joints in aluminium alloys

A start was not made on investigating the fatigue strength of welded joints in aluminium alloys until relatively late (first publication 1947). At that time the high tensile aluminium alloys used in the aircraft industry, which were subjected to fatigue loading, were riveted without exception. In the meantime, inert gas welding for aluminium alloys had been developed. Fatigue tests were then carried out on welded joints to an appropriate extent, corresponding to the increasing use of aluminium alloys in welded structures. The fatigue strength of welded joints in aluminium alloys is approximately 1/3 to 2/3 that of comparable welded joints in structural steels, with the advantage of a specific weight which is lower by 2/3, which makes these alloys interesting for lightweight design (e.g. vehicle bodies and ship superstructures, mobile jibs and bridges).

Just as for structural steels there is a connection between the (substitute) endurance limit, σ_{A0} (usually for $N = 10^7$), and the tensile strength, σ_U, in each case for smooth specimens, $\sigma_{A0}/\sigma_U = 0.5$ for low tensile alloys, decreasing to 0.3 for high tensile alloys. The measures which increase the static strength (alloying, work hardening, and cold and warm age hardening) are therefore only of limited effectiveness under fatigue loading, even in smooth specimens. The same applies to notched specimens to an increased extent. In aluminium alloys the loading frequency influences the fatigue strength. Raising the frequency has a strength increasing effect (presupposing that the temperature remains constant), markedly in the finite life range (up to factor 1.5 at $N = 10^5$ for $f = 1000$cps compared with $f = 10$cps), and less markedly close to the endurance limit (up to factor 1.1). The influence of corrosion, which is time dependent and therefore most effective in the high cycle and low frequency range, is superimposed on the influence of frequency. The endurance limit is reduced to less than a half in the presence of seawater.

In order to determine the reduction factors for welded joints made from aluminium alloys, Table 3, the test results published by Hertel,[11] Gurney,[2] Schütz/Winkler,[70] Haibach/Atzori,[71] Buray,[72] and Kosteas/Graf,[73,74] were used as a basis. These results refer to high tensile alloys such as AlMg 5 (not age hardenable, $\sigma_U = 270$ N/mm^2, $\sigma_{0.2} = 150$ N/mm^2), AlMgSi 1 (warm age hardened, $\sigma_U = 320$ N/mm^2, $\sigma_{0.2} = 260$ N/mm^2) and AlZnMg 1 (warm age hardened, $\sigma_U = 360$ N/mm^2, $\sigma_{0.2} = 280$ N/mm^2) which are in widespread use nowadays. The fatigue strength of these materials as sheet metal is given in a relatively uniform manner by Schütz and Kosteas: $\sigma_P = 155\text{-}170$ N/mm^2, $\sigma_{A0} = 95\text{-}100$ N/mm^2 for $N = 2 \times 10^6$. The pulsating and alternating fatigue

Table 3. Survey of the reduction factors, γ, for aluminium alloys with tensile pulsating fatigue strength σ_p = 155-170 N/mm^2 and alternating fatigue strength σ_{A0} = 95-100 N/mm^2, failure probability P_f = 0.1.

Type of weld	Loading on base material	Reduction factor γ
Butt weld	$\sigma_\perp$	0.45-0.75
Double bevel butt weld	$\sigma_\perp$	0.2
Fillet weld (in cruciform joint)	$\sigma_\perp$	0.3-0.5
Edge and side fillet weld (in strap joint)	$\sigma_\perp$ or $\sigma_\parallel$	0.15-0.2
Transverse stiffener	$\sigma_\perp$	0.4-0.7
Butt and fillet weld (on an I section girder)	$\sigma_\parallel$	0.45-0.7
Longitudinal stiffener	$\sigma_\parallel$	0.15-0.2

strengths for welded joints refer to these values for the parent materials, when stating the reduction factor, γ. This factor is transferable unchanged to N = 10^7, whilst the reference stresses decrease by the ratio 1:1.2. The reference values for sectional bar metal are lower (by a factor of 0.8) than those for sheet metal.

By analogy with structural steel, increased static strength does not generally produce an improvement in the fatigue strength of welded joints. The aluminium alloys, considered above have already been graded as high tensile, as they are highly alloyed and work hardened, or cold or warm age hardened, which is partly reversed by welding. However, the corresponding reference stresses for low tensile alloys are not much lower.

Because of the somewhat smaller reduction factors, γ, and the considerably smaller reference stresses the welded joints in aluminium alloys only achieve half the fatigue strength of the welded joints in structural steels with the same dimensions. However, the welded joints in aluminium are two-thirds lighter with the same dimensions because of the difference in densities, so the fatigue strengths with reference to weight are higher than for the corresponding welded joints in steel. As structures made of aluminium alloys work out to be lighter, the mass dependent loadings are also lower.

The following details are worth noting in comparison with welded joints in structural steel. The high level of reduction for the butt joint is striking, and this cannot usually be improved by machining flush. Machining lays pores open which, as surface notches, have a particularly detrimental effect (γ = 0.6). Therefore, machining is not worthwhile. Fillet welds are produced with deep parent metal penetration and therefore low root notch effect because of the more extensive fusion of the parent material. The double bevel butt joint performs badly in comparison with the simple butt weld and the edge fillet weld for some unexplained reason, so that the correctness of the γ value is doubtful. The differing reduction in strength for single and double sided transverse stiffeners as a result of differing notch effect is more obvious than for structural steel. Strap joints, with edge fillet welds and side fillet welds, and longitudinal stiffeners show a particularly marked reduction.

For welded joints in aluminium alloys, too, the influence of shape proves to be dominant with respect to fatigue strength. However, the reduction factors shown in Table 3 do not correlate with notch severity with sufficient consistency. The higher notch and mean stress sensitivity of aluminium alloys and their welded joints, which is occasionally put forward, cannot be inferred as a generalisation from the publications above. The greater reduction in some cases must be explained by greater thermal distortion and by thermal softening caused by welding.

Evaluation of the relevant literature by Atzori/Attoma[75,76] leads to conclusions which deviate in part from the above. No serious difference in the reduction factors was established between structural steels and aluminium alloys, Fig.73. Machining is worthwhile. The relative values of the fatigue strengths were determined for $P_f = 0.5$ (instead of for $P_f = 0.1$) and are normalised by the fatigue strength of the unmachined transverse butt weld (instead of unwelded parent material) which hence includes notch effect and thermal softening. Both pulsating and alternating fatigue strength values are covered.

Haibach/Atzori[71] specify a normalised S-N curve for the high cycle strength and endurance limit range, with a quantified scatter band for welded joints made of aluminium alloys, Fig.74, derived from the published results on non-work hardenable alloys, in particular AlMg 5. It can also be considered valid for other aluminium alloys, welded joints and limit stress ratios. The further decrease in fatigue strength for $N \geqslant 2 \times 10^6$ compared with the corresponding diagram for welded joints in structural steels is noteworthy. The substitute

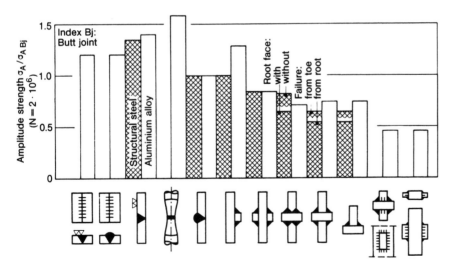

73 Endurance limit for welded joints made of structural steel and aluminium alloy, with reference to the endurance limit of the unmachined transverse butt weld, after Atzori and Dattoma.[76]

endurance limit is evaluated at the standard value $N = 2 \times 10^6$. In addition, the exponent k of the straight S-N curve for finite life fatigue strength is increased. An even higher value (k = 5.75) was determined for the work hardening alloy AlZnMg 1. More recent information on the fatigue strength of welded aluminium structures and joints can be taken from Ref.36.

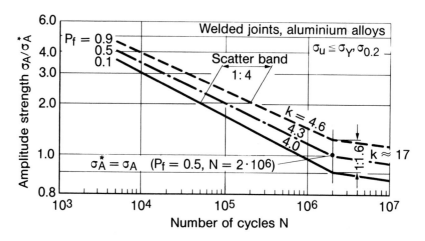

74   Normalised S-N curve with scatter bands for welded joints made of aluminium alloys, any notch class and limit stress ratio, after Haibach and Atzori.[71]

5

Fatigue strength of welded components, design improvements

5.1   General aspects and welded components in structural steel engineering (bar structures)

Now that the fatigue strength of simple welded joints, subjected to different types of loading, has been described, this chapter considers more complex components which recur frequently and which are jointed by seam welds. Reliable statements can no longer always be made in the form of reduction factors, partly because of a lack of test results and partly because of the range of dimensions, loading and support. In the latter case, qualitative statements on design measures which lead to a major increase in the fatigue strength* of the welded component are all the more desirable. Therefore statements are made, varying from case to case, on design improvements for welded components in which fatigue strength is the main consideration. These statements are then summarised in a separate section at the end of the chapter. The criterion of strength assessment and the guideline for design improvement is the distribution of structural and notch stresses. Stress peaks must be avoided or reduced to an endurable level.

Reference to design systematics, which would really be desirable, has to be dispensed with so that the treatment is not unduly long.

Bar structures predominate in structural steel engineering, *i.e.* in building, bridge, crane and hangar constructions, as well as in vehicle and machinery frames. Above all, railway bridges, cranes and crane runways as well as vehicle frames are subjected to fatigue loading. The members of these structures are primarily subjected to axial forces (uprights, rods in a latticework), bending moments and transverse forces (girders, bars in a framework), and secondarily to torsional moments. See section 6.4 with regard to stud welded setbolt dowels in steel-concrete composite girders, which are not dealt with in this chapter.

---

*   Where the term strength is used below, this refers to fatigue strength.

5.1.1   *Gusset plate on tension and compression members*

The fatigue strength of tension or compression members with welded joints can be derived from the data in section 2.1.2; with regard to frame members, see sections 5.1.11 and 5.1.12. Reduction factors are stated in Fig.75 for the non-load-carrying gusset plate as a type component 'plane re-entrant corner with weld end', from which it follows that high fatigue strength is only achieved with a relatively large rounding radius in the corner combined with machining.

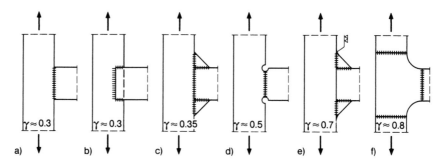

75   Gusset plate on a tension member, design variants, reduction factors, γ, partially after Neumann.[3]

5.1.2   *Butt joint on I section girders*

Here 'butt' refers to a jointing of girders without a change in direction, which can be achieved both by butt and fillet welds, Fig.76. The rolled girder without a joint (a) is selected as the reference component ($\gamma = 1.0$). The rolled girder joint with butt welds (b) suffers a considerable reduction of fatigue strength because welding through the wall thickness leaves impurities in the root, the surface of

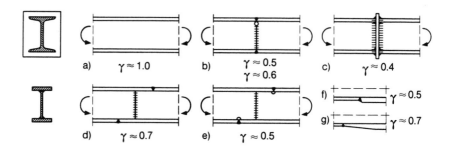

76   Butt joint in I section girder, reduction factors, γ, partially after Neumann.[3]

the weld is not well shaped and several weld ends must be accepted. An additional reduction occurs in low quality rolled steels (unkilled steels) as a result of welding over zones of segregation. Drilling out the zones of segregation removes this source of cracking but is in itself connected with high notch stresses, so that no improvement is achieved. The rolled girder joint with fillet welds (c) results in a greater fatigue strength reduction than the corresponding joint with butt welds because it has a more severe notch effect. Moreover, it is more expensive to produce and is liable to fractures involving lamellar tearing, if lamination is present in the cross plate. However, it is unavoidable as a bolted assembly joint. Amongst butt jointed girders, type (d) with welded-over web-to-flange fillet welds only shows a slight reduction compared with the unjointed welded girder. Staggered butt welds produce a gain in fatigue strength compared with those which are not staggered. Providing a cope hole in the web at transverse welds in the flange (e) produces no advantage, just as it does not for the rolled girder because of the severe notch effect connected with the hole, although it does avoid welding in an overhead position, which is disadvantageous, and the flange welds can have a pass deposited on the opposite side to increase fatigue strength (overhead then of course). If the flanges are butt welded the joint can be combined with a transition of thickness (f, g) which is, however, connected with an additional reduction, unless the bevelling is rather flat. Compressive residual stresses, which increase fatigue strength, are produced in all these joints by welding the flanges first, and then the web. It should be noted as a comparison that the riveted joint, which was used in former times, did not have the fatigue strength of the best of these welded joints.

Comparable circumstances are present in butt welded joints in box, channel section and tubular girders.

A look at the historical development of sheet metal butt joints at the stage of transition from riveting to welding is instructive. Butt joints of this type were at first completely covered by pad plates, then later only at the weld ends ('safety plates'), Fig.77. The fatigue strength is very low, as a result of the severe notch effect, although the cross section and the section modulus are increased. The inclined weld, which behaves in a somewhat more favourable manner than the transverse weld, has also failed to become generally accepted because of the loss of material when cutting the sheet metal to size.

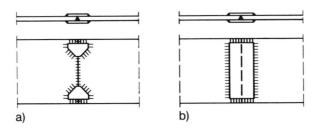

a)   b)

77   Covering of butt joint, out of date.

5.1.3   *Web-to-flange weld on I section girders*

Slender I section girders subjected to bending are made economically from plate steel, whereby the chords or flanges are connected to the web via double fillet welds. The fatigue strength of girders welded in this way is lower than the corresponding strength of rolled girders (crack initiation from the starting point of the weld), Fig.78. The double fillet weld (a) (a single sided fillet weld can also be used in some circumstances) gives adequate fatigue strength compared with the more expensive types (b) to (d). Only version (e), a sectional steel to web plate combination, achieves considerably higher fatigue strength as a result of the position of the weld in the area subjected to lower bending stress. In contrast, intermittent welds have a considerable strength-reducing effect because of the notch effect of the weld ends, which can be minimised by a staggered arrangement (on opposite sides of the web) (indispensable when there is loading with concentrated transverse forces, *e.g.* crane runways and bridge girders with directly connected rails or sleepers). Spot welded girders subjected to bending (g) also have relatively low fatigue strength. The longitudinal weld was superior to the riveted longitudinal joint (h) from the beginning with regard to cost and fatigue strength. In the presence of transverse bending force, in addition to the bending stress, $\sigma_{\parallel}$, a shear stress, $\tau_{\parallel}$, and a transverse stress, $\sigma_{\perp}$, are also acting in a double fillet weld, the combined effect of which on the strength can be determined in accordance with the von Mises distortion energy hypothesis. The double bevel butt weld has a relatively high strength in this case. The data above on the double fillet weld in the I section girder subjected to bending can be transferred to similar types of double fillet welds, *e.g.* on longitudinally stiffened plates in shipbuilding.

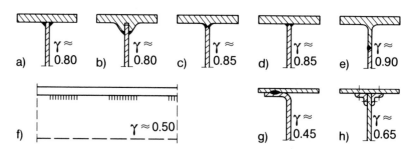

78   Web-to-flange weld on I section girder, design variants, reduction factors, γ, partially after Neumann.[3]

5.1.4   *Cover plates on I section girders*

Cover plates (also called doubler plates) serve to strengthen the cross section of the chords of trusses, or of the flanges of I section girders, if the standard section available is not adequate, or if reinforcement is only necessary for some length of the girder. The cover plate can be narrower or wider than the flanges, Fig.79.

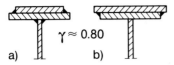

a)   b)

79   Cover plate on I section girder, reduction factor, γ.

Although the reinforcing effect of the narrower plate is considerably less than that of the wider one, the former is preferable in the presence of fatigue loading because the notch effect of the ends of the plate can only be reduced adequately in this version by design and manufacturing measures, Fig.80. The cover plate end without transverse weld (a) is not generally permissible because of the risk of crevice corrosion. The version with transverse weld (b) without any other measures gives a severe notch effect and a corresponding reduction of fatigue strength. This version is used when the end of the plate can be placed in an area with a low bending moment. Adequate strength at the plate end is achieved by giving the corners a rounded shape, bevelling the plate and grinding the weld toe (c). Making the plate ends pointed (d) is not advantageous. The end of the wider plate (e) gives particularly low fatigue strength, to which reference has already been made.

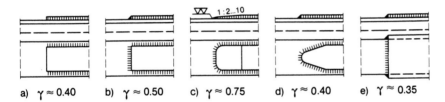

a) γ ≈ 0.40   b) γ ≈ 0.50   c) γ ≈ 0.75   d) γ ≈ 0.40   e) γ ≈ 0.35

80   Cover plate end, design variants, reduction factors, γ, partially after Neumann.[3]

The section modulus of an I section girder subjected to bending can also be increased by separating the web of a rolled section in a toothed manner and joining them together again in a staggered fashion, instead of using cover plates. This is known as a 'honeycomb girder', as in Fig.81. The notch stresses, which arise at the large openings thus created, require consideration, especially in the presence of loading with transverse forces.

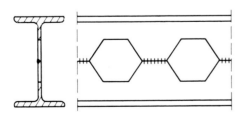

81   Honeycomb girder.

5.1.5   Stiffeners on I section girders

Stiffeners in the form of transverse ribs on I section girders, Fig.82, are necessary to maintain the cross sectional contour, especially in the presence of high local loading, such as when concentrated transverse forces are acting. It should be remembered with regard to fatigue strength when designing structures, that the transverse welds (a) or the end of the short longitudinal welds (c, d) represents a marked notch, so that in (railway) bridge construction the stiffener occasionally only butts against the transverse tension flange without a welded joint (with or without a base plate) (b). On the other hand, allowing the stiffener to come to an end above the tension flange does not produce sufficient sectional stiffening and is prone to failure when the flanges are subjected to eccentric transverse forces.

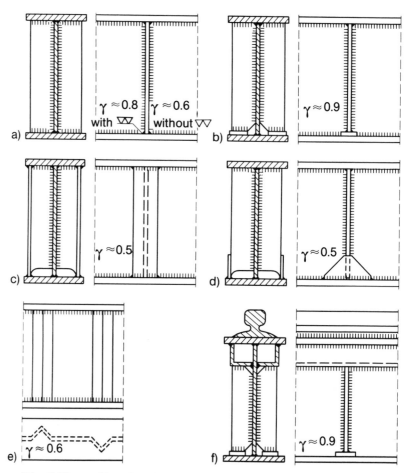

82   Stiffener of I section girder, design variants, reduction factors, γ, partially after Neumann[3] and Stüssi and Dubas.[10]

Cutting out the inner corners, which is frequently performed (triangle or quarter circle cutouts, Fig.83), comes from the era of rolled sectional girders made of unkilled steel, in which the segregation zones in the transition area from web to flange had to remain uninfluenced by welding because of their tendency to crack. Later the inner corners were cut out on similar web-to-flange welded girders made of segregation-free structural steel to avoid defects caused by bad fitting and welding as well as unfavourable residual stresses at the inner corner with the weld crossing. The relatively high notch stresses of the transverse weld on the side of the web subjected to transverse force bending of the girder are also avoided. Therefore the fatigue strength of girders with stiffeners with cut out inner corners is substantially improved (in contradiction to the fatigue test results in Ref.77). Transverse corrugations in the web (e) (in Fig.82) instead of transverse ribs also perform relatively well. A short longitudinal weld (c, d) instead of a transverse weld can also produce an advantage with regard to fatigue strength. The T section, which is resistant to buckling, is preferred as a stiffener in the presence of large transverse forces.

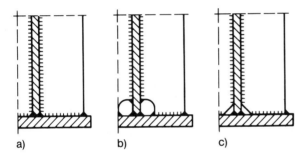

83 Stiffener with and without cutout at inner corner.

The reduction factors given in Fig.82 are also applicable to the transverse wall of box girders.

Crane runway girders have not only to be transversely stiffened, the upper chord must also be designed to be torsionally rigid (f). Otherwise the transverse forces on the girder and the transverse bending moments in the flanges which occur with the high number of cycles of the individual crane wheels, cause early crack initiation.

5.1.6   *Box girders and deck plates*

Box girders of varying sizes and applications are shown in Fig.84. The design for vehicle frames (a) can tolerate the (through-welded) butt welds only, which are relatively favourable as regards the notch effect, especially with respect to the shear flow by torsional moments superimposed on the bending moments. The design for statically loaded compressive uprights (b) is completed using fillet welds. In the design for box girders in arched bridges, the use of double bevel butt welds is preferable, as they are more favourable as regards notch

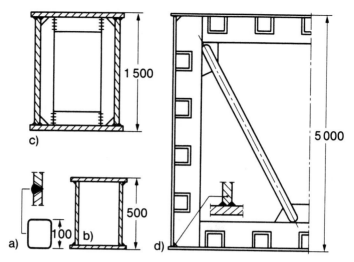

84 Box girder subjected to bending, design variants, partially after Neumann[3] and Stüssi and Dubas.[10]

severity. This design therefore has to be accessible from inside and has the transverse walls to retain the cross sectional contour. The design for girder bridges with wide spans (d) is stiffened in small cells to reduce the risk of buckling whilst the cross sectional contour is secured by diagonal frame members. It is particularly stiff against torsion and therefore stable with respect to torsion buckling.

The deck plate of girder or truss bridges is fitted with longitudinal stiffener ribs or rectangular or trapezoidal hollow section stiffeners in the direction of traffic to transfer the loading on the deck to the transverse girders. To avoid fit-up problems and weld crossings, the longitudinal stiffeners are combined with cutouts as in Fig.85 (also usual in shipbuilding). Trapezoidal hollow section stiffeners for railway bridges which are subjected to high levels of cyclic load are designed as in Fig.86 to minimise the notch effect.[78]

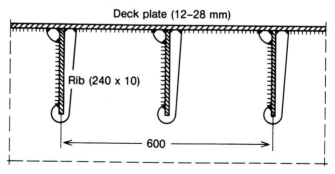

85 Longitudinal stiffeners of bridge deck plate combined with cutouts, after Neumann.[3]

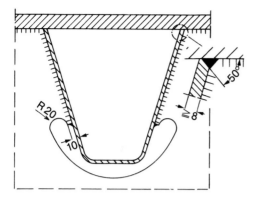

86 Longitudinal through stiffeners of railway bridges combined with
cutouts, after Haibach and Plasil.[78]

5.1.7   *Brace plates on I and channel section girders*

Gusset and brace plates are attached to the sides of I section girders (mainly
subjected to bending load) as mounting plates or as connecting members
against buckling, Fig.87. The reduction of fatigue strength caused by the
welded joint is considerable, unless the weld end toe is machined to a rounded
shape. If torsion is superimposed, there is an additional reduction.

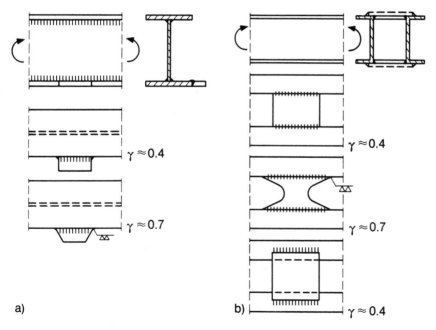

87 Gusset and brace plates of I section girder (a) and channel section girder
(b), design variants, reduction factors, γ, partially after Neumann.[3]

5.1.8   *Joints between I section girders*

Inseparable joints between girders and uprights, which are treated in structural theory as rigid to bending, must be reinforced adequately, and in the case of high quality versions, they also need to be given a rounded profile to avoid local overstressing (with regard to fatigue) and early loss of strength (with regard to collapse) after the plastic limit load has been exceeded, as shown in Fig.88 for corner joints. The corresponding separable joint is shown additionally. The corresponding T and cross joints can be designed as a symmetrical extension of the shape of the corner joints.

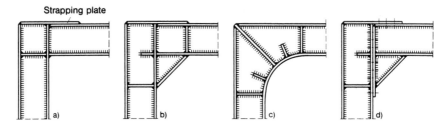

88   Corner joint of I section girder, design variants (a) to (d).

Joints between transverse and longitudinal girders are designed as separable (to be treated as bending free or bending rigid support in structural theory) or inseparable (to be treated as bending rigid support in structural theory), Fig.89. The transverse girder with separable joint is given a cutout with a machined rounded notch (machining is used to remove the micronotches caused by flame cutting) at the chord of the longitudinal girder. The inseparable transverse girder joint is relatively expensive and less fatigue resistant because of the upper transverse butt joint with the cope hole in the web. The corresponding double sided joints can be designed as a symmetrical extension of the shape of the single sided joints in Fig.89 (the separable joint then has a strapping plate).

The same considerations as above apply to ridge corner joints, Fig.90.

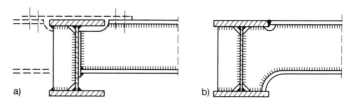

89   Transverse-to-longitudinal girder joint, separable (a) and inseparable (b) versions.

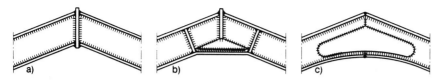

90   Ridge corner joint, design variants (a) to (c).

5.1.9   *Openings in the web of box and I section girders*

To reduce weight or to allow pipes or cables to pass through, large openings may be made in the web of box and I section girders. Openings of this type are connected with high levels of notch stress. The cross section is weakened, even if the openings are made in the area of the neutral zone (neutral with regard to bending stress). The distribution of transverse shear stress which has a maximum in the neutral zone, is extremely disturbed. The stress concentration at the edge of the opening can be reduced by reinforcing the edge whereby the rule of thumb can be applied that the material which corresponds to the size of the opening is distributed appropriately around its edge. Reinforcement produces a considerable reduction of notch stress. However, the notch stresses rise again if the reinforcement is too strong. Therefore the reinforcement has to be neither too weak nor too strong. Reinforcing a web plate hole by using a section of pipe is shown in Fig.91. Further details on openings are to be found in section 5.3.9. For determining the notch stress concentration at the non-reinforced or reinforced edge of the opening, see Ref.79, 80.

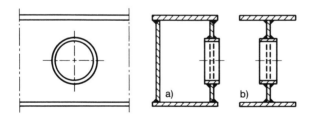

91   Web opening reinforcement: box girder (a) and I section girder (b).

5.1.10   *Special cruciform joints*

The special cruciform joints (c), (d) in Fig.92 are applied to structural steel engineering besides the conventional types (a), (b). They have the advantage that their dimensions are adjustable to the loading conditions considered and their shape is suitable for optimisation (see section 8.2.6). Additionally they are not at risk of failure by lamellar tearing. The fatigue strength reduction factors stated in Fig.92 have been determined for structural steel St 52-3 subjected to alternating load ($R = -1$) the reference stress being $\sigma_{A0} = 165$ N/mm^2 for $P_f = 0.5$, see Ref.81.

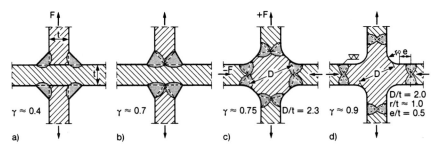

92   Cruciform joint: conventional types (a), (b), special types (c), (d); reduction factors, γ, with reference to $\sigma_{A0} = 165$ N/mm^2 for St 52-3, after Kaufmann.[81]

5.1.11   *Joint design, force introduction and stiffening of channel and angle section bars*

More or less thin walled channel section bars are preferred for use in framework structures, in vehicle construction in particular. Tubular constructions are only usual for frames which are expressly torsionally stiff. Vehicle frames and their members are subjected to loading from bending and torsion. The torsional loading on the frame arises from axle distortion and eccentric loading of the cargo area. Additional torsional loading on the frame members arises as a result of transverse forces acting outside the centre of shear. The critical points in frames of this type are the joints between side and cross members, the points where external forces are introduced, and the stiffeners. Welded components of this type can be so critical as regards fatigue that screwed or riveted joints are often preferred.

The centre of shear must be treated with particular care in the structural design of welded joints in frames of this type.[82,83] The centre of shear is the point in the cross section around which the moment of the internal shearing forces which arise in the presence of external transverse force disappears. In cross sections which are not point symmetrical the centre of shear does not coincide with the centre of gravity. Thus, in the channel section it is situated behind the back of the web and in the angle section it is situated in the intersection of the two section lines. The bending remains free of torsion only if the transverse forces are acting in the centre of shear. The introduction of transverse forces in the centre of gravity as a result of not knowing the interrelationship described above, causes torsion which reduces strength, Fig.93. As an example of the application of the above, Fig.94 compares favourable and unfavourable designs for the joint between bearing block and side rail in an agricultural trailer. Figure 95 gives a further example of the use of the unfavourable (at the centre of gravity) and favourable (in the centre of shear) introduction of transverse force into the channel section cross member of a T joint. In addition to the introduction of force in the centre of shear, positioning the fillet weld in the neutral axis of the cross bar subjected to bending, the unrestrained bending elongation and tensile transverse contraction and the free ability of the flanges to warp (described below) also have favourable effects. To supplement the above, a corresponding comparison is shown in Fig.96 for a T joint made of

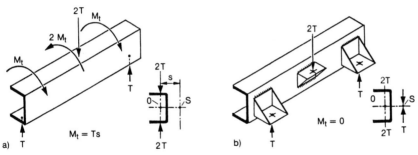

93 Introduction of transverse forces into channel section bar with torsion (a), and free of torsion (b). Centre of gravity, S, and centre of shear, T, torsional moment, M_t, with transverse force in the centre of gravity, rotation of the bar cross section around the centre of shear.

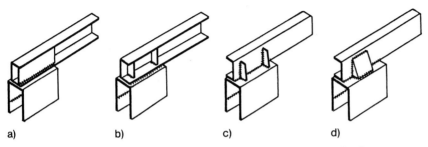

94 Bearing block to side rail joint on agricultural trailer: disadvantageous (a, b) and advantageous (c, d) design variants, after Kloth et al.[82]

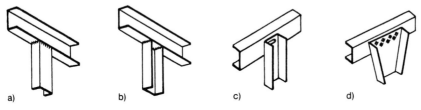

95 Introduction of transverse force into channel section bar: disadvantageous (a, b), and advantageous (c, d) design variants, after Kloth et al.[82,83]

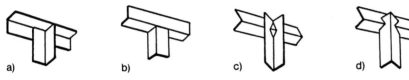

96 Introduction of transverse force into angle section bar: disadvantageous (a, b) and advantageous (c, d) design variants, after Kloth et al.[82,83]

angle section bars in which an exceedingly large gain in fatigue strength is achieved without increasing the manufacturing costs. In addition to the strength increasing aspects, which have already been mentioned, the section modulus of the cross bar is increased by being placed on edge. In addition, angle section bars are free from warping when subjected to torsion.

Moreover, in the structural design of welded joints in frameworks, the normal stress caused by suppressed warping should be given particular attention.[82,83] The proportion of the longitudinal displacement, which is not plane, of the cross section points in the presence of torsion in the bar, is designated warping. This arises in a particularly distinct manner in thin walled channel section bars (just as in I section girders), Fig.97(a). If warping of the cross section is restrained or suppressed by stiff end plates, unexpectedly high normal stresses occur in the direction of the bar (torsion bending stresses), Fig.97(b), which reduce the fatigue strength of these joints considerably. Unfavourable and favourable frame corner designs, reinforced and unreinforced, are compared in Fig.98. The non-warping angle section has a favourable effect as a transverse member as has the channel section with the flange ends cut off (to avoid torsion bending stress peaks caused by suppressed warping at the flange ends). Jointing in the neutral zone of the side rail via the centre of shear of the cross member is particularly advantageous.

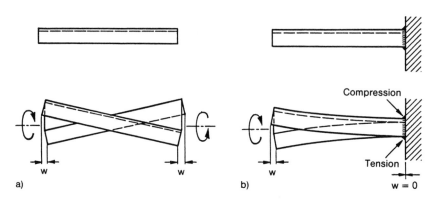

97   Torsion bending of a (thin walled) channel section bar: warping deformation, w, (a) and torsion bending stress (b).

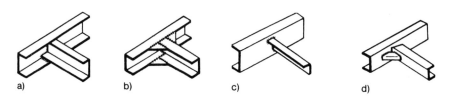

98   Frame corner joint: disadvantageous (a, b) and advantageous (c, d) design variants, after Kloth *et al*.[82,83]

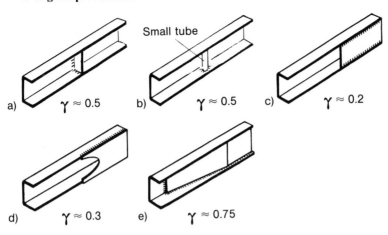

Small tube

a) $\gamma \approx 0.5$ b) $\gamma \approx 0.5$ c) $\gamma \approx 0.2$

d) $\gamma \approx 0.3$ e) $\gamma \approx 0.75$

99 Stiffening of channel section bars: advantageous (a, b), disadvantageous
(c, d) and very advantageous (e) design variants, reduction factors, γ, for
endurable torsional moment, partially after Kloth *et al.*[82,83]

The aspect of torsion bending stresses dominates the design of stiffeners in
channel section bars, Fig.99. Stiffener plates (a) or tubes (b) vertical to and
between the flanges have a warping restraint effect ('coupling against warping')
which leads to considerable torsion bending stresses and therefore to a
reduction of fatigue strength. The least favourable are stiffeners using a flange
connection plate (c, d). Here there is a sudden transition from the open cross
section, which is not torsionally stiff, to the closed cross section, which is
torsionally stiff (a factor of 500 for torsional stiffness is a frequent value in
practice), and which coincides with the weld end notch in (c) and with the
machined smooth profile in (d). There is a reduction of fatigue strength to less
than 0.25 compared with the smooth open or closed channel bar. The
advantageous gradual transition of torsional stiffness (e), in contrast, permits
the very favourable reduction to 0.75 to occur.[82,3]

5.1.12 *Conventional latticework node joints*

A system of rods connected by hinge joints (originally via bolts and rivets) is
designated a 'latticework', such as was used first in bridge girders. When
welding techniques came on the scene latticework node joints were welded
more and more, only the assembly joints were still bolted. Bending moments
were also transmitted via the welded joints; however, mostly as a secondary
effect. If the ability to carry bending moments is used purposefully, the
designation 'framework' is more accurate (*e.g.* for the Vierendeel truss
consisting of upper and lower chords plus vertical posts). Latticeworks made
from I, channel, T and angle section bars are designated 'conventional', unlike
the tubular latticework dealt with in the next section. Fully welded latticeworks
are not without their problems, not only for assembly reasons, but also with
regard to fatigue strength (secondary stresses and residual stresses). In a
partially welded latticework the gusset plates are permanently attached to the
chord members only.

Lattice girders subjected to bending (*e.g.* in bridges) consist of tension and compression chords which are joined together (vertically) via posts and (obliquely) via braces. Lattice compression rods (*e.g.* masts) also consist of chords and braces. Light latticeworks are designed with a single 'wall' (only one system plane of elements, *e.g.* upright I as chord with one web plane), heavy latticeworks with two walls (two system planes of elements, *e.g.* horizontal I as chord with two flange planes). The usual chord and brace (including upright) sections are listed:

— Chord sections: T, I, ⊓, H, ⊐⊏, ⊏⊐, ⊓⊓,
— Brace sections: T, ∧, ⊐⊏, <>, H, ⊦⊣.

With regard to the design of the node joints, those latticeworks which are mainly subjected to static loading, in roof trusses, uprights and masts are not problematic (a high notch effect is permissible), whilst the latticeworks of bridges, cranes and conveyors, which are subjected to fatigue loading, are not without problems (a high notch effect is not permissible). The node joints of the latticework, which are subjected to fatigue loading, should enable a steady flow of force to take place and therefore need to be well rounded at all notches and can only really tolerate butt and double bevel butt welds. Member joints, which are eccentric to the system plane, should be avoided. The fatigue resistant

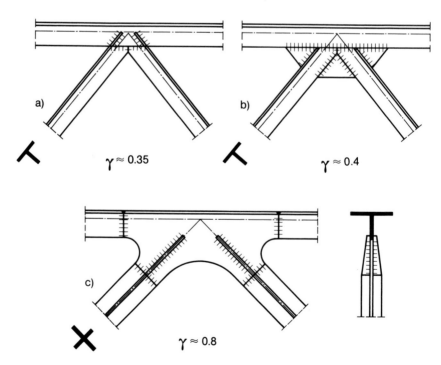

100   Light latticework node joints: disadvantageous (a, b) and advantageous (c) design variants, after Stüssi and Dubas.[10]

lattice node joint is an expensively manufactured flat component with bar connection stubs and large radii between them. Fatigue resistant lattice node joints have a streamlined shape and are expensive. Statically adequate dimensioning of the weld lengths, weld thicknesses and cross sections must be ensured, in addition to design measures for improved fatigue resistance.

The node joint of a light latticework, made of T section members, is represented in Fig.100. Version (a) should be loaded only statically, version (b) can be subjected to superimposed fatigue loading, version (c) with the rounded gusset plate can also be subjected exclusively to fatigue loading. The butt weld is preferred in all versions. Eccentric joints are avoided. The flange extension ribs in (a) can be omitted in (b) (as it has adequate weld lengths without ribs). The node joint of a heavy latticework made of horizontal I section girders is represented in Fig.101(a). As the weaker brace members must be of the same width as the chord to make the butt joint of the flanges possible, they are produced by using welded webs (instead of less advantageous brace plates). As a result of the strong notch effect, only loading which is mainly static is permissible. The design suitable for fatigue loading (b) uses upright I section girders with special node joint elements. Version (c), which is particularly favourable as regards notches, with only one post butting on to the chord vertically (Vierendeel joint) has not, however, become popular.

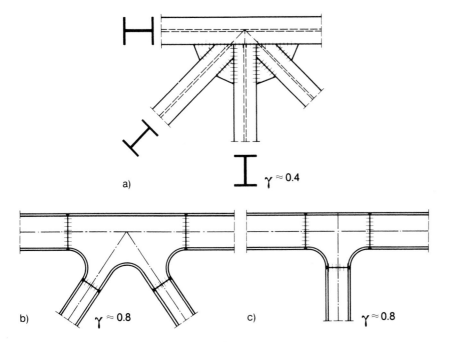

101   Heavy latticework node joints: disadvantageous (a) and advantageous (b, c) design variants, after Stüssi and Dubas.[10]

5.1.13   Tubular latticework node joints

Latticeworks made purely of tubular steel, or in combination with conventional open section bars as chords, have become popular, especially in the construction of offshore oil drilling rigs. The higher price of tubes is a factor against tubular lattices, but this is counterbalanced if the loading is mainly static as a result of more advantageous compression member behaviour; in the presence of fatigue loading, this factor is more important because of the very low fatigue strength of the welded tubular joints. Tubular latticeworks can be produced using tubes with a circular or rectangular cross section. The cheaper circular tube requires more expensive machining of the edges of the joint, corresponding to the line of intersection of the tubes which are butted together. The more expensive rectangular tube can simply be sawn to produce the edge for the welded joint. The plane tube arrangements most frequently found, which are subjected to axial or bending load in this plane, are represented in Fig.102 and 103. They are designated by the capital letters of the alphabet to which their arrangement is similar. The tube joints can be reinforced by box or plate gussets or by rings. However, increasing the thickness of the wall in the area of the joint is often the more economical method. In space frames of roofs and scaffolds separable connectors are frequently used instead of welded joints.

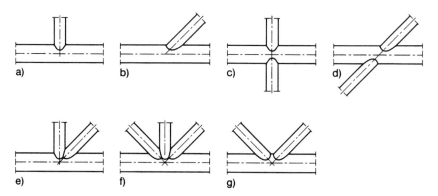

102   Tube arrangement in node joint with brace tubes subjected to axial loading: T, Y, double T, TY (= N), KT and K types, after Wardenier.[86]

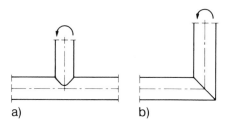

103   Tube arrangement in joints subjected to bending load: T and L type.

Fatigue cracks in welded tube joints can be initiated at various points of the length of the weld from one of the two weld toes or from the weld root. They can propagate in the brace or chord tube, or penetrate the weld. The multiplicity of possible fractures makes the assessment of fatigue strength difficult.

Welded tube joints have relatively low fatigue strengths[2,84-88] reduction factors $\gamma = 0.45\text{-}0.30$ and less (with reference to the fatigue strength of the unwelded brace tube), and are allocated to the notch classes with severe to very severe notch effect ($\gamma = 0.4\text{-}0.2$). High thickness of the chord tube and partially overlapping sectional areas of the braces in the joint have a favourable effect. After more precise strength and stress analysis the tube joints can in fact be allocated to notch classes with differing levels of reduction, depending on geometrical parameters, type of load and quality of manufacture. Differing endured nominal stresses, which are open to a normalised representation in the S-N and fatigue strength diagrams, would be a part of this (see sections 3.1.1 and 2.3.1). Reference stresses suggested as an alternative such as 'punching shear' in the line on which the brace tube butts against the chord tube or circumferential stresses in the chord tube, determined in a simplified way, may only improve the prediction of strength in combination with the above-mentioned nominal stresses for the brace tube.

The highest structural stresses ('hot spot stress', see section 7.7) have been established as the basis of a uniform method of evaluation of test results. For the tube joints which have been investigated, these are the highest surface stresses in the area of the joint between the brace and the chord, immediately in front of the weld toe, after subtracting the increase of the notch stress in accordance with an established procedure. These stresses can be measured by means of strain gauges or be calculated by the finite element method. As the structural stresses characterise the failure initiating notch stresses to a limited extent only, there were still uncertainties in the evaluation of the test results.

The highest structural stress at the crack initiation point, referenced to the nominal stress in the brace tube at the point where the system line passes through the surface of the chord, gives the structural stress concentration factor K_s (see equation (44)) which is dependent on the dimension ratios (those which are independent of each other) of the tube joint, Fig.104, (the following assumes circular tubes, symmetrical braces and uniform brace cross sections):

$$K_s = f(\alpha, \beta, \gamma, \tau, \theta, \varepsilon) \tag{25}$$

The geometrical parameters have the following definitions:

$$\alpha = \frac{2L}{D} \tag{26}$$

$$\beta = \frac{d}{D} \tag{27}$$

$$\gamma = \frac{D}{2T} \tag{28}$$

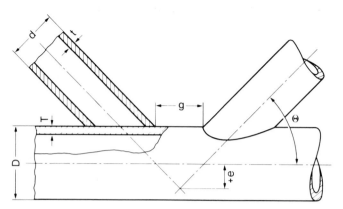

104   Geometrical parameters of tubular node joints.

$$\tau = \frac{t}{T} \tag{29}$$

$$\varepsilon = \frac{2e}{D} \tag{30}$$

with length of chord between two nodes, L; diameter of chord tube, D; diameter of brace tube, d; thickness of chord tube, T; thickness of brace tube, t; angle between chord and brace, θ; eccentricity of the intersection of the brace centre lines, e.

Occasionally the gap, g, and the overlap, q/p, which are particularly influential as regards strength, are used additionally. They can be calculated from the above-mentioned parameters, Fig.105.

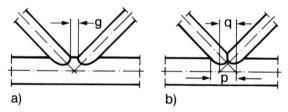

a)   b)

105   Gap, g, and overlap, q/p, for tubular joint.

A particularly simple basic form of the function, f, in equation (25) follows as a linearised appproximation to the logarithms of the above-mentioned geometrical parameters (therefore it is only sufficiently accurate in a limited range of the parameters):

$$K_s = k\alpha^{n_1} \beta^{n_2} \gamma^{n_3} \tau^{n_4} (\sin\theta)^{n_5} \varepsilon^{n_6} \tag{31}$$

where k and n_1 to n_6 are constants.

The large number of approximation formulae, which are actually proposed[86] use the above-mentioned set-up with slight modifications. The accuracy of these approximations is questionable in some cases.

If tension, in-plane bending and out-of-plane bending are acting simultaneously, superimposition of the relevant K_s values is only permissible if the maximum structural stress values occur at the same point and in the same direction.

K_s values of the order of magnitude of 10 or 20 can occur on welded tubular joints with a corresponding very large reduction of the fatigue strength. As a result of the unusually strong local increase of structural stress, which covers a fairly large area, it was possible to prove that there was a correlation between the static strength and the fatigue strength of the tubular joint. It is therefore recommended to use the guidelines for fatigue resistant design in the presence of static loading too.

It was possible to draw the practical conclusions below from fatigue tests on tubular joints in addition to the classifying framework which is expressed by the structural stress concentration factors. The local deflection of the tube wall of the chord caused by the brace tube has a dominating effect on the strength. Therefore a greater thickness of the chord tube, a tangential force flow into the chord tube, and overlapping brace ends have a strength increasing effect; a V butt weld or single bevel butt weld is preferable to a fillet weld; additionally the dimension ratios should be favourably chosen.

Node joints of tubes with a circular cross section require careful groove preparation at the ends of the braces, usually on specialised machines, to produce gap-free fit-up for a fillet weld, or a fit-up with a V or single bevel groove for a butt weld. Simpler preparation is required for joints of tubes with a rectangular cross section. A fillet weld is only permissible when the butting angle is greater than 60°, otherwise a butt weld with V or single bevel groove should be selected, Fig.106.

In addition to node joints, butt joints in the chord tubes also occur in practice, mainly to produce adequate chord lengths. The chord joint can be produced either as a butt weld joint or a fillet weld joint with a crosswise intermediate plate. The butt weld joint can be subjected to higher levels of loading. Moreover, there is a risk of lamellar tearing with the fillet weld joint. Rectangular tubes are worse than tubes with a circular cross section, because of the less favourable notch conditions at the corners of the rectangular cross section.

Reference should be made to section 7.7 with regard to the dimensioning of tubular node and butt joints on the basis of the structural stresses.

A design element occasionally seen in practice is the tubular strap joint, consisting of a slotted or flattened tube end with a gusset plate attached into or on to it, Fig.107. Tubular strap joints have relatively low fatigue strengths.[89] The most advantageous is the flattened tube end with a butt joint to the gusset plate (economical but limited by the deformability of the tube).

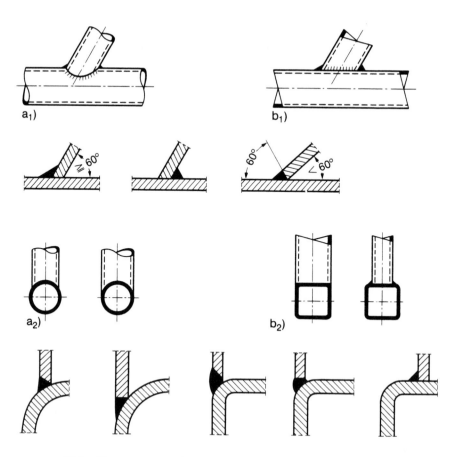

106   Groove preparation and weld position at node joints of circular tubes
(a_1, a_2) and rectangular tubes (b_1, b_2) in transverse view (a_1, b_1) and
cross section (a_2, b_2).

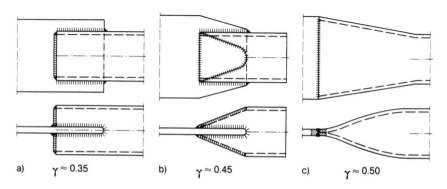

107   Tubular strap joint, design variants, after Wellinger *et al.*[89]

5.2   Welded components in shipbuilding (bar to plate structures)

The main components of a ship's hull are extensive plates and shells with load bearing and stiffening bars inside. For operational reasons, oblique and arched surfaces are more frequent than in structural steel engineering, to which similarities exist on the other hand from a design point of view (therefore see also section 5.1). The ship's hull is subjected to fatigue loading by torsional and bending moments caused by sea waves. Vibrations excited by the propulsion system are superimposed. In addition, variable external pressures from the water and internal pressures from the cargo caused by the loading and unloading cycles have to be taken into account. Comparisons of designs with regard to fatigue were carried out in Japan in particular.[90,91] The values given below for the increase of structural stress characterised by the structural stress concentration factors K_s or K_{s0} are based on measurements on the component (usually reduced in size) using strain gauges, on measurements on photoelastic models and on computations by the finite element method. They are generally checked with regard to the endurable structural stresses by fatigue testing. However, the values given below should be used with care in so far as the additional dependence on the dimensional situation is only covered by the special cases investigated.

5.2.1   Corner of double bottom (bilge corner)

The transition between the bottom of a ship and its sidewall, designed as double plates as in Fig.108 (typical of container ships) is subjected to loading from axial force, bending moment and transverse shear force, put into effect in the tests, to which reference will be made below, by the diagonal tensile force in Fig.108, which causes a specific combination of the stated forces in the corner cross section.

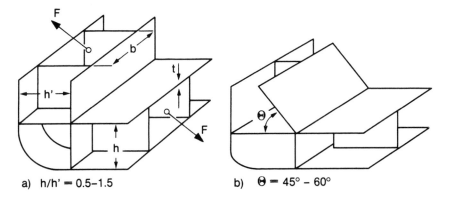

a)   h/h' = 0.5–1.5   b) Θ = 45° – 60°

108   Double bottom corner joint (bilge corner), design variants (a, b), subjected to diagonal tensile force, F, in the fatigue tests, after Iida and Matoba,[90] and Matoba et al.[91]

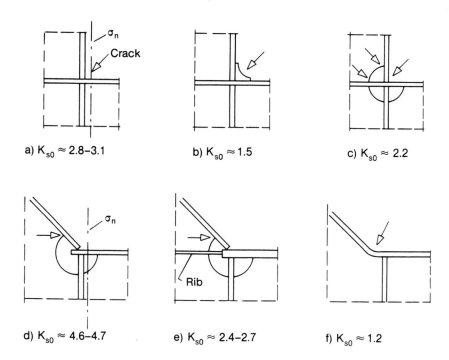

a) $K_{s0} \approx 2.8\text{-}3.1$ b) $K_{s0} \approx 1.5$ c) $K_{s0} \approx 2.2$

d) $K_{s0} \approx 4.6\text{-}4.7$ e) $K_{s0} \approx 2.4\text{-}2.7$ f) $K_{s0} \approx 1.2$

109   Double bottom corner joint, design variants (a) to (f), structural stress
concentration factors, K_{s0}, after Iida and Matoba,[90] and Matoba
et al.[91]

The structural stress concentration factor is represented as:

$$K_s = K_{s0}K_{cw} \qquad (32)$$

subdivided into the factor K_{s0} for a full load carrying width of the component
and the factor K_{cw} as a result of the reduced load carrying width of the
component.

The range of the width between webs $b/h = 0.4\text{-}0.6$ produces $K_{cw} = 1.2\text{-}1.8$.
A smaller width between webs is therefore advantageous from the point of view
of stress concentration.

The basic design shapes with right angled or oblique shaped inner corners
are compared in Fig.109. The crack initiation points are indicated by arrows.
The fatigue strength can be estimated via K_{s0}. The following design features
have a favourable effect:[90,91]

—   rounded corner with cover plate without a welded joint;
—   rounded gusset plate in a right angled corner;
—   rib as a continuation of the bottom plate with an oblique angled corner.

Compared with the right angled corner, the oblique angled corner does not
produce any obvious advantage in the range of dimensions investigated.
Designing a rounded cutout in the web at the corner is not a favourable measure
(compare section 5.1.5).

5.2.2 *Corner of sidewall frame to double bottom*

The corner joint between the sidewall frame and double bottom is differently designed (compared with the solutions in the previous section), if free standing sidewall girders butt against the double bottom at a right angle or obliquely, Fig.110, (typical of cargo vessels). This component was investigated subjected to bending from transverse force, T. The basic design shapes are compared in Fig.111. The following design features have a favourable effect if there is an oblique angled joint:[90,91]

— small oblique angle;
— reinforcement by a thickened gusset;
— base plate or collar plate in the corner;
— adjusted thickness of cover plate strip;
— tapering of the cover plate strip;
— dovetail joint for cover plate strip.

In the case of a right angled joint, Fig.112, the girder extending into the double bottom is sufficient.[90,91] Using a dovetail joint for the cover plate strip produces no advantage.

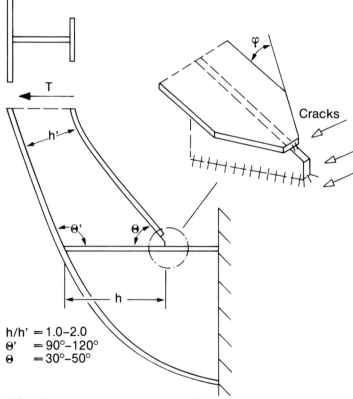

h/h' = 1.0–2.0
θ' = 90°–120°
θ = 30°–50°

110 Framework corner joint, sidewall frame to double bottom, subjected to transverse force, T, in the fatigue tests, after Iida and Matoba,[90] and Matoba *et al.*[91]

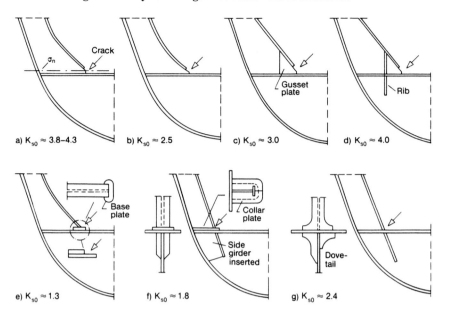

111   Framework corner joints, with oblique angle, design variants (a) to (g), structural stress concentration factors, K_{s0}, after Iida and Matoba,[90] and Matoba *et al.*[91]

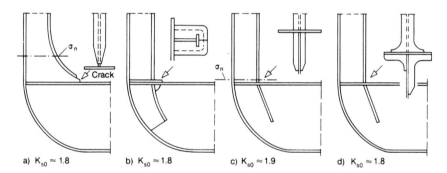

112   Framework corner joints, with right angle, design variants (a) to (d), structural stress concentration factors, K_{s0}, after Iida and Matoba,[90] and Matoba *et al.*[91]

5.2.3   *Corner joint of bar stiffeners*

The corner joint of bar stiffeners of the plates and shells of a ship's hull can be designed in several different ways. The joint was investigated[90,91] subjected to diagonal tensile loading. The result is shown in Fig. 113. The reinforcing plate or rib on the gusset plate is introduced to reduce the risk of buckling and vibration. As a result of the additional notches it has an unfavourable effect on the fatigue strength. Giving the ends of the gusset plate a smooth profile is recommended with regard to the fatigue strength. K_{s0} is surprisingly favourable for the joint as

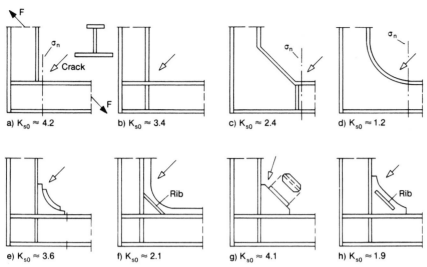

113   Corner joint of bar stiffeners subjected to diagonal tensile force, F, in the fatigue tests, design variants (a) to (h), structural stress concentration factors, K_{s0}, after Iida and Matoba,[90] and Matoba *et al.*[91]

in Fig.114, which is only partially connected showing a relatively weak reinforcement and which has a relatively low static load carrying capacity. An explanation of this can be seen in the position of the weld, close to the neutral zone of the stiffening bars.

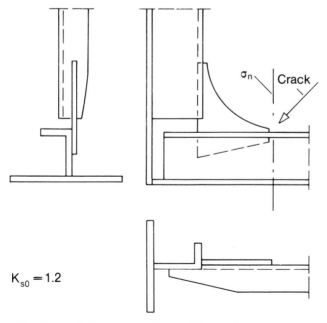

114   Corner joint of bar stiffeners, lightweight design, structural stress concentration factors, K_{s0}, after Iida and Matoba,[90] and Matoba *et al.*[91]

5.2.4   *Longitudinal assemblage*

A further critical welded component of a ship's hull is the joint of a longitudinal partition bulkhead (alternatively longitudinal girder) ending at a transverse partition bulkhead under continuation of the deck plate, Fig.115 (longitudinal assemblage). The most important loading is the longitudinal tensile or compressive loading of the cover plate. The result of systematic fatigue tests with stress calculations and measurements on components made of higher tensile shipbuilding steel (DH 36) is represented in accordance with Ref.92. The reduction factor, γ, which is shown, is the ratio of the endurance limit for pulsating load of the welded component (failure probability $P_f = 0.1$) to the endurance limit for pulsating load of the mill finished structural steel St 37 (240 N/mm²). The structural stress concentration factor, K_{s0}, shown is the mean value from calculation and measurement. The strain range $\Delta\varepsilon = 0.01$ (corresponds to 210 N/mm² under uniaxial stress conditions) is stated as being the local endurance limit ($P_f = 0.5$) for pulsating load.

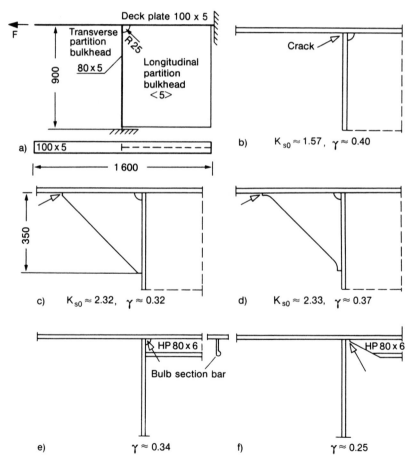

115   Longitudinal assemblage (a), design variants (b) to (f), structural stress concentration factors, K_{s0}, and reduction factors, γ, after Petershagen.[92]

5.2.5 *Cutouts and cope holes*

In ship structural design, many cutouts occur of various size, number and shape which have to fulfil different tasks. The shape, design and dimensioning of these cutouts is accomplished on the basis of the structural stresses at the non-reinforced or reinforced edge which must remain lower than local strength values. Both engineering formulae and finite element methods are available for assessment of the structural stress concentration.[79,95,209]

A special type of cutout is the web hole in the double bottom transverse girder which serves for continuing the longitudinal ribs, Fig.116. Its structural stress state has been repeatedly evaluated[95-97] for tensile load, transverse load and bending moment in the girder superimposed by external hull pressure.

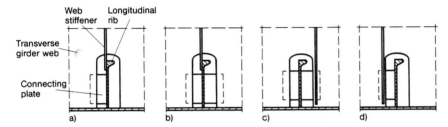

116 Web hole in double bottom transverse girder for continuing the longitudinal ribs, design variants, after Fricke.[96]

Intermittent welds are combined in shipbuilding with cope holes at the weld ends which close the weld to exclude crevice corrosion. Such holes are also to be found at the flange butt joint of I section girders. Test specimens made of higher tensile steel, as in Fig.117, produced the reduction $\gamma \approx 0.33$. A different investigation on a girder subjected to bending produced structural stress concentration factors, K_s, as in Fig.118, maximum value of structural stress at the upper or lower notch, relative to the nominal bending stress at the same point. The values, K_s, are dependent on the shape of the cope hole and the ratio of the thickness of the plates, amongst other things. The relatively high K_{s1} values of the design in the middle can be explained by the proximity of the weld ends. Cracks propagate into the web and flange plate in the case of intermittent welded web-flange joints in girders without cope holes. They propagate into the flange plate only in the case of a joint with cope holes.[95]

Further design details from shipbuilding are investigated in Ref.93, among them the ends of cover strips welded on to a tensile plate.

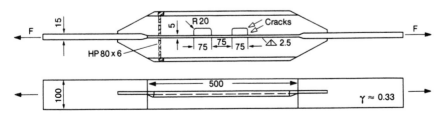

117 Test specimen with two weld end cope holes, reduction factor, γ, after Petershagen and Paetzold.[92]

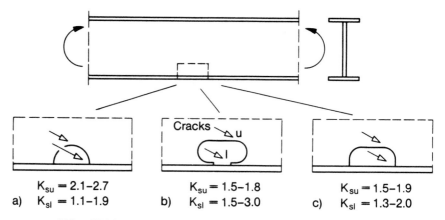

$$K_{su} = 2.1\text{--}2.7 \qquad\qquad K_{su} = 1.5\text{--}1.8 \qquad\qquad K_{su} = 1.5\text{--}1.9$$

a) $\quad K_{sl} = 1.1\text{--}1.9$ b) $K_{sl} = 1.5\text{--}3.0$ c) $K_{sl} = 1.3\text{--}2.0$

118   Weld end cope holes, design variants (a) to (c), structural stress
concentration factors, K_{s0}, for upper (index u) and lower (index l)
notch, after Matoba *et al.*[91]

5.2.6   *Hatch and deckhouse corner*

The fatigue problem at hatch corners has become particularly well known as a
result of the cases of brittle fracture (starting from fatigue cracks) in US Navy
supply ships (Liberty Ships) during the Second World War. The inadequate and
the improved designs are compared in Fig.119. The following design measures
have a favourable effect:

—   giving a more generously rounded profile to the corners (however, this
    increases the length of the vessel);
—   increasing the plate thickness at the corners;
—   removing the weld from the corner position.

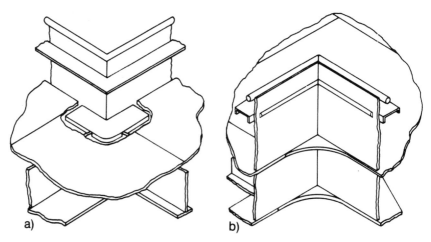

119   Hatch corner in a supply ship, inadequate (a) and improved (b)
design variants.

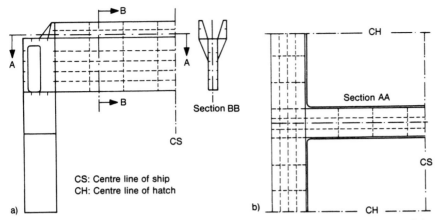

120   Hatch corner in a container ship: cross section (a) and plan view (b) of
ship, after Hapel and Just.[94]

The fatigue problem at hatch corners has become even more important in
later times, especially in container ships, for which it is well known that the
decks are designed more or less open for loading and unloading. The doubled
sidewalls in this type of vessel are only connected by several box-type
transverse ties, Fig.120. If the ship's hull is subjected to torsional moments
(when running obliquely in a heavy sea) warping occurs on the open hull of the
ship. This is restrained by the transverse ties which are loaded by a transverse
bending force. The transverse bending force in the ties causes a critical stress
concentration in the hatch corner. The latter is intensified by the fact that the
bending moment mainly runs via the cover plates in the tie to sidewall
connection area and the vertical walls of the box-type tie have little load
carrying effect. The increase of structural stress (including the notch effect of
the rounded corner) referenced relative to the nominal bending stress in the
connection cross section (calculated without the rounded corner) is composed
of the structural stress concentration factor, K_{s0}, of the plane model of the
transverse tie and the structural stress concentration factor, K_{ch}, of the reduced
height, the effective load carrying height, of the vertical walls of the box-type
ties in the connection.

$$K_s = K_{s0}K_{ch} \tag{33}$$

The structural stress concentration factor, K_{s0}, was determined
systematically both by functional analysis and finite element methods and by
photo-elasticity for the model shown in Fig.120 subjected to varying load cases
dependent on the dimension ratios r/h and b/h.[94] With r/h $\approx$ 0.2, the factor K_{s0}
$\approx$ 1.4 was produced for existing ship designs. The factor, K_{ch}, is dependent on
the ratio of the section moduli of the cover plate and the whole box (values in
practice 0.4-0.8) and on the design details of the joint at the vertical box walls
(e.g. corner reinforcing gusset plates or increased wall thickness). Values of K_{ch}
which are relevant to real conditions lie close to 2.0.

The transition corners from the deck girders to the deckhouse wall are notch
areas which are critical as regards fatigue. The notch effect can be reduced by
design measures (gusset plate with rounded profile or rivets instead of welding).[98]

5.3   Welded components in tank, boiler and pipeline construction (circular shell and plate structures)

Circular shells and plates predominate in tank, boiler and pipeline construction. A collection of design examples for welded joints in this field can be found in Ref.99. The primary loading is internal pressure which may pulsate as a result of start-up and shutdown cycles. Superimposed on the latter are temperature stresses caused by temperature differences in the shell or plate (usually up to 100°C, thermal shock in special cases), which can also occur in a pulsating manner. Stress analysis of this type of component by measurement and calculation is highly developed. The shape-dependent, inhomogeneous membrane and bending stresses, *i.e.* the structural stresses, are generally sufficiently well known, as are the structural stress limits which can lead to the initiation of fatigue cracks. As the wide range of knowledge on the distribution of structural stress in this technical field is dealt with in detail in Ref.9, the representation here can be restricted to a few basic facts and questions concerning the strength of the welded joints. Large scale tanks, which are similar from a design point of view, are not covered here since they are mainly subjected to static loading.

5.3.1   *Pipes and tanks with longitudinal and spiral welds*

Cylindrical pipes can be rolled without seams or, for large diameters, be butt welded via longitudinal or spiral welds. First the fatigue strength of the seamless rolled (and therefore mill finished) pipe subjected to internal pressure is of interest. The fatigue strength of welded pipes is compared with the latter to indicate the reduction of fatigue strength in each case. The endurance limit of the seamless pipe with striation and lack of roundness caused by rolling ($\Delta d/d$ $\leqslant 0.1$ in accordance with normal terms of delivery) is reduced corresponding to the fatigue notch factor $K_f \approx 2.5$ compared with the polished smooth specimen. The influence of the roughness caused by rolling is included in this factor, but the increase of circumferential stress on the inside of the (ideally round) thicker walled pipe is taken into account separately.[9,27] Thus for seamless steel pipes made from unalloyed and alloyed steels with $\sigma_U =$ 400-700 N/mm², the pulsating fatigue strength is $\sigma_P =$ 140-200 N/mm². The same factor applies for high frequency resistance welded pipes with longitudinal seams. In contrast, values of $\sigma_P =$ 80-115 N/mm² are usual for submerged-arc welded pipes with longitudinal or spiral welds which correspond to a reduction compared with the seamless pipe of $\gamma \approx 0.6$. The relevant lower limit curves for endurable stresses (also in the fatigue strength for finite life range) corresponding to the German standard for steel pipes, are specified in section 7.6.2. The stress increasing influence of the lack of roundness can approximately be covered by calculation also,[100] which then permits correspondingly higher endurable stress.

Whilst no difference between longitudinal and spiral welds was established in the data above from fatigue tests on tubular tanks,[100] in the subsequent investigations[101] a difference was established. It was found that the spiral weld (inclination towards longitudinal direction about 35°) endures a number of cycles about ten times greater under simple circumferential stress, σ_{ci}, without axial stress, σ_{ax}, compared with superimposed circumferential and axial stress ($\sigma_{ci} = 2\sigma_{ax}$), as it occurs in tubular tanks. Identical numbers of cycles were established for both loading cases for the longitudinal weld. The explanation is derived from investigations into notch stress in the oblique weld (see section 8.2.6) carried out[101] with $K_f = 1.5$ for the weld notch and $K_f = 2.0$ for the shape deviation of the pipe at the weld (ridge formation). The differing strength behaviour of longitudinal and oblique welds in the presence of simple circumferential stress is connected with a differing fracture position. In the longitudinal weld the fracture runs in an unchanging direction in the weld toe or weld root notch, in the spiral weld it runs in the weld area in steps across the weld. In both cases the fracture surface tends to be perpendicular to the direction of the first principal stress (therefore axial).

5.3.2   Pipes and tanks with circumferential welds

The reduction factors for butt and fillet welds in accordance with sections 2.1.2 and 2.1.4 for simple loading are directly applicable to circumferential butt and fillet welds in pipes and tanks, Fig.121, because there are no weld ends. And so is the equivalent stress formula as in equation (5), which is taken from a relatively old Swiss specification for the design of vessels and pipelines in the presence of multiple loading of the weld (see also tubular butt joint in structural steel engineering, section 5.1.13). The butt weld is made with or without an inserted ring. The inserted ring (a) which shapes the root is removed after welding. The inserted ring (b) which remains in position and is at risk of crevice corrosion (b) is used if a smooth inner surface is to be achieved. The joint, which produces the least fit-up problems by using a fillet weld, (c), can withstand less loading and is also subjected to crevice corrosion. The bulge has a strength reducing effect if the pipe is subjected to longitudinal loading as a result of the increase in the structural stress.

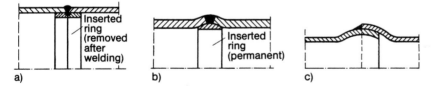

121   Pipe joint: butt weld with ceramic backing ring (a), butt weld with welded-on backing ring (b), and fillet weld (c).

5.3.3   Pipe wall

The temperature gradient through the wall thickness is the principal factor for design and dimensioning (if necessary against high temperature plastic fatigue) of the (gas-tight) pipe walls used in high performance boilers ('finned pipes'), Fig.122.

122   Pipe wall of a boiler.

5.3.4   *Pipe bends*

Pipe bends are either manufactured smooth or (to increase their flexibility) with peripheral corrugations. In practice they are predominantly loaded by internal pressure and bending moments. When subjected to bending in particular, the inside of the bend experiences excessive structural stresses compared with the outside (for explanation and calculation see Ref.9). Circumferential corrugations, used to increase the bending flexibility, have superimposed shell bending and notch stresses in their apex. When subjected to internal pressure, no reduction has been established in the fatigue strength of pipe bends compared with straight pipes, provided that the lack of roundness, $\Delta d/d$, remains below 3%. However, a reduction does occur if the lack of roundness is greater. A smooth pipe bend is clearly superior to a corrugated pipe bend with regard to fatigue strength when subjected to bending load. As the hoped for bending flexibility of the corrugated pipe bend is only established to an inadequate extent, there is actually no advantage in this design. If there are welds in the bend area, their notch effect or fatigue strength reduction is superimposed.

5.3.5   *Pipe branches*

Pipe branches, particularly those in high pressure pipelines, may have high structural stresses and therefore correspondingly increased notch stresses as a result of the large unsupported wall areas of the branch under internal pressure. Webs and collars are welded on to the outside of high pressure pipe branches in the axial and circumferential direction to reduce these. The same effect is achieved by a shaped plate welded into the branch in the symmetry plane.

5.3.6   *Tubular bellow compensators*

Tubular bellow compensators are used to accommodate longitudinal or transverse displacements, caused by heat, which take place in pipelines, largely without constraint (axial compensator or flexible compensator). The bellow rings are inserted into the compensator using welded joints. High radial bending stresses can occur in the plane radial surfaces of the bellow rings, in addition to the circumferential stresses caused by internal pressure, as a consequence both of internal pressure and of axial tension or axial bending. The strength limit is introduced as an endurable strain range. Approximation formulae are available for practical calculations.[9,27]

5.3.7 *Pipe and vessel flanges*

Pipe flanges and, by analogy, vessel flanges consist of the cylindrical pipe or shell end, the flange in a narrower sense, acting as a ring plate, and, if necessary, a conical transition ring. Figure 123 shows welded joints of various types and positions on this component, *i.e.* versions for low, medium and high levels of loading. The flange is subjected to loading in a complicated manner by internal pressure, axial forces, bending moments, sealing and screw forces. The (nominal) stress assessment should be carried out for several prescribed sections.[9] The weld toe notches require particular attention with regard to notch stresses. It is advantageous to remove the weld from the transition area. There are no systematic investigations on the fatigue strength of welded flanges available.

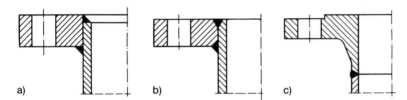

123 Pipe and vessel flange for low (a), medium (b) and high (c) levels of fatigue loading.

5.3.8 *Vessel nozzles and pipe branches*

Pipe and vessel nozzles are the most critical welded elements in pipe and vessel construction as regards fatigue strength. Their design and dimensioning must be undertaken with particular care.

Analysis of structural and notch stresses reveals the behaviour described below (see also Ref.103). Cutting a circular opening necessary for transfer of gas or liquid introduces high stress on the edge of the opening. For comparison a plate with a hole can be used as a basis, subjected to the principal stresses $\sigma_{ax} = 0.5\sigma_{ci}$ for a cylindrical vessel, or subjected to the principal stresses $\sigma_{me} = \sigma_{ci}$ for spherical vessels, which leads to the stress concentration factors $K_k = 2.5$ or 2.0. These stress concentration factors are increased by the extent of the weakening of the circumference by the hole, if a vessel is considered instead of the plate. In addition, a notch transverse bending effect occurs dependent on the curvature of the shell, t/R, and on the nozzle-shell diameter ratio, r/R, which can result in multiples of the above mentioned stress concentration factors. The notch stress maximum for the nozzle in a cylindrical vessel is found in the axial apex of the hole on the inside of the shell; for the nozzle in a spherical vessel it occurs over the whole inner edge.

There is therefore a need to reinforce the edge of the hole. This takes place to a limited extent via a weld jointed nozzle pipe, which penetrates the hole or butts against it, possibly also by means of a ring shaped plate placed on the vessel shell reinforcing it in the area of the nozzle. Even if the pipe penetrates the wall of the vessel, the reinforcement is more efficient on one side (in the direction of the outside of the vessel, whilst the notch stress peak is on the inside of the vessel) and only a narrow section of the pipe end close to the hole has an adequate strengthening effect. Thus the reinforcement can only reduce, not remove, the stresses on the edge of the hole. The stress peak on the inside edge of the hole still remains, although it is less distinct. Welds increase the notch effect if they coincide with the above mentioned critical point or line. The more effective measure, namely a reinforcing ring on the inside of the vessel, is usually not possible for manufacturing and operating reasons. If, on the other hand, a particularly strong reinforcement is selected on the outside, the notch effect in the transition area near the reinforcement on the outside of the vessel can exceed the notch effect on the inside. The former has its highest value on the outside in circumferential direction for cylindrical vessels, the latter on the inside in axial direction. Reinforcing rings are also fitted without a weld on the outside to avoid the above mentioned reinforcement notch effect, if they can be positioned effectively (possible for nozzles on a spherical vessel). Welded-on reinforcing rings and collars are limited to special cases.

Therefore the structural and notch stress states at nozzles require particularly careful analysis in the presence of fatigue loading, if necessary also using expensive measuring and calculation procedures. Here a nozzle reinforcement, which is adequately dimensioned from a static strength point of view should be used as a basis. Structural stress concentration factors K_s = 2.0-3.3 are determined in calculations for the particularly well designed nozzles in nuclear reactor construction. More normal nozzle designs have a higher factor.

The increases of stress considered so far apply to structural stresses. The actual notch stresses arising from shape transitions, corner or rounding, or from the toe or root of the weld, which is applied there, are to be superimposed on the structural stresses. As a result of the much smaller notch dimensions compared with the nozzle dimensions, the corresponding structural and notch stress concentration factors should be superimposed by multiplication as a first approximation.

In addition to the above mentioned considerations on the increase of notch and structural stress, aspects concerning the risk of corrosion and the flow interference have to be given particular attention. Open slits make crevice corrosion possible and should be avoided. Nozzle pipes, which penetrate the vessel wall, cause dead areas of flow in which corrosive materials tend to be deposited. The nozzle shape illustrated in section 8.2.5, optimised as regards the notch effect, has probably not become popular because of the dead area.

Some basic nozzle design variants (including pipe branches) are shown in Fig.124, in which only one of several possibilities is illustrated in each case as regards weld position, type of weld and groove shape and, of course, only one specific combination of dimensions. The nozzles (a, b), which butt against the vessel or are inserted, offer little reinforcement acting on the wrong side of the wall and high notch stresses arise in the acute angled transition areas.

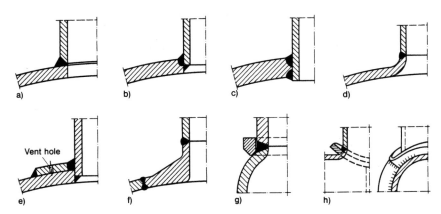

124 Nozzle design variants (including pipe branches): butted or inserted
nozzle pipe (a, b), penetrating nozzle pipe (c), necked out vessel wall (d),
annular plate reinforcement (e), section shaped nozzle ring (f),
unwelded reinforcing ring (g), collar reinforcement (h).

Penetration of the vessel wall by the nozzle pipe (c) improves the reinforcing
effect strongly and on the right side of the wall, but is unfavourable as regards
flow and corrosion factors. Necking out of the vessel wall (d) (or of the nozzle
pipe) does not offer much reinforcement, but, on the other hand, reduces the
notch effect strongly and moves the weld outside the zone of high notch
stresses. An annular plate placed around the nozzle pipe on the vessel wall (e) is
effective as regards reinforcement, but is combined with relatively high notch
stresses on the face and nozzle pipe sides. Adjusting the annular plate to shape
is also expensive in labour. A section shaped ring, (f), fitted using butt welds, is a
particularly advantageous solution which is, however, correspondingly
expensive. A design with a collar welded on (h) is effective as regards
reinforcement but is also notch intensive; moreover, for sufficiently good
support, the collar needs to go round the main body (sphere, pipe or tank) as
well as the nozzle pipe. An alternative design in the case of bifurcated pipe
branches is a sickle shaped internal web plate. A reinforcing ring (g) applied
without welding is only suitable for nozzles on spherical vessels because only
here is the joint suitable for the necessarily planar reinforcing ring. It represents
an example worth noting for the avoidance of the high notch stresses of a
transverse fillet weld by dropping the weld completely (comparable with a
transverse stiffener in bridge construction, which is not connected to the
tension chord, section 5.1.5). A further effective measure for equalising notch
stresses is represented by necking out the cylindrical vessel wall in an ellipse
(major ellipse axis in the circumferential direction).

A fatigue fracture can run in various ways depending on the shape,
dimensional and notch conditions concerned, additionally modified by
welding defects and imperfections of manufacture. It can start on the outside or
inside of the vessel or nozzle pipe at various points in the transition area from
nozzle pipe to vessel. It can propagate radially or tangentially to the edge of the

nozzle pipe into the wall of the vessel, into the wall of the nozzle pipe or into the weld seam. It can therefore be difficult to predict initiation and propagation of a crack in individual cases and to estimate the fatigue strength.

Pulsating internal pressure tests on hollow cylinders with penetrating nozzle pipes lead to the S-N diagram shown in Fig.125 in which endurable circumferential stress amplitudes are plotted as a function of the double ratio of dimensions $(t_1/t)/(d_1/d)$. The curves in the diagram can also be applied to nozzle pipes which butt against the cylinders if the weld root on the inside of the pipe is ground down and the root gap there is removed. Reduction factors $\gamma = 0.6$-0.7 should be taken into account for welded through and necked out versions, in relation to the curves which are quoted. A corresponding reduction

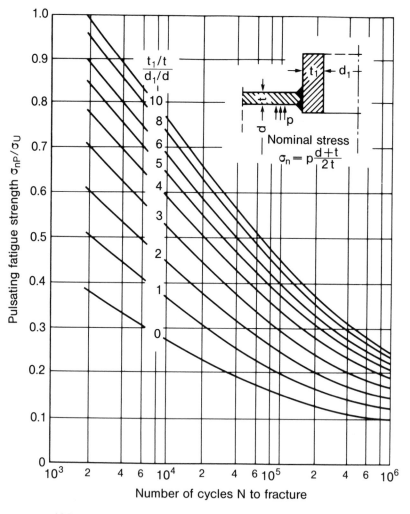

125 S-N diagram for nozzles with high quality welded joints subjected to internal pressure, after Schwaigerer.[9]

of strength has to be taken into account with bending moments transferred via the nozzle pipe which are mostly caused by thermal expansion. The values quoted should be reduced at elevated temperature in the ratio of the high temperature yield strength to the low temperature yield strength. Reference 9 is recommended with regard to details of the fatigue strength assessment by calculation.

In a different representation[104] the reduction for frequently used nozzles is quoted uniformly as $\gamma = 0.5$ with reference to $\sigma_p = 240\,N/mm^2$, with a steeper slope to the fatigue strength for finite life for higher tensile steel and a flatter slope for lower tensile steel (differentiated grading of the reduction in Ref.102). Finally, plotting the endured local strain amplitude against the number of cycles (strain S-N curve)[27] arose as a result of the attempt to represent the cyclic strength results for nozzles in a more concise manner.

In pulsating internal pressure tests on eccentric (more tangential than radial) nozzle pipes welded in from the outside with single sided V butt welds (good design for draining)[105], Fig.126, the highest structural stresses were generated on the nozzle pipe flank side as membrane stress tangential to the nozzle pipe, or as bending stress radial to the pipe on its face side, in each case with fatigue cracks normal to them. Structural stress concentration factors, measured with 3mm strain gauges were relatively favourable at between 2.0 and 3.0. A surface and weld factor of 1.5 was introduced to correlate the pulsating fatigue strength values for the nozzle with those for polished specimens.

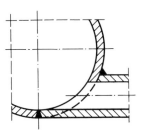

126   Eccentric nozzle pipe.

5.3.9   Reinforcement of cutouts

As has already been explained for the nozzle, every cutout, every opening of circular or general shape, requires an edge reinforcement to reduce the relatively high notch stresses on the edge of the cutout[209] and to increase the rigidity of the component. Referring to Fig.127, annular plates (a), pipe sections (b), angled rings (c), and necked out rings (d), are the usual edge reinforcements, which are mostly weld jointed to the component. The dimensions of the edge reinforcement can be selected as a first approximation in such a way that the volume of material corresponding to the cutout is distributed evenly and symmetrically around the edge. If the edge reinforcement is too weak, the notch stresses of the cutout are only reduced insignificantly. If the reinforcement of the edge is too strong, the edge acts like a relatively rigid inclusion with corresponding other notch stresses. There is an optimum between these two cases with the smallest possible notch stress (about half of the notch stress without reinforcement). The shape of the cutout can be selected to be advantageous, depending on the basic stress state. Cutouts with reinforced edges completely free from increase of notch stress ('neutral holes') can be derived in theory but cannot be used in practice. Unsymmetrical reinforcements, particularly reinforcements on one side, are only partially effective and cause superimposed notch bending stresses. If only the part of the cross section edge which is subjected to the highest notch stresses is reinforced or if longitudinal ribs are welded on tangentially to the edge of the cutout and close to it, care should be taken to ensure that the reinforcement or rib begins and ends in an area of low stress.

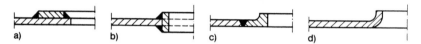

127   Reinforcement of cutouts: ring plate (a), pipe section (b), angled ring (c), necking out (d).

5.3.10   Block flange

Block flanges are compact ring bodies weld jointed into or on to shells or plates, Fig.128. They offer a connection facility via a bolted joint including the machined jointing plane and usually enclose a circular hole. Designs (d) with a large rounding radius in the transition area and welds outside the notch area have the highest level of fatigue strength (presupposing a sealing root pass). Versions (a, b, c) with fillet welds and with and without a centering shoulder have a lower strength. A further aspect of design is the risk of bulging at the unfused slit face at the weld root during annealing or drilling (solution: avoid unfused slit face or drill ventilation holes as in (a)). The unfused slit face should always be sealed to avoid crevice corrosion.

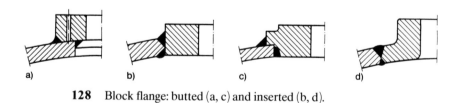

128   Block flange: butted (a, c) and inserted (b, d).

5.3.11   Pipe to plate joints

A plate with many welded-in pipe ends in a regular arrangement is a typical component of heat exchangers. It is subjected to plate bending stresses as a result of the pressure difference between the outside and inside surface as well

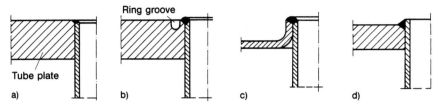

129 Pipe-to-plate joint: sealing welds (a, c) and load bearing welds (b, d).

as the axial force of the pipes (their thermal expansion). To promote defect-free root formation and to avoid crevice corrosion, the tubes are fitted into the bores as tightly as possible and are connected to the base plate via fillet or edge welds, Fig.129. Shapes (a) and (c) are suitable for sealing, but larger axial forces can be carried in shapes (b) and (d) with fatigue resistance. The ring groove in (b), which takes up more space, is intended for relieving the stresses in the weld.

5.3.12   *Vessel ends*

Vessel ends can be designed flat, tapered, dished, tapered with varying gradients, dished according to differing profiles (*e.g.* torispherical or compound curve end). In the central part of the ends a manhole can occur which is necked in or out. Thus the ends consist of a flat, tapered or slightly curved central part (if required, with manhole) and a cylindrical shell section with undefined length. The dished end can be connected to the cylinder by a sharply curved but smooth transition ring. Alternatively there can be a kink where the end and the cylinder butt together; overlapping joints are also possible, Fig.130. Types (a), (e) and (f) are unsuitable in the presence of fatigue loading,

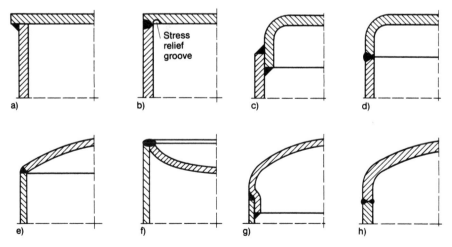

130   Vessel end: flat or dished, with or without a curved transition, with butt or fillet weld.

types (b), (c) and (g) are suitable subject to certain limitations, types (d) and (h) are well suited to the purpose. In type (b) the groove is supposed to relieve the notch stress at the weld root which is not really the case (see section 8.2.8). Type (g) can be considerably more loaded than type (c) because of more favourable force flow and the slit face outside the main flow.

The vessel end supports the internal pressure of the vessel, which represents a pulsating load in operating conditions which are variable with time; the effect of this pulsating load on fatigue strength should be estimated on the basis of the structural and notch stresses which occur in the vessel end. Membrane and bending stresses in the circumferential and meridional directions, which are particularly high at the transition from the cylinder to the end, Fig.131, arise as structural stresses. They can be analysed[106] in accordance with the well known theories of axisymmetric membrane and bending shells, in the transition area in accordance with the 'stepped body model', separated into ring-shell elements with element forces which are referenced to the meridional or peripheral length unit. The radial membrane tensile stresses in the end are supported by circumferential pressure in the transition area. During this process the transition area is displaced inwards, so that meridional bending stresses too are generated in it (tensile inside, compressive outside). Circumferential tension is again prevailing in the cylindrical part of the vessel. The increase in structural stress on the inside of the shell at the transition is characterised in the example

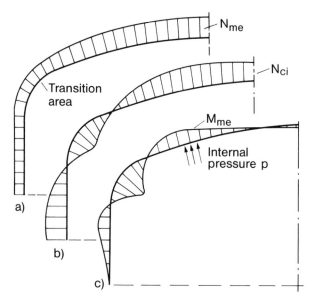

131   Internal forces in the curved transition between end and cylinder of the vessel subjected to internal pressure: normal force in meridional (a), and circumferential (b) direction, bending moment in meridional direction (c), in each case referenced to the peripheral or meridional length unit and independent of thickness, after Eßlinger.[106]

here by the factor 2.5. It decreases with an increased radius of curvature at the transition. In the case of a flat plate end, usually connected to the cylindrical shell without a smooth transition (welded or screwed joint), bending stresses prevail in the end plate.[9] The shell and plate thicknesses t_{sh} and t_{pl} should be adjusted to one another to achieve an advantageous distribution of (structural) stress ($t_{pl} \approx 1.5\, t_{sh}$). The actual notch stresses, which assume high values if the transition has a small radius of curvature, especially if there is a kink without a transition radius, are superimposed on the structural stresses. If there is a welded joint in the transition, additional notch stresses occur at the weld toe and root.

The most critical point with regard to fatigue strength is the smooth transition or the kink which corresponds to it. The highest structural and notch stresses occur on the inside of the transition in the meridional direction. The least favourable case is the flat, tapered or dished end with a sharp bend and a weld in the bend (single bevel butt weld, fillet weld or edge weld, the latter being at risk of crevice corrosion additionally). It is necessary to use a rounded transition, which is butt welded outside the transition, if possible with a capping pass, in the presence of fatigue loading. The stress concentration factor (structural and notch stress) for the rounded transition which is given in accordance with Ref.9 is dependent on the ratio end height, h, to end diameter, d, (in accordance with measurements on manufactured vessels and some results of calculations, where the reference stress is the axial stress pd/4t), Fig.132. The diagram can only offer a rough estimate and should be

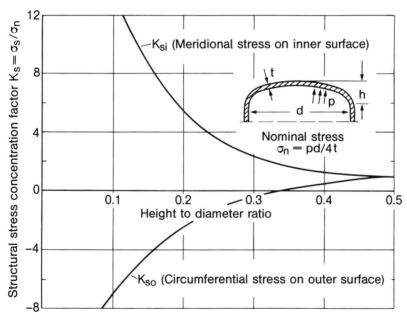

132 Structural stress concentration factor, K_s, of the transition inside (index i) and outside (index o) for dished vessel ends, mean values of scatter band, measurement on manufactured vessels, after Schwaigerer.[9]

supplemented by a stress analysis in the individual case. Structural stress concentration factors calculated using finite element methods for small off-centred axial nozzles and large centred ones in the transition area (radius of curvature 0.1 times end diameter) of dished ends are given in Ref.107. Here multiplicative superimposing of the decoupled influencing parameters thickness-to-diameter ratio of the end, relative eccentricity of the nozzle and thickness of end/thickness of nozzle ratio are assumed. The circumferential groove suggested for flat ends without a radiused transition, which does not have the intended substantial stress relieving effect on the weld root (see section 8.2.8), may only improve the fatigue strength if a painstaking design is undertaken with checking of the structural and notch stresses. The overlapping joint with edge fillet welds is disadvantageous with respect to stress concentration because of the large changes in cross section, but advantageous with respect to the root position in the thickened cross section of the material. This design solution, too, should be checked on the basis of the structural and notch stresses in the individual cases.

It should be noted, with regard to the published stress patterns and stress concentration factors for vessel ends, that either experimental measurements are used as a basis (structural and notch stresses superimposed if the radius of curvature is large enough) or (axisymmetric) analysis which, if ring shell elements are used ('stepped body model') does not model the notch stresses.

So far, the notch stresses at the weld toe and the weld root have not been covered either by measurement or by calculation. Therefore particular care is recommended in evaluating stress results, especially as only isolated fatigue test data are available.

5.3.13   *Vessel ends with manhole*

Vessels ends with a manhole, which may be necked-in or out, reveal an increase of structural and notch stress on the outside of the transition from the end to the neck additionally to the stress increase on the inside of the transition from the end to the shell. The membrane stresses in the end are supported by circumferential compressive stress in the necked-in or necked-out section, whereby bending stresses too are caused (as in the shell to end transition),[107] Fig.133. The increase of stress is greater for a necked-in design than a necked-out one, presupposing that the geometrical parameters are the same. When plotting the stress concentration factor (maximum equivalent stress σ_{eqmax}, referenced to the nominal stress $pr/2t$ in the end, taken to be a spherical membrane) over the diameter of the hole, d, referenced to $\sqrt{rt}$ with radius of curvature, r, of the end, a narrow scatter band is produced according to Ref.9 with stress concentration factors between 4 and 6 (rising with $d/\sqrt{rt}$).

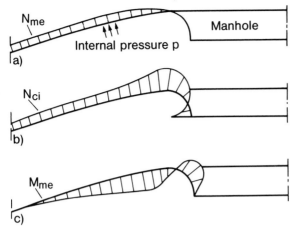

133 Internal forces in the vessel end with necked-in manhole, vessel subjected to internal pressure: normal force in meridional (a) and in circumferential (b) direction, bending moment in meridional direction (c), in each case referenced to the peripheral or meridional length unit and independent of thickness, after Eßlinger.[106]

5.3.14 *Fire flue base*

Fire flue bases are designed flat (less rigid) or dished (more rigid) with necked-in openings to the fire flues. Apart from internal pressure they are also subjected to loading from the axial force of the fire flue pipes as a consequence of their uneven thermal expansion. The necked-in section also gives increased structural and notch stresses here as does the inside of the shell-to-base transition. A design detail is considered on the basis of a notch stress analysis in section 8.2.8.

5.3.15 *Vessel caps*

Dish-shaped vessel caps, used as bolted closures for pressure vessels, consist of a dished end with a welded-on thicker flange ring on which the bolts are acting and which bears the sealing ring over all its surface or on one side only over a narrow area, Fig.134. High notch stresses arise at the transition from the end cap to the flange ring. In addition a meridional structural stress peak occurs in the smooth part of the end close to the flange ring, which can be estimated in accordance with Ref.9 (factor 2-7 referenced to pr/2t, higher for a narrow one sided seal than for a broad seal over the whole surface).

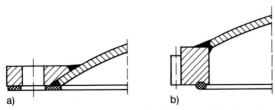

134 Vessel caps, welded joint between dished end and flange ring.

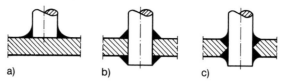

135   Bolt, butted with fillet weld (a), penetrating with double fillet weld (b), and penetrating with double bevel butt weld with fillet weld (c).

5.3.16   *Welded-on bolts*

A welded-on bolt, either butting on to or penetrating a surface and jointed by using a fillet weld, Fig.135, is a relatively high quality design. The penetrating type (b) is able to withstand higher loads than the version which butts against the surface (a) (as a consequence of doubling the weld and removing the risk of lamellar tearing). A double bevel butt weld with double fillet weld (c) is able to withstand the highest loading. A stud welded bolt (see section 6.4) has the least load bearing capacity.

5.3.17   *Supporting brackets, supporting stays, bearing rings, saddle bearing, load bearing eyes*

The above-mentioned supporting members of vessels should be designed in such a way that the supporting forces are acting more in a tangential direction on the vessel wall than in a radial one (to reduce local bending of the wall) and so that they are introduced over a long, well curved joint line enclosing a large support area (a bearing ring with box section or a supporting box is advantageous) and finally so that they do not restrain the deformation of the vessel caused by internal pressure in a disadvantageous manner. Reinforcing plates between the supporting members and the vessel wall are to be recommended.

5.4   Appropriate design with regard to fatigue strength

Welded structures and their components should be designed in such a way that the required fatigue strength, service life and safety are achieved with the lowest possible expense. An appropriate design with respect to fatigue strength is characterised by a design process with a prescribed budget, which aims at an optimum of strength, life and safety to an equal extent in all components. The aspects of strength are stressed here as against other aspects such as ease of manufacture, testing and maintenance which occasionally compete with it.

   The questions of shape design, both of the structure as a whole and its division into elements to be jointed (global structure) and of the positioning and shape of the welded joints (local structure), are of primary interest with regard to fatigue strength.

5.4.1   *Methodological basis*

The most dangerous limit states as regards strength, besides instability, are fatigue fracture and brittle fracture. Although the latter should be considered to be an independent limiting condition of static strength, as final fracture it is also an important constituent of fatigue failure both in test specimens and in the structure presupposing that the final fracture is brittle to a greater or lesser extent. On the other hand, it can be seen from surveys of brittle fractures in practice that a major percentage of such fractures are initiated by fatigue cracks in the structure.

Fatigue and brittle fractures are initiated at discontinuities of shape, at notches and cracks, that is in areas with excessively high local elastic stress, at stress peaks (structural stresses, notch stresses and crack stress intensity). A relatively small amount of material at the position of the stress peak can have a decisive (negative) effect on the strength of the whole structure. On the other hand, fatigue or brittle fracture can be delayed or avoided if the stress peak is reduced or removed by design measures. Even then, if ductile material is used and reduction of the stress peak should be achieved by yielding, design improvements are effective because the ductility of metallic materials is limited, and as this ductility is more and more exhausted at the point where there is a stress peak, the risk of brittle fracture grows. Avoidance of local structural and notch stress peaks by design measures is by far the most effective means of increasing the strength and life of the structure. The effectiveness is so great that a removal of stress peaks of the structure, when subjected to the basic load cases, often without very precise knowledge of the actual loads, is sufficient to ensure that the structure does not fail in practice.

Structural stress peaks occur where there are discontinuities in the global structure (joint lines, corners, stiffeners, cutouts), notch stress peaks where there are discontinuities in the local structure (transition in cross section, weld toe, weld root, weld ends, weld ripples, gaps, slits, weld nugget edges, welding defects and cracks). The global structure is defined by the dimensions which are laid down in the design drawings. Only the type, positioning and thickness of the welds are established as regards the local structure. Further data for the stress analysis of the local structure must be measured on the weld or must be set in some other meaningful manner.

The global structure is completely under the designer's control, the local structure only partially, the manufacturing engineer is responsible for the rest.

5.4.2.   *General guidelines*

The general guidelines listed below can be stated for the appropriate design of welded structures with respect to fatigue strength in conformity with the design examples which have been given:

—   moderate abrupt changes in stiffness in the presence of tension, bending and torsion by long and rounded transition areas;

- use butt or single and double bevel butt welds in preference to fillet welds;
- use double sided in preference to single sided fillet welds;
- use lap joints with aligned plates and offset laps only in preference to offset plates;
- place weld, particularly weld toe, weld root and weld end in an area of low stress (*e.g.* the neutral zone in the presence of bending load, sections with a small bending moment, points where the notch stress passes through zero on the edge of a hole, zones outside transition areas and corners), separate notch effects instead of superimposing them;
- relieve stress on weld toe, weld root and weld end notches by inserting milder notches before or after them in the force flow;
- reduce bending moments by narrower support spacing in the presence of transverse bending force;
- reduce torsional moments by introducing transverse force at the centre of shear;
- reduce deformation restraint in the presence of tension and bending by using small stiffening lengths;
- avoid warping restraint in the presence of torsion by means of flange ends which are cut off or bevel jointed, and by using bars with cross sections which are free of warping;
- reduce local bending when forces are transmitted into thin walled components by designing weld lines which enclose a large area or by increasing wall thickness locally;
- position beads on thin plate areas to reduce bending effect;
- avoid cutouts if they disturb the force flow; cutouts at corners of stiffeners transverse to the force flow are efficient;
- use bolted or riveted joints, forged or cast connectors, instead of welded joints at particularly critical points, especially when this can be combined with advantages in assembly;
- close root gaps if there is a risk of corrosion.

Structural stress peaks can be determined by measurement or calculation to check the quality of the design. Measurement of the structural stress peaks on a prototype (brittle lacquer and strain gauges) or on a model (photo-elasticity) is an essential development process for designs subjected to fatigue loading (*e.g.* aircraft, vehicles, agricultural machinery, engines, turbines). The strain measured at the weld toe is a reliable evaluation criterion for welds, provided that failure initiates at the weld toe and not at the weld root (see section 8.3.1). Calculation has been used increasingly in recent times instead of measurement, especially based on finite element methods.

6

Fatigue strength of spot, friction, flash butt and stud welded joints

6.1   Spot welded joints

6.1.1   *Design aspects and fatigue phenomena*

In the spot welded joints subjected to fatigue loading considered below (pressure and fusion spot welding) it is a precondition, as for the previous seam welded joints subjected to fatigue loading, that dimensioning, design, material, manufacturing and testing aspects relevant to static loading have already been fulfilled.

The dominant influence of the local stress parameters (structural stress, notch stress and crack stress intensity) on fatigue strength also applies to pressure or fusion spot welded joints. Design parameters, which influence the structural and notch stresses, include plate thickness, plate width, spot diameter, number of spots, spot arrangement, distance between spots (the pitch) and the cross sectional shape of the plate (*e.g.* with beads or folded edges). The loading of the spots in the structure, if the design is advantageous, should mainly be tensile shear because of the higher fatigue strength in this load case. However, cross or peel tension may be superimposed as a secondary effect.

Details are given in Chapter 10 on the structural stressing mechanism, the notch effect and the crack-like effect of the sharp slit inherent to the geometry of spot welded joints (the local approach). The critical crack initiation area is the slit-shaped edge of the joint face of the weld spot (the edge notch). Moreover, hot cracks and bonding defects cannot be completely excluded here. The crack is initiated (and propagated) outside the heat affected zone and not at the edge notch (and only then if there is substantial hardening in the heat affected zone and loading in the lower cycle fatigue range.

Welding defects within the weld spot generally have little effect on fatigue strength and can be disregarded. This has been proved, for example, by the extreme measure of drilling out the whole centre area whereby the fatigue strength of the weld spot remains nearly unchanged.

The following typical crack initiation and propagation mechanism was detected[108] in the high cycle fatigue range when subjected to tensile shear

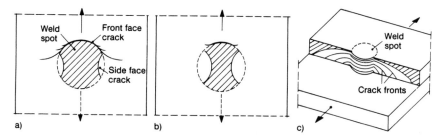

136   Crack initiation and propagation in the spot welded joint, as a plate fracture primarily at the front face (a), as a joint face fracture, primarily at the side face (b), and front face fracture surface (c), after Overbeeke and Draisma.[108]

loading. Two basically different fracture phenomena are observed, the plate fracture and the joint face fracture, Fig.136. In the plate fracture, which occurs with larger spot diameters, small cracks appear transverse to the tension in the plate at several points of the front face edge of the weld spot. Whilst they propagate over the thickness of the plate, shear cracks initiate at the side face edges of the weld spot, accelerated by the diversion of the force flow caused by the transverse crack. In the joint face fracture, which occurs with smaller spot diameters, shear cracks are initiated on the flank side edges of the weld spot which propagate over the joint face, whilst at the same time transverse cracks appear at the front face edge of the spot, which can also be explained by the change in the force flow. It is supposed that as a result of the severe notch effect in spot welded joints the crack initiation phase is small relative to the crack propagation phase.

6.1.2   *Fatigue test specimens*

Fatigue tests on spot welded joints have been carried out using mainly single face, tensile shear specimens subjected to pulsating loading. Some tests have been carried out with cross tension and peel tension specimens also, see Ref.113 and Fig.137. The fatigue test with tensile shear and cross tension specimens (German standard DIN 50124 and DIN 50164) has been submitted to the ISO for standardisation. The DVS Merkblatt 2709[114] refers to further tensile shear specimens used in the aircraft and space industries involving multiple shear face specimens, multispot joints and strapped joints. As no practicable simple correlation between fatigue strength and joint dimensions is possible as a result of the complexity of stress distribution and crack propagation, the fatigue strength of the joint is characterised by the endured load amplitude on the specimen (which is identical to the load amplitude on the weld spot for single spot, single face specimens), not by the endured nominal stress. The cyclic load for fatigue fracture considered here is dependent on so many dimensioning, design, materials and manufacturing parameters that in practice often only the component fatigue test with the actual combination of parameters is sufficiently informative.

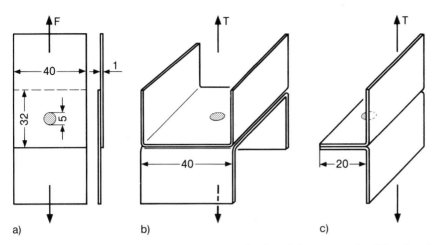

a)   b)   c)

137   Single spot test specimens: tensile shear (a), cross tension (b) and peel tension (c).

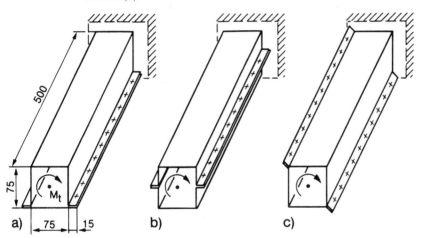

a)   b)   c)

138   Multispot test specimens: different hat profile combinations, after Eichhorn and Schmitz.[137]

Industrial practice (*e.g.* in the automotive industry) favours the fatigue test on the more component-like multispot hat profile specimen subjected to torsional loading (and sometimes to internal pressure too), Fig.138, to determine the effectiveness of various materials, methods of manufacture, and design parameters via the relevant S-N curves. Another object of this kind of testing is to determine the endurable weld spot forces which are needed for evaluation of relevant finite element results from structural stress and weld spot force analysis. For that purpose, the torsional loading which generates shear forces in the weld spots (in longitudinal direction of the specimen) is supplemented by internal pressure loading, which generates peel tension at the weld spots. The shear force, F_s, at the weld spot is derived from the shear flow in the cross section according to Bredt's formula (with torsional moment, M_t; weld spot pitch, l_s; area, A^*, circumscribed by the hollow section):

$$F_s = \frac{M_t l_s}{2A^*}$$ (34)

The shear force in the weld spot is therefore proportional to the distance between the spots (the pitch) and independent of plate thickness. If a single spot is omitted, the shear force in adjacent spots rises by a factor of approximately 1.5. Therefore, successive shear fractures of all the weld spots can result. The peel and cross tension force at the weld spot in the specimen subjected to internal pressure depends on the geometrical details of the considered cross section. Endurable forces of that kind, therefore, cannot be generalised. The limiting damage criterion for the hat profile S-N test is the decrease of stiffness of the specimen caused by large cracks.

6.1.3   General evaluation of fatigue strength data

The dependence of fatigue strength on the number of cycles (S-N curve) and on the mean load (expressed by the limit load ratio R = P_l/P_u), which is typical of spot welded joints, is shown in Fig.139, a test result[115] which was selected because of its careful statistical evaluation (AlMgSi 1 plate, 1.5mm thickness, two spot tensile shear specimen). A more recent investigation on the fatigue strength of a spot welded aluminium alloy (AlMg 5) can be found in Ref.116. The scatter band of the S-N curve shows the decline reaching beyond N = 2 × 10^6 cycles which is typical for aluminium alloys. It can be harmonised with the normalisation for seam type welded joints (gradients and scattering correspond to a considerable extent). The fatigue strength curves dependent

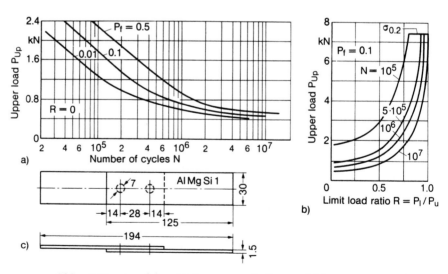

139   S-N curves (a) and fatigue strength diagram (b) for two spot specimen (c). Aluminium alloy AlMgSi 1, failure probability P_f = 0.01, 0.1, 0.5, after Buray.[115]

on the limit load ratio show the typical steep decline for severely notched welded joints, starting from the static ultimate load, $R = 1$, limited by the yield stress $\sigma_{0.2}$ in the diagram. The pulsating fatigue strength $(R = 0)$ is very low, generally only 10-15% of the statically endurable tensile shear force for the joint, equivalent to a reduction factor $\gamma \approx 0.1$ compared with the pulsating fatigue strength of the plate without the weld spot, therefore clearly below the worst seam type welded joint. The path of the $N = 10^7$ curve between $R = 0$ and $R = 0.75$ corresponds to a constant load amplitude independent of R. Therefore spot welded joints are better suited to higher static preload and for lower numbers of cycles (fatigue strength for finite life range, especially low cycle fatigue strength range).

The above statements are valid not only for aluminium alloys but for spot welded steels also. It can be seen from an older survey[117] that a reduction factor $\gamma = 0.1\text{-}0.5$ (with reference to $\sigma_P = 240\,N/mm^2$) can be expected, double face shear higher than single face shear, thin plate higher than thick plate (relative to the pitch), small pitch higher than large pitch (relative to the spot diameter). Missing or inadequately fused spots in a multispot specimen can also reduce its fatigue strength considerably. Thus an additional reduction of up to one third can be determined for a single row, single shear face spot welded joint.[118]

6.1.4   *Detailed consideration of fatigue strength*

The fatigue strength of spot welded joints is now considered in more detail on the basis of further investigations.[108-112, 117-135, 342, 343, 351-353]

It has been revealed by comparative tests on spot welded joints in low tensile steel compared with higher tensile steel that high cycle fatigue strength values remain uninfluenced or are lowered by the higher static strength. The same applies to joints between the two materials.

This can be seen from several investigations[119, 122, 123, 126, 127] on the fatigue strength of low and higher tensile steel spot welded joints. For example, the alternating and pulsating fatigue strength of joints made from higher tensile steel ZSte 380 $(\sigma_{0.2} = 320\,N/mm^2)$ is slightly lower than the fatigue strength of equal joints made from low strength steel St 1403 $(\sigma_{0.2} = 150\text{-}200\,N/mm^2)$.[119] This does not mean that higher tensile steels are completely unsuitable for cyclic loaded spot welded components. They offer an advantage where static prestresses are high, where low cycle fatigue is considered or where loading sequences occur with some (rare) extremely high load peaks.

The question can now be answered whether the results of fatigue tests with spot welded joints can be allocated to the normalised S-N curve for seam welded joints, Fig.40 or 74. Whereas the allocation was successfully performed in Ref.123 with test results from multispot tensile shear specimens made from ferritic and austenitic steels (only minor deviations in curve gradient and scatter band occurred), the possibility of allocation is denied in Ref.119, 120 on the basis of test results from hat profile specimens made from low and higher tensile steels and subjected to torsional and internal pressure loading. The

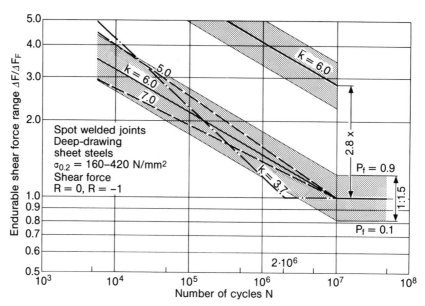

140   Endurable shear force range for weld spots in steel plates of 1mm
thickness, S-N curves and service life curve with scatter band,
P_f = 0.1-0.9, proposed on the basis of test results.

endurance limit is achieved only for N $\geqslant$ 10^7 (instead of N $\geqslant$ 2·10^6) which is
well established in severely notched specimens. On the other hand, the S-N
curve gradient exponent, k, is increased (k = 5.0-7.0 instead of k = 3.7) *i.e.* the
S-N curve runs flatter. The deviation can be explained to some extent by the
higher number of cycles for the endurance limit, Fig.140. The scatter bands of
the different lines (only one scatter band is actually drawn) are overlapping.
    The shear force endurance limit ΔF_F = 2F_{AF} of the weld spot between 1mm
thickness steel plates, N = 10^7, R = 0 and R = $-$1, P_f = 0.5, is given in Ref.119,
120, 122 as ΔF_F $\approx$ 1070, 1530 and 750-1100N respectively. There is no
explanation available for these differences as far as the same steel is concerned,
e.g. St 1403. In any case, the endurable weld spot force ΔF_F $\approx$ 1000N is an
acceptable value for design force derivation. This is confirmed by other
investigations.[118,125] The S-N curves in Fig.140 with the endurance limit at N =
10^7 which are plotted dimensionless with respect to ΔF can thus be interpreted
also as endurable forces with the dimension kN for plate thickness t = 1mm.
Concerning the aluminium alloy AlMg 5, the endurable force ΔF_F $\approx$ 620N for
the endurance limit at N = 10^7 can be derived from investigation.[116] No
generally applicable endurance limit load can be stated for cross tension, even
for a given plate thickness, because the endurable transverse force depends on
the support spacing of the transverse bending effect. In any case this force is
substantially lower than the endurable shear force, *e.g.* ΔT_F $\approx$ 200N for the
standardised cross tension specimen with plate thickness t = 1mm. It is
proposed to define the permissible forces at less than 25% below the endurable
forces for P_f = 0.5.

The above endurable weld spot forces are valid for the plate thickness t = 1mm and weld spot diameter d = 5mm usually combined with this plate thickness. It has been found by several investigators that the endurable shear force is approximately proportional to the plate thickness. The endurable transverse force seems to increase more steeply with plate thickness. With two different thicknesses combined in one joint, the endurable force is between the respective values but closer to the value of the smaller thickness. The influence of the weld spot diameter is not so strong. The endurable shear force is approximately proportional to the square root of the diameter. The diameter has very little influence on the endurable transverse force in cross tension or peel tension.

The endurable total load increases with the number of weld spots in a multispot tensile shear specimen with a spot row or a spot field but the endurable force per spot decreases because of uneven load distribution on the weld spots. The spots at the end of the row or at the corners of the field are subjected to higher forces than the others.

A further influencing parameter on fatigue strength relevant to automotive engineering is the superimposition of cyclic shear and cyclic transverse loading (in-phase and out-of-phase) and the superimposition of cyclic load and static preload conditions. Some variants of this kind have been investigated[119] in the form of a superimposition of relevant torsional moment and internal pressure loadings. The strength reduction was 20-25% under these conditions.

The service fatigue strength of spot welded joints subjected to variable amplitude loading was investigated[119,120] using a Gaussian random load sequence (see Ref.133 also). The endurable maximum load amplitudes of the spectrum are higher by a factor of 2.5-3.0 with reference to the endurable load amplitude in the constant amplitude test. The curve gradient exponents are similar, $\bar{k} \approx k$, or a little higher. The life curve with the load increase factor 2.8 and the curve gradient exponent $\bar{k} = 6.0$ is plotted in Fig.140. The calculation[119,120] of the damage accumulation according to the simple or modified Miner rule gives the total damage S = 0.5-1.5.

The temperature dependence of the fatigue strength of the spot welded joint corresponds to that of the base material, i.e. a decrease in strength with increase in temperature, whereby the crack initiation point can be moved to a position in front of the heat affected zone.

Spot welded specimens in the as-welded condition have been compared with stress relieved specimens by Overbeeke and Draisma.[111,112] Crack closure effects raising the fatigue strength occurred for R ⩽ −1 in the case of as-welded specimens and for R ⩽ 0.1 for stress relieved specimens.

Increasing the size of the annular bonding area with the same weld nugget area (via process parameters, especially electrode pressure and welding time) improves fatigue strength. Further factors which can increase fatigue strength considerably include static or impact type compression, post-weld compression by the electrode or hydraulic compression of the spot and/or the heat affected zone, prestraining to two thirds of the ultimate tensile shear force and post-weld heat treatment. This is because of strain hardening, more favourable residual stresses and removal of microseparations. As a result of an optimum combination of the possible strength increasing manufacturing measures (to be defined by experiment for each individual case) an increase in fatigue strength of one third of the ultimate tensile shear force is possible, equivalent to an increase in the reduction factor to $\gamma \approx 0.25$.

When the strength values obtained from the standardised flat specimens are transferred to a construction joint with folded edges in the vicinity of the weld spot, an increase of fatigue strength can be expected as a result of reduced weld spot slanting (increase of fatigue strength 50% according to Ref.124).

As spot welded joints are particularly liable to corrosion from the inaccessible crevice-like joint surfaces, the stated strength values can frequently not be achieved in practice or only with special corrosion protection. A decrease in fatigue strength by crevice corrosion was found[135] for weld spots in low and higher tensile steel increasing from humid air via dipping into pure water, spraying with salt water to dipping into salt water. The corrosion was more severe for larger gap widths.

Spot welded joints of galvanized steel sheets have a sufficiently high fatigue strength compared with non-galvanized material if relevant manufacturing measures are taken. Oil films in the vicinity of the weld spot increase fatigue strength (in the finite life range of steels) by reducing the rate of crack propagation.

The dependence of fatigue strength on the frequency content of the load cycles stated in Ref.131 has not been confirmed by other investigators.[120]

The inadequate state of knowledge regarding fatigue strength of spot welded joints, and their sensitivity to imperfections of manufacture and to corrosion, limits their use in automotive, railway and aircraft engineering to joints with low reliability requirements.

6.2   Weldbonded joints

Weldbonding of steel and aluminium plates has been an important advance, particularly with regard to fatigue loading combined with corrosion. Here the crevice-like gaps around the weld spots are filled by an adhesive which can be applied before or after spot welding. Adhesive with a low viscosity is applied in the joint crevice after welding (using gravity and capillary action). Alternatively, paste-type adhesive can be applied immediately before welding, which is expelled from the spot contact surfaces during welding and before bonding by electrode pressure.

Weldbonding has the following advantages compared with spot welding as regards strength (in order of importance):

—   high corrosion resistance (no crevice or friction corrosion);
—   increased fatigue strength (up to two thirds of the ultimate tensile shear force, equivalent to $\gamma \approx 0.5$);

— increased static strength;
— increased energy absorption when subjected to impact;
— increased damping when subjected to vibration.

Compared with adhesive bonding, weldbonding has the advantage of suppressing continuous creep in the adhesive layer so that the creep strength of the joint is strongly increased.

Using high strength construction adhesives (*e.g.* epoxy resins) several times the strength of a spot welded joint can be achieved.[136-138] This is because of the suppression of single plate bending and secondary cross tension force at the weld spot (the fracture is shifted from the spot edge to the outside edge of the adhesive layer) and by the prevention of crevice corrosion. Despite the unquestioned advantages of weldbonding with respect to service fatigue strength it is not used in mass production because of disadvantages in manufacture.

6.3   Friction welded and flash butt welded joints

Friction welding of rod-type components (of similar materials) and subsequent removal of the upset metal achieves endurance limit values which (with increased scattering) lie above those of the base materials and which can also be maintained in practice with appropriate checking for freedom from cracking.[139] Test results are available for alternating torsional loading in addition to pulsating tensile loading and rotating bending loading. The fatigue crack initiates at geometric notches independent of the location of the friction face in notched components (*e.g.* pinion shaft, steering knuckle, steering shaft). The fatigue strength depends mainly on the notch effect.[140]

An increase in endurance limit is also observed[117] in the joint area of flash butt welded joints after the weld flash has been removed because of a rise in hardness in this zone. The fatigue strength of the mill finished base material is achieved without difficulty.

6.4   Stud welded joints

The arc pressure welded joint between stud and base plate, Fig.141, is mainly used in structural steel engineering (head stud in steel-concrete composite girders) as well as in shipbuilding, pipe, tank and boiler construction. As regards fatigue strength, the notch effect of the joint is the dominant factor,[141,142] ahead of the influence of quality of manufacture and residual welding stresses. Fatigue cracks initiate at the sharp notch under the bulge and propagate in the stud if stud tension runs straight through (a) (rare in practice). If there is tension or bending in the base plate (b) (including stud transverse force (c)) they propagate in the base plate, approximately normal to the principal loading direction, adjacent to the stud weld base at the front face. The reduction factor for the endurable bending stress is approximately $\gamma \approx 0.5$ for stud and

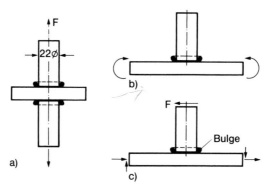

141 Stud welded joints, subjected to loading by running-through tension (a), base plate bending (b), and stud transverse force (c).

base plates made of low tensile structural steel St 37. Higher tensile steel produces only slightly higher values.

Fatigue strength tests on hollow and solid shear studs, which were connected to the I section girder by circumferential fillet welds and are subjected to shear loading individually in the test, produced an average $(P_f = 0.5)$ endurable range of shear stress for the stud of $\Delta\tau = 70\,\text{N/mm}^2$ at $N = 2 \times 10^6$ with the following additional specification: given value for $R = 0$, value for $R = -1$ somewhat higher, gradient exponent of the S-N curve, $k = 5.4$, no endurance limit in the range investigated, $N < 10^7$. The value $\Delta\tau_F \approx 56\,\text{N/mm}^2$ for $P_f = 0.1$, referred to the pulsating shear strength of structural steel St 37, produces the reduction factor $\gamma = 0.4$. If the fillet weld is adequately dimensioned, the fatigue crack propagates in the girder flange at the front face at first and finally undermines the stud weld base in a trough shape. In a composite girder bending test, which is closer to reality than individual stud loading, the fatigue fracture propagates into the web without undermining the stud and therefore shearing it off. The ranges of shear stress endured by the composite girder are somewhat higher, so that here $\Delta\tau_F = 70\,\text{N/mm}^2$ or $\gamma = 0.5$ characterises the lower limit of the scatter band, $P_f = 0.1$.

7

Design codes, assessment of nominal and structural stress

7.1   Codes based on science, empirical knowledge and tradition

Codes on dimensioning, design and manufacture should record the generally accessible and recognised state of technical knowledge and ability so that they can be used as a basis for design and manufacture with adequate reliability and economy. Codes describe what the majority of authorities on the subject consider to be appropriate in practice. The representation is extremely simplified without taking too much account of scientific plausibility. Thus a recognised methodology can formally appear although its content is based on purely empirical knowledge. This partially explains the multiplicity of permissible stresses and safety factors. In fairness to this lack of scientific consistency, it should be mentioned that it does avoid scrupulous precision, which can be a hindrance in practice. However, assessment of fatigue strength according to codes only provides a rough reference point for adequate operational reliability. Only a small part of the extensive practical knowledge from relevant industrial activities is laid down in the codes.

Observation of codes is a legal requirement stipulated by building inspection boards and authorising boards (industrial regulations, relevant laws, decrees, *etc*). At the same time, the codes are used as the basis of delivery contracts. Therefore they not only have a technical, but also a legal character. Only verifiable information is suitable for codes.

Codes also contain conventions and traditions therefore those which have arisen separately in different countries or technical fields, are difficult to integrate at a later date. Successful harmonisation presupposes the existence of previously co-ordinated basic standards.

Designing largely without reference to codes is permissible in certain cases. Special strength assessments are possible even when the codes have to be applied. Some important technical fields possess no code on design, dimensioning and manufacture, for example the automotive and aircraft industries. In tank, boiler and pipeline engineering, as in shipbuilding, bridge building and structural steel engineering, spectacular accidents have led to codes being enshrined in the law at an early stage. However, here too, the general regulations, especially as regards fatigue strength remain in need of further inquiry on the basis of the phenomena put forward in section 1.1 (multi-parameter problem and scattering).

The general data given below on the content of codes with regard to dimensioning for fatigue loading should provide a general survey, put the permissible stresses into concrete terms so that rough estimates are possible, and give guidelines for assessing quality of manufacture. Reference is made to the specific code concerned with respect to the details.

7.2   Dimensioning for fatigue in accordance with the codes, assessment of nominal stress

The core of the codes as regards 'dimensioning' is generally the assessment of nominal stress. Nominal stresses in all parts of the structure, determined by simple technical methods of calculation for specified loading cases, remain smaller than the material dependent stated permissible nominal stresses σ_{per}. The latter are smaller by a safety factor, S, than the endurable nominal stresses, σ_{en}, the magnitude of which is dependent on the type of failure (here fatigue: endurance limit, fatigue strength for finite life, or service fatigue strength) and, in the case of fatigue, on the notch class and on the quality of manufacture. Originally, there was only one single permissible stress for each material whilst special types of failure, notch classes and qualities of manufacture were taken into account by increasing the nominal stresses in the structure (*e.g.* factor γ for fatigue in the old German design specifications for riveted railway bridges, BE 1934 and BE 1951). For fatigue in the finite life range, the safety factor can refer to endured number of cycles (safety factor S*) instead of endured stresses (safety factor S) (*e.g.* tank and boiler construction) which then requires a conversion or a graphic transformation via the (normalised) S-N curve.

Thus the basic form of the assessment of nominal stress is

$$\sigma_n \lessgtr \sigma_{per} \tag{35}$$

$$\sigma_{per} = \frac{\sigma_{en}}{S} \tag{36}$$

or alternatively

$$\sigma_{per} = \sigma_{en} \text{ for } S^* \times N \tag{37}$$

Stress assessment with reference to the endurance limit is a part of all codes relevant to fatigue. The permissible nominal stresses with reference to the endurance limit are given in a fatigue strength diagram (dependent on limit stress ratio, on mean stress or lower stress) for current structural steels and types of joint (notch case including quality of manufacture) in the form of various curves. Recently (IIW design recommendation[143]) permissible stress amplitudes have been recommended for welded constructions which are not stress relieved in which the size of the stress amplitudes is no longer dependent on the mean stress (but permissible upper stress never higher than statically permissible stress). This is based on the corresponding strength behaviour of thick walled, as-welded components. Assessment with reference to the endurance limit presupposes a cyclic design load of constant amplitude. If a large number of smaller load amplitudes also occur or if the component does not need to be fatigue resistant with reference to the endurance limit but only to the fatigue strength for finite life, this leads to over-dimensioning.

The permissible stresses listed above can largely be raised where, as has just been stated, a stress assessment with reference to fatigue strength for finite life or service fatigue strength is possible on the basis of the actual loading conditions. The majority of codes relevant to fatigue include the latter. First of all, fatigue design for finite life is a matter of the permissible stress with reference to fatigue strength for finite life under load amplitudes which continue to be constant, represented as S-N curves, which are mostly equidistant and parallel to each other, for the different materials, joint types and qualities of manufacture.

Secondly, fatigue design for finite life concerns the permissible stress with reference to service fatigue strength for load spectra *i.e.* for variable load amplitudes (whereby the permissible stress refers to the largest amplitude of the load spectrum). Corresponding service life curves are also stated as being mainly equidistant and parallel, or, as a simplification, only the damage calculation according to Miner's rule is required, based on the above-mentioned S-N curves.

The basic types of code dimensioning, which have been described, are further differentiated in the following way (extent dependent on code concerned):

— load case specifications dependent on the field of application;
— strength hypothesis for multi-axial stresses (von Mises' distortion energy hypothesis, Tresca's principal shear stress hypothesis or exceptionally the principal tensile stress hypothesis);
— largest principal shear stress as the greatest difference between two principal stresses in the presence of principal stresses which vary with time to differing extents and are phase-displaced when produced by periodic loading;[161,162]
— temperature dependence of the permissible stresses if temperatures are increased considerably (up to 600°C);[162]
— increase of structural stress at the crack initiation point, amount estimated for specifying the notch class or more precisely measured or calculated for comparison with permissible structural stress values without specification of notch class;[156-159]

— factors for structural stress increases produced by geometrical imperfections in manufacture (*e.g.* lack of roundness, linear and angular misalignment, angular distortion);
— reduction or quality factors for the weld seam to characterise the reduction of the permissible nominal stress in the base material in the presence of a weld;[163]
— surface factor for roughness or rolled skin on the base material;[160,163]
— additions to the wall thickness for corrosion and wear;
— additions to the nominal wall thickness with regard to permissible wall thickness tolerances;
— assessment of welding defects.[177]

In components and constructions with cycles more in the low cycle fatigue strength range, what is important is the number of cycles, above which dimensioning with respect to fatigue strength should be undertaken, which goes beyond dimensioning with respect to static strength. Static dimensioning with reference to the yield limit, according to the code, also covers low cycle fatigue strength up to $N = 10^3$-10^4 cycles, the lower value applying to severely notched welded joints, disadvantageous as regards fatigue, the higher value applying to smoothly notched welded joints, which are advantageous as regards fatigue (thus according to Ref.160; up to $N = 2 \times 10^4$ according to Ref.152 independent of the notch severity).

7.3   Survey of the design codes

The most important design rules in the FRG for constructions (including welded ones) with a specific assessment of fatigue strength (in addition to the assessment of static strength) are compiled below into a preliminary review. Concerning comparable design rules in other countries the proceedings[35,36] are recommended. This review concentrates exclusively on seam type welded joints. Spot welded joints are not permissible on components subjected to cyclic loading where there are reliability requirements. Rules are quoted first which have the character of fundamental standards, whereby a distinction is made between calculation standards and quality assurance standards.

The design recommendation IIW-693-81[143] is intended as a suggestion for a basic standard for design and dimensioning of welded structures in steel, which are subjected to fatigue loading. The standard DIN 18800 *Steel structures — dimensioning and design* is being introduced in the FRG to correspond with the Eurocode 3[150] and ECCS-TC6[151] (corresponding GDR specification TGL 13500.[145] The standard DIN 8563 *Quality assurance of welding work*[146] is valid as a national basic standard for manufacturing requirements, especially for welded structures subjected to fatigue loading too. A corresponding draft ISO 6213 is being discussed at an international level.

Assessment of fatigue strength as in rule DS 804[147] is necessary for welded steel railway bridges. Highway and road bridges are considered to be mainly statically loaded structures. An assessment of fatigue strength is not usually required (according to Ref.185 only one tenth of traffic load cycles have a fatigue effect), but the design recommendations for fatigue resistant structures are to be observed, particularly if there is loading from rail vehicles. In the same way, structural and waterway steel engineering do not generally require an assessment of fatigue strength, despite occasional occurrence of failures (*e.g.* in sluice gates). The DASt Richtlinie 011[148] is valid for steel structures made from high tensile fine grain structural steel, and includes an assessment of fatigue strength. The standard DIN 18808[149] supplemented by Eurocode drafts[150] is applicable to tubular joints subjected to fatigue loading.

The crane standard DIN 15018[152] supplemented by crane track standard DIN 4132[153] contains the most elaborate assessment of fatigue strength of the codes for welded steel structures including service fatigue strength. '*Principles for dimensioning large-scale equipment in opencast mining*'[154] was based on this standard. Such equipment is mainly produced by welding nowadays as structures subjected to intense fatigue loading. The structural steel engineering standard TGL 13500[145] which includes an assessment of fatigue strength, applies in the GDR.

An assessment of fatigue strength in accordance with German Lloyd's design specifications[155] is to be established for seagoing vessels and drilling platforms. The specification was also originally based on DIN 15018. Depending on the country concerned, the differing rules[156-159] which include an assessment of structural stress, are applicable to tube girders and tube joints of drilling platforms.

The AD-Merkblatt S1[160] of the TRB-Code (technical rules for pressure vessels) and the Merkblatt 301[162] of the TRD-Code (technical rules for boilers) specify the number of cycles up to which a separate assessment of fatigue strength is not necessary in tank, boiler and pipeline engineering. The assessment of fatigue strength itself is embodied in AD-Merkblatt S2,[161] in TRD-Merkblatt 301, appendix 1[162] and in the steel tube standard DIN 2413.[163] The regulations in the ASME code[164] or the KTA code[165] are valid for components of the primary circuit in nuclear reactors, which also introduce a roughly estimated local increase of structural stress (at nozzles and other openings). In the GDR assessment of fatigue strength is included in the TGL 22160[166] for pipeline dimensioning and TGL 32903[167] for tank dimensioning.

In rail vehicle engineering DS 952,[168] supplemented by the international quality standard UIC 897-13,[169] is binding for the German Railways and their suppliers. The DVS-Merkblatt 1612[170] with its extensive list of design shapes is more generally applicable as are Merkblatt 415 and 416[171,172] of the German Steel Advisory Centre. The automotive industry manages successfully without a code, just assessing service fatigue strength by testing. In the GDR the code TGL 14915[173] for strength calculations in welded components, is valid for the whole of the machine construction industry, including automotive engineering.

The aircraft industry, like the automotive industry, manages largely without set regulations concerning fatigue strength. However, there is a generally accepted design document available in the form of the 'Aircraft engineering manual for structural dimensioning' (HSB),[174] which is constantly expanded. In the machine building industry the GDR standards 19340, TGL 19333 and TGL 13350 are available, regulating the assessment of fatigue and service fatigue strength.[175] The comparable guidelines[25,32] are recommended in the FRG.

An Italian standard[75] for welded aluminium structures is in preparation which covers engineering sectors. The guideline documents[177-183] on the acceptability of detected defects and imperfections i.e. the assessment of fitness for purpose (see section 9.5) have a strong impact on all engineering fields mentioned above.

7.4   Design loads, load spectra, loading classes

Design loads for fatigue loading (load cases, load amplitudes, number of cycles, load spectra) are only rarely quantified in the codes because they are dependent on the particular conditions of use of the structures concerned and cannot be generalised or only in a limited way. A detailed analysis of the travel, loading, working, lifting or other operational processes on the structure including the test cases, which vary with time and of the superimposed environmental influences (e.g. wave movement, wind strength, gusts, roughness of the ground) mostly measured as accelerations, and their transformation in the vibrating structure, into loads or stresses, which are variable with time, must be supplied by the structural engineer to enable an assessment of fatigue strength to be carried out. In contrast, the more extensive comparability of the test results speaks in favour of standardised load spectra. Only a few general references can be made here (see the list of technical fields given in section 7.3 above). Some recent investigations into design loads can be found in Ref.176.

Service load diagrams are specified in railway bridge construction which contain the axle loads and wheel base e.g. for a motor coach, goods train, or passenger train with locomotive engine. Starting from the number and mixture ratios of the types of trains on a route section (traffic density) the load spectra at the joints relevant to dimensioning can be established for all joints via the statics beam model of the railway bridge. The possibility of trains meeting on the bridge is to be included. Starting from these load spectra, a uniform basis of reference is taken into consideration[184] for dimensioning the structure with regard to service fatigue strength. The interrelationship between single and multilevel loading of bridges has been investigated on the basis of Miner's rule by Herzog.[185]

In crane construction, the equal value of fatigue damage, expressed by an equidistantly parallelised family of standardised fatigue life curves, for increased load amplitude, and for increased number of cycles in the spectrum, is used to establish loading classes which connect the two stated quantities via a

Table 4. Loading classes for cranes, after DIN 15018[154]

Severity of operation	Frequency of operation			
	OO	IO	CO	HO
VO	B1	B2	B3	B4
LO	B2	B3	B4	B5
MO	B3	B4	B5	B6
HO	B4	B5	B6	B6*

* Repetition of B6 without verification by test results[55,56]

single parameter and, in this form, determine the maximum permissible spectrum stress. The distribution of load amplitude in the spectrum, *i.e.* the spectrum shape is specified according to the severity of the crane operation (spectrum shape coefficient, p): very light operation (VO), rarely full load, p = 0; light operation (LO), occasionally full load, p = 1/3; medium operation (MO), frequently full load, p = 2/3; heavy operation (HO), constantly full load, p = 1.0. The total number of cycles of the spectrum *i.e.* the spectrum size is specified according to frequency of operation: occasional operation (OO), 2 × $10^4 \leq N \leq 1 \times 10^6$; interrupted operation (IO), $1 \times 10^5 \leq N \leq 6 \times 10^5$; continuous operation (CO), $6 \times 10^5 \leq N \leq 2 \times 10^6$; hard continuous operation (HO), $2 \times 10^6 \leq N$. Severity and frequency of operation establish the loading classes B1 to B6 in accordance with Table 4. The loading classes for the various types of crane are specified in Ref.152.

In shipbuilding and drilling platform construction a straight line spectrum of the wave heights is used as a design basis, of which, according to the code, only 1% lead to loadings which are relevant to service fatigue strength, represented by a straight line spectrum of the load amplitudes,[155] Fig.142.* The actual

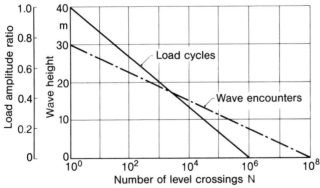

142 Load spectrum for dimensioning of ships and drilling platforms, after Ref.155.

᠄ Instead of 'Level crossing number, N' 'Number of cycles, N' is common in practice in representations of load spectra; this corresponds to the conversion into the block program test.

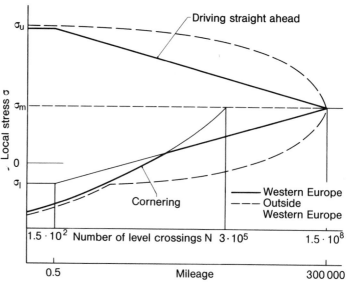

143 Stress spectrum[133] for dimensioning of truck axle components, after Ritter and Lipp.[202]

percentage and the generally deviating spectrum shape are dependent on the wave lengths, on the type and size of the structure as well as speed and direction of cruising relative to the wave movement. Ships can avoid extreme loadings to a certain extent, drilling platforms cannot. Corresponding to this, ships receive the rating B2 and drilling platforms to class B3 for a medium sea state. In addition, the loading and unloading cycles of the ship have to be taken into account.

In tank, boiler and pipeline construction the number of start-up and shut-down operations which occur during the whole life of the equipment are decisive for assessment of fatigue strength. The relevant internal pressure is of pulsating type. In addition the temperature loading cycles, which occur because of the temperature difference in the wall, are to be considered, for which a distinction should be made between warm and cold starts. Additional loads applied via nozzle pipes should be taken into account in special cases. Assessment of static strength based on maximum load is often sufficient for many current tanks. The number of cycles necessary for assessment of fatigue strength is specified by code.[160] In primary circuit components of nuclear reactors normal and abnormal operational situations, test cycles and definite fault situations are to be covered precisely.

In automotive engineering there are obligatory design loads specified for trailer couplings only. Design loads and safety factors are specified for rail vehicles (*e.g.* buffer impact forces) for dimensioning by calculation and acceptance based on testing. In both cases these are quasi-static design loads. Figure 143 shows additionally for road vehicles the suggested (but voluntary) stress spectrum measured in service at the critical point of an axle component which can be used as a basis for dimensioning in trucks and similar commercial

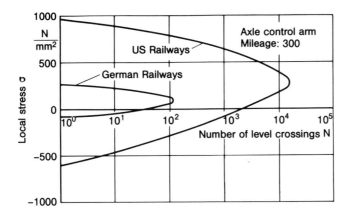

144 Stress spectrum on the axle control arm of a rail vehicle
undercarriage, distance covered 300 miles.

vehicles. Driving straight ahead produces a straight line spectrum under the
favourable road conditions in Western Europe, in the less favourable
conditions outside Western Europe it produces the 'fuller' normal distribution
spectrum. A total distance covered of 300 000 miles is used as a basis, which
corresponds to a number of cycles $N = 1.5 \times 10^8$ at an average speed of 30mph
and an axle natural frequency of 5cps. Only approximately 1% of the vehicles
actually have to sustain the dimensioning spectrum (required failure
probability $P_f = 0.001$). Cornering (only the load-relief on the inner wheel is
recorded) produces a relatively small number of particularly large load
amplitudes. The design spectrum for wheel and axle components of
commercial vehicles is further developed in Ref.203 including the separation
of the spectrum for driving straight ahead (96% of the mileage), the spectrum
for cornering (4% of the mileage) and an additional spectrum for braking
(1 normal braking per mile, 1 emergency braking per 30 miles), the absolute
and relative amplitudes of these spectra being dependent on the type of vehicle
and the considered point of stress evaluation. They are established for an
occurrence probability of 0.01 (referring to the total number of operating
trucks), the same being the case for the spectrum in Fig.143.

The stress spectrum on the axle control arm of a rail vehicle undercarriage is
shown in Fig.144 for design purposes, which also comes to a lower level in
Europe than abroad, caused by varying rail and permanent way qualities.

In aircraft engineering, the spectra of the ground to altitude pressure cycles,
of the take-offs and landings, of the loading from gusts and from taxiing on the
runway are used as the basis for dimensioning. A standard (double) spectrum
for the design of wings is shown in Fig.145. The pressure waves emitted by the
jet streams (sonic pressure), which can cause damage to the structure close to
the jet (sonic fatigue) have not been evaluated as load spectra up to now.

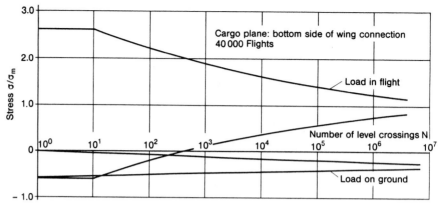

145 Stress spectrum for dimensioning of cargo plane components, wing connection, after Ref.174.

7.5   Notch classes and quality of manufacture

The peculiarity of the dimensioning rules for welded structures subjected to fatigue loading is the assignment of the individual joint to notch classes, for which differing S-N curves and service life curves are specified. The latter are generally (in special scales) linearised, parallelised and graded equidistantly with regard to level of stress and number of cycles. Thus welded joints, which are graded in accordance with geometrical shape, type of weld, type of loading and quality of manufacture, are assigned to a series of standardised S-N curves and service life curves of permissible or endurable stress values. The allocation takes place on the basis of the results of relevant fatigue tests. The conventional designation 'notch case' or 'notch class' is only correct to the extent that varying fatigue strength is caused by varying notch effect.

In fact the above is only the case as regards design details, not regarding some aspects of manufacture and material. A rough estimate for the local increase of structural stress is only allocated to the notch class in the codes in a very few cases and rarely the notch class is replaced by the local increase of structural stress. In particularly high quality components (*e.g.* primary circuits of nuclear reactors) the division into notch classes is disregarded because only designs which are favourable as regards notch effect are permissible (indicated in Ref.165 by 'stress indices' which range from 2.0 to 3.3 for nozzles).

The standardised S-N and service life curves of endurable or permissible stresses are differentiated according to position and number depending on the individual code. The varying number of notch classes is particularly obvious. The considerable scattering of the test results supports a small number of classes. The more concise dimensioning in the individual case because of the finer grid for the permissible stresses supports a larger number of classes. The following numbers of notch classes for welded joints (without base material) are found in the codes: five in DIN 15018 (K0 to K4) as well as in German Lloyds' code, six in DASt 011, nine in DS 804 (KII to KX), ten in IIW-693-81, five in the older issues of DVS 848, now no longer valid and seven in the valid DS 952 (B to H), three in TRB-AD S1 and S2 (K1 to K3), six in the proposal for aluminium structures.[75] In addition, the differences in today's assessment of the influence of the mean stress are obvious. In future, in bridge construction and shipbuilding the dependence on the mean stress of the permissible stress amplitude is to be dispensed with, based on the high residual welding stresses in these structures. However, in contrast, the mean stress is an influencing parameter which has to be taken into account in tank, boiler and pipeline engineering because of the components which are often relieved of residual welding stresses.

Which welded joints are allocated to which notch classes depends on the code, as the various technical fields give preference to different joints. To illustrate two notch class systems, classification as in DIN 15018 is illustrated in Tables 5-10 and as in TRB-AD S1 and S2 in Tables 11-13, which together cover the majority of welded joints which occur frequently. A complementary notch class system for tube joints in latticework (tubes with circular and rectangular cross section) has been established.[150,151]

Tables 6-13 reveal that in addition to design shape and loading type, inspection and manufacturing conditions are also decisive for allocation into notch classes. However, assessment of inspection and manufacturing results (quality assessment) with regard to notch or strength classes is still a largely unsolved problem and now stands in the centre of efforts being made with regard to the codes.

The minimum requirements on the quality of manufacture involving fusion welding are quantified in DIN 8563.[146] They are classified into four (butt weld) or three (fillet weld) requirement classes. The manufacturing requirements are subdivided into external and internal findings. The external findings include the characteristics: excessive weld reinforcement, incompletely filled groove, weld thickness deficiency, plate edge misalignment, excessive asymmetry of fillet welds, undercuts and toe grooves, open end of weld craters, visible pores, visible slag inclusions, fused-on weld spatter, arc strikes, excessive penetration, root concavity and lack of penetration. The internal findings include the characteristics: cavities, solid inclusions, lack of fusion, lack of penetration, cracks and end crater pipes. The requirement classes (also called 'assessment classes') establish the permissible findings from non-destructive inspection within narrow limits (assessment 'high quality') or less narrow limits (assessment 'low quality'), four classes for butt welds (designated AS, BS, CS, DS), three classes for fillet welds (AK, BK, CK). Slightly modified data apply to thin sheet steel (plate thickness 0.5-3.5mm) and to aluminium alloys. The particularly high quality butt weld, which is ground flush, lies outside the above-mentioned assessment classes. The assessment classes are to replace the previously usual division into 'standard quality' and 'special quality'. Reference should be made to DIN 8524[39] as regards the designation and explanation of welding defects.

Table 5. Crane standard DIN 15018. Explanation of the notch classes and quality characteristics in Tables 6-10

Type of weld	Weld quality	Weld execution	Symbol (examples)	Inspection method for defect-free weld execution	Abbreviation
Butt	Special	Root machined, capped, ground flush in direction of stress, no end craters		Non-destructive inspection of the weld on 100% of weld length *e.g.* by radiography	P 100
	Standard	Root machined, capped, no end craters		As for special quality but only for tensile loading with $\sigma_{max} \geqslant 0.8 \times \sigma_{per}$	P 100
				Non-destructive inspection of the other most important welds in random sample tests for at least 10% of the weld length of each welder *e.g.* by radiography	P
Double bevel butt with superimposed fillet welds	Special	Root machined out, welded through, weld toe without notches machined if necessary.		Non-destructive inspection of the plate subjected to tension transverse to its plane for lamination and microstructural damage in the weld area *e.g.* by ultrasonics	D
	Standard	Width of residual groove at the root up to 3mm or 0.2 times thickness of the welded-on component. The smaller value is to be used			
Fillet	Special	Weld toe without notches, dressed if necessary			
	Standard	-			

To simplify the text in Tables 6-10 the statement 'fillet weld' in the column 'Description and representation' also applies to the double fillet weld if both symbols are given. If a double fillet weld is required for a notch class then this is stated both in the column 'Description and representation' and 'Symbol'.

Italics in the column 'Description and representation' in Tables 6-10 indicate whether the weld, the continuous part influenced by the weld, or both, are classified in the notch class concerned. This makes classification into different notch classes possible for the weld and for the component.

Table 6. Crane standard DIN 15018, notch class K0, slight notch effect

Order no.	Description and representation		Symbol
011	*Parts connected* with *butt weld* of special quality, transverse to loading direction		P 100 P 100
012	*Parts* of different thicknesses *connected* with a *butt weld* of special quality, transverse to loading direction with asymmetrical joint and bevel ⩽ 1:4 if supported or with symmetrical joint and bevel ⩽ 1:3	Bevel ⩽ 1:4 Bevel ⩽ 1:3	P 100 P 100
013	*Welded-in gusset plate* with *butt weld* of special quality, transverse to loading direction		P 100 P 100
014	Transversely *connected web plates* using *butt weld* of special quality		P 100 P 100
021	*Parts connected* with *butt weld* of standard quality, longitudinal to loading direction		P or P 100 P or P 100
022	*Web plates* and *chord sections* made from section or bar steel excluding flat bar, steel *connected* with *butt weld* of special quality		P or P 100 P or P 100
023	*Parts connected* with *double bevel butt weld* with superimposed *fillet welds*, longitudinal to loading direction		K

Table 7. Crane standard DIN 15108, notch class K1, moderate notch effect

Order no.	Description and representation		Symbol
111	*Parts connected* with *butt weld* of standard quality transverse to loading direction		P or P 100 P or P 100
112	*Parts* of different thicknesses *connected* with a *butt weld* of standard quality, transverse to loading direction, with asymmetrical joint and bevel ≤ 1:4 if supported or with symmetrical joint and bevel ≤ 1:3		P or P 100 P or P 100
113	*Welded-in gusset plate* with *butt weld* of standard quality, transverse to loading direction		P or P 100 P or P 100
114	Transversely *connected web plates*, using *butt weld* of standard quality		P or P 100 P or P 100
121	*Parts connected* with *butt weld* of standard quality, longitudinal to loading direction		
123	*Parts connected* with *fillet weld* of standard quality, longitudinal to loading direction		
131	*Continuous part*, on to which parts are welded with continuous double bevel butt weld with superimposed fillet welds of special quality, transverse to loading direction		

Table 7. Contd

Order no.	Description and representation		Symbol
132	*Continuous part*, on to which discs are welded with a double bevel butt weld with superimposed fillet welds of special quality, transverse to loading direction		
133	*Compression members* and *web plates*, on to which transverse bulkheads or stiffeners with cut off corners are welded using double fillet welds of special quality: grading into this notch class is only valid for double fillet welds		
154	Web plates and curved chord plates, connected with *double bevel butt weld* with superimposed *fillet welds* of special quality		

Table 8. Crane standard DIN 15018, notch class K2, medium notch effect

Order no.	Description and representation		Symbol	
211	*Parts* made of section or bar steel, excluding flat bar steel, *connected* with *butt weld* of special quality, transverse to loading direction			P 100 P 100
212	*Parts* of different thicknesses *connected* with *butt weld* of standard quality, transverse to loading direction with asymmetrical joint and bevel ≤ 1:3 if supported, or with symmetrical joint and bevel ≤ 1:2	Bevel ≤ 1:3 Bevel ≤ 1:2		P or P 100 P or P 100
213	*Butt weld* of special quality, and *continuous part*, both transverse to loading direction, at chord plate crossings with welded-in plate corners; weld ends machined notch free			P 100 P 100

Table 8. Contd

Order no.	Description and representation		Symbol
214	*Welded-on parts* on junction plates using *butt weld* of special quality, transverse to loading direction		P 100 P 100
231	*Continuous part*, on to which parts are welded using a continuous double fillet weld of special quality, transverse to loading direction		
232	*Continuous part*, on to which discs are welded with double fillet weld of special quality, transverse to loading direction		
233	*Chord and web plates*, on to which transverse bulkheads or stiffeners with cut-off corners are welded with double fillet weld of special quality, transverse to loading direction		
241	*Continuous part*, to the edge of which parts are welded, which are bevelled on the ends or are given a smooth profile, using a butt weld of standard quality, longitudinal to loading direction; weld ends machined notch free		
242	*Continuous part*, on to which parts or stiffeners which are bevelled or given a smooth profile, are welded; end welds executed in the range $\geqslant 5t$ as double bevel butt weld with superimposed fillet welds of special quality		End welds only
244	*Continuous part*, on to which a chord plate is welded with bevel $\leqslant 1:3$; end weld executed as a fillet weld of special quality, with a $= 0.5t$ in the range marked $\geqslant 5t$		End welds only

Table 8. Contd

Order no.	Description and representation		Symbol
245	*Continuous part*, on to which bosses are welded with fillet weld of special quality		
251	Parts connected by cruciform joint with *double bevel butt weld* with superimposed *fillet welds* of special quality, transverse to loading direction		D
252	Parts connected with *double bevel butt weld* with superimposed *fillet welds* of special quality, bending and shear loading		D
253	Parts connected with *double bevel butt weld* with superimposed *fillet welds* of special quality, between chord and web, subjected to concentrated loads in the web plane transverse to weld		
254	Web plates and curved chord plates connected with *double bevel butt weld* with superimposed *fillet welds* of standard quality		K

Table 9. Crane standard DIN 15018, notch class K3, strong notch effect

Order no.	Description and representation		Symbol
311	*Parts connected* with *butt weld* from one side on root backing, transverse to loading direction		V
312	*Parts* of different thicknesses, with symmetrical joint and bevel ⩽ 1:2 if supported, or with symmetrical joint and bevel ⩽ 1:1, *connected* with *butt weld* of standard quality, transverse to loading direction		P or P 100 X P or P 100

Table 9. Contd

Order no.	Description and representation		Symbol
313	*Continuous part* and *butt weld* of standard quality, both transverse to loading direction at crossings of chord plates with welded-in plate corners; weld ends machined notch free		P or P 100 P or P 100
314	*Tubes connected* by backed *butt weld*, not capped		
331	*Continuous part*, on to which parts are welded with double fillet weld of standard quality, transverse to loading direction		
333	*Chord and web plates* on to which transverse bulkheads or stiffeners are welded transverse to loading direction with uninterrupted double fillet weld of standard quality; grading into notch class is only valid for fillet weld region		
341	*Continuous part*, on to the edge of which parts with bevelled ends are welded longitudinal to loading direction; weld ends machined notch free		
342	*Continuous part*, on to which parts or stiffeners with bevelled ends are welded longitudinal to loading direction; the end welds are executed as double fillet welds of special quality in the range marked $\geqslant 5t$		End weld only
343	*Continuous part*, through which a plate with bevelled ends or ends with a smooth profile penetrates and is welded; end welds executed as double bevel welds with superimposed fillet welds in the range marked $\geqslant 5t$ and machined notch free		End weld only
344	*Continuous part*, on to which a chord plate with $t_o \leqslant 1.5t_u$ is welded; end welds executed as fillet welds of special quality, in the range marked $\geqslant 5t_o$		End weld only

Table 9. Contd

Order no.	Description and representation		Symbol
345	*Parts,* on the ends of which joint straps of $t_o \leqslant t_u$ are welded using fillet welds of special quality; end welds executed as fillet welds of special quality, in the marked area; if only one side of the parts is covered by a joint strap the eccentric force effect is to be taken into account		End weld only
346	*Continuous part,* on to which *longitudinal stiffeners* are welded with intermittent double fillet weld or using cutouts with double fillet weld of standard quality between; grading into the notch class is valid for the weld between the end welds		
347	*Continuous part,* on to which rods of section or bar steel are welded with a closed circumferential fillet weld of special quality		
348	*Tube rods* connected with fillet weld of special quality		
351	Parts connected by cruciform joint with *double bevel butt weld* with superimposed *fillet welds* of standard quality, transverse to loading direction		
352	Parts connected with *double bevel butt weld* with superimposed *fillet welds* of standard quality, bending and shear loading		
353	Parts connected with *double bevel butt weld* with superimposed *fillet welds* of standard quality, between chord and web, subjected to concentrated loads in the web plane transverse to weld		K
354	Web plates and curved chord plates connected with *double fillet welds* of standard quality		

Table 10. Crane standard DIN 15018, notch class K4, very strong notch effect

Order no.	Description and representation		Symbol	
412	*Parts* of different thicknesses with asymmetrical supported joint without bevel, eccentrically *connected* with *butt weld* of standard quality, transverse to loading direction			P P
413	*Parts connected* with *butt weld* of standard quality, transverse to loading direction at crossings of chord plates			P P
414	*Flanges* and *tubes connected* with two *fillet welds* or with *single bevel butt weld* with superimposed *fillet weld*			
433	*Chord* and *web plates* on to which transverse bulkheads are welded with continuous single sided fillet weld of standard quality, transverse to loading direction			
441	*Continuous part,* on to the edge of which parts, which butt at right angles are welded longitudinal to loading direction			
442	*Continuous part,* on to which parts or stiffeners, which butt at right angles are welded longitudinal to loading direction with double fillet weld of standard quality			
443	*Continuous part,* which is penetrated by a plate at right angles and which is welded with double fillet weld of standard quality			

Table 10. Contd

Order no.	Description and representation		Symbol
444	*Continuous part* on to which a chord plate is welded with fillet welds		
445	*Superimposed parts* with holes or slots in which fillet welds are executed		
446	*Continuous parts,* between which a brace plate is welded with a fillet weld of standard quality or butt weld		P or P 100
447	*Continuous parts,* on to which *rods* are welded with fillet welds		
448	*Tube rods* connected with fillet weld		
451	*Parts connected* by cruciform joint with *double fillet weld* of standard quality or single sided *single bevel butt welds* with superimposed fillet weld on root backing plate, transverse to loading direction		D
452	Parts connected with *double fillet weld* of standard quality subjected to bending and shear		D
453	Parts connected with *double fillet weld* of standard quality, between chord and web, subjected to concentrated loads in web plane transverse to weld		

Table 11. Pressure vessel code TRB-AD S2, notch class K1

Item no.	Representation	Description	Preconditions
1		Longitudinal or circumferential weld for pressure bearing walls of the same thickness	Welded from both sides
2		Longitudinal or circumferential weld for tank walls of different thicknesses	Welded from both sides
			Welded from both sides; edge offset the same inside and outside
3		Welded-on parts without introduction of additional forces or moments	Welded through from one or both sides, external findings as in DIN 8563, part 3, group BK, exclusive of the characteristics: groove overfill and underfill and uneven legs
4		Nozzle with butting pipe	Welded to be fully load bearing (without residual gap), nozzle pipe drilled out or root ground
5		Nozzle with inserted or penetrating pipe	Welded through from one or both sides
6		Block flange	Welded through from one or both sides

Table 12. Pressure vessel code TRB-AD S2, notch class K2

Item no.	Representation	Description	Preconditions
1		Longitudinal or circumferential weld for pressure bearing tank walls of the same thickness	Welded from one side; if equivalence with seam welding from both sides is provided, the connection can be allocated to notch class K1
2		Longitudinal or circumferential weld for tank walls of different thicknesses	Welded from both sides
	≤30°		Welded from both sides; edge offset is the same inside and outside
3		Welded-on parts without introduction of additional forces or moments	Welded through from both sides, external findings as in DIN 8563, part 3, group BK, exclusive of the characteristics: groove overfill and underfill and uneven legs
4	≤60°		Welded through from both sides, external findings as DIN 8563, part 3, group BK, exclusive of the characteristics: groove overfill and underfill and uneven legs
5		Flat end of tank with shell or double shell, welded-on parts with the introduction of additional forces or moments, reinforcing ribs	Welded through from both sides, external findings as in DIN 8563, part 3, group BK
6		Welded-on flange	Welded from both sides, AD-Merkblatt B8, Table 1 weld execution 3-5

Table 13. Pressure vessel code TRB-AD S2, notch class K3

Item no.	Representation	Description	Preconditions
1	≤30°	Longitudinal or circumferential weld for pressure bearing tank walls of uneven thickness	Welded from one or both sides
2		Welded-on parts without introduction of additional forces or moments	Welded from one or both sides
3		Flat end of tank with shell or double shell, welded-on parts with the introduction of additional forces or moments, stiffening ribs	Welded from both sides
4		Block flange	Welded from both sides
5		Welded-on flange	Welded from both sides, AD-Merkblatt B8 Table 1, weld execution 1 and 2
6		Nozzle with butting pipe	Weld not fully load bearing
7		Nozzle with inserted or penetrating pipe	Weld not fully load bearing

Table 14. Rail vehicle code, extract from DVS Merkblatt 1612, assessment class as per DIN 8563 and permissible stress as per DS 952

Type of weld. Application	Description of weld. Manufacture	Recommended scope of inspection	Decisive characteristic (DIN 8563)	Assessment class	Type of stress	σ_{per} as per curve
Butt weld between parts of same thickness	Root cap welded and machined notch free	Up to 100% non-destructive		AS	σ τ σ_{eq}	B^+ G B^+
	Root cap welded	Up to 100% non-destructive	Excessive weld convexity, underfilling, edge misalignment	AS	σ τ σ_{eq}	D^+ G D^+
		Up to 10% non-destructive		BS	σ τ σ_{eq}	D G D^+
		Visually inspected		CS	σ τ σ_{eq}	D^- G D
	Root not cap welded	Visually inspected	Excessive weld convexity, root concavity, root notch, lack of fusion at weld root	CS	σ τ σ_{eq}	$E1$ H $E1^+$

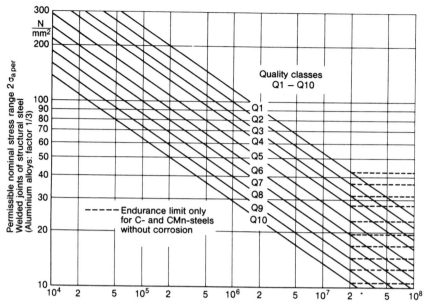

146 S-N curves for quality classes, welded joints, structural steel and aluminium alloys, after Ref.177.

It is to be regretted that in the classification of manufacturing quality described above, according to assessment classes, there are frequently no quantitative data on the permissible imperfections and defects, that the effect of these imperfections and defects on the fatigue strength is very variable, depending on the characteristic (for example the notch effect of the shape and dimensional deviations on the one hand, and defects on the other is very different) and that the extent of inspection to be prescribed for the assessment class concerned is not specified. Therefore the various engineering sectors are currently applying regulations which supplement and extend DIN 8563. Fatigue tests are being run in shipbuilding with the aim of allocating characteristic-dependent notch classes within each assessment class.[188, 189] It has been established to date that imperfections of manufacture do not reach the notch effect of the most unfavourable design details (notch case allocation K2 to K4). A particularly thoroughly produced (with regard to the extent of the inspection and the evaluation of the inspection results) guideline[154,190] has been introduced for large scale equipment in brown coal opencast mining (excavators and live booms). The scope includes visual inspection and checks on shape and dimensions, surface crack detection, radiographic and ultrasonic inspection. The findings are evaluated on a scale of 0-5. The list of permissible welded joints and their weld details is given in greater detail than in DIN 15018. In structural steel engineering, the possibilities of weld quality assessment in accordance with DIN 8563 are being considered with some reservations.[193] In rail vehicle engineering according to DVS Merkblatt 1612,[170] the permissible stress is dependent on the notch class, type of joint and manufacturing quality, whereby the assessment class is quoted as an additional factor, Table 14.

However, the change in the permissible stress is not quantified if there is a deviation from the assessment class given. However, it does show the correct way forward from the point of view of method, which can be substantiated in future by further data. A more recent comparative investigation by Neumann[191] considers relevant DIN, ISO and TGL standards especially with respect to the situation in the GDR. A proposal for a quantitative defect evaluation has been made by Schultze[192] for high strength steels.

According to the British proposal for assessing defects[177] the nature and size of the defect (shape imperfections and dimensional deviations are not covered) determine the quality class. An S-N curve is allocated to each of the total of ten quality classes, Fig.146, (partially identical to the notch class S-N curves). Thus the possible unfavourable superimposing of notch and quality classes is excluded, which is not always justified.

7.6   Permissible stress, safety factor, multiaxiality hypothesis

The permissible stresses with respect to fatigue are derived from the endurable stresses (for $P_f = 0.1$ or 0.5) by introducing the safety factor with reference to stress, $S = 1.3\text{-}2.0$, or the safety factor with reference to number of cycles, $S^* = 4\text{-}20$. To simplify the representation, the same safety factor, S, is used for mean stress and stress amplitude in most cases. Permissible stresses for fatigue resistant welded joints are given in the relevant codes. There are formal differences in the assessment method between bridge, crane and shipbuilding engineering on the one hand (assessment of nominal stress in the conventional simple form) and tank, boiler and pipeline engineering on the other hand (assessment of nominal stress with differentiation towards the assessment of structural stress).

7.6.1   Bridge, crane and shipbuilding industries

First of all, let us consider the permissible stresses with respect to fatigue strength for infinite life (endurance limit) in welded joints which are specified dependent on the material (structural steels St 37 to St 52, high tensile structural steels), on the limit stress ratio, R, and on the notch class of the joint. The safety concepts of some of the codes considered below are illustrated in Fig.147. According to DIN 15018,[152] $S = 4/3$ against endurable stresses for $P_f = 0.1$, and according to DS 804,[147] $S = 1.65$ against endurable stresses for $P_f = 0.5$. The permissible stresses in Eurocode 3[150] are based on endurable stresses for $P_f = 0.5$ against which a safety margin of two standard deviations of the scattering test results is observed. The stress versus failure probability curve in Fig.147 is based on the scatter range $T_\sigma = 1{:}1.5$ of the normalised S-N curve for seam welded joints in the endurance limit range. Apparently, the permissible stresses in DIN 15018 and DS 804 include the failure probability $P_f \approx 0.001$, those in Eurocode 3 $P_f \approx 0.023$.

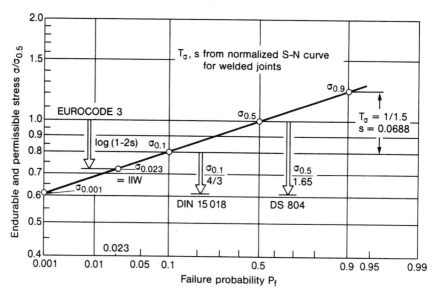

147 Illustration of the safety factor concept behind the permissible stresses of different codes, after Olivier, Köttgen and Seeger.[231]

The permissible stress values stated in DIN 15018,[152] based on the latter in German Lloyd's specification[155] and stated independently in table form in DS 804,[147] are represented graphically in Fig.148. The standard DIN 15018 is based on a section by section linearised curve representation in the fatigue strength diagram according to Smith, DS 804 is based on a similar one in the diagram according to Haigh. Despite having the same basis of test results, the curve pattern is only identical to the extent of a rough approximation and neither is there any correspondence with the family of curves for welded joints suggested by Stüssi[15,49] and Neumann.[3,16]

All the curves in Fig.148 and the curves for the permissible stress amplitude preferred in future are limited in the area of high mean stress by the statically permissible stresses. The diagram for St 52 or a similar one, expanded upwards in the area of high tensile mean stress, can be used for weldable high tensile fine grained structural steels depending on the individual yield limit which is present. The DASt-Merkblatt 011[148] specifies permissible stresses for the fine grained structural steels StE 460 and StE 690, dependent on notch class and limit stress ratio. Compared with the endurable stresses, the stated permissible stresses contain the safety factor $S = 1.33$ both as regards the endurance limit (for $P_f = 0.1$) and with regard to the yield limit σ_Y (in the load case HZ, primary and secondary loads, as in DIN 4100[194] corresponding to $\sigma_{per} = 0.75\sigma_Y$). The safety factor with regard to the endurance limit for $P_f = 0.5$ is $S = 1.63$. The failure probability resulting from the use of this safety factor is $P_f = 0.01$ (can be calculated from the scatter band of the normalised S-N curve). The safety factor with regard to the yield limit in the assessment according to load case H, only primary loads, is $S = 1.5$ corresponding to $\sigma_{per} = 0.67\sigma_Y$.

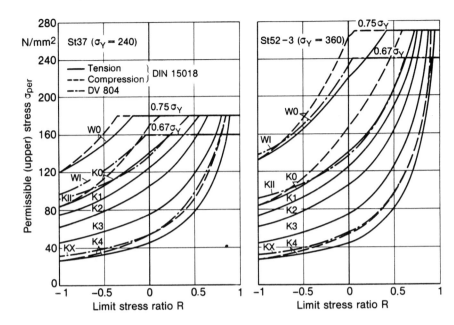

148 Permissible stresses, structural steels St 37 and St 52, base material W0 and WI, welded joint notch classes K0 to K4 or KII to KX, as in DIN 15018[152] or DS 804[147] respectively.

The explanation below is given for the fatigue strength curves in Fig. 148. W0 in DIN 15018 and WI in DS 804 designate unnotched base material; K0 to K4 in DIN 15018 and KII to KX in DS 804 designate the notch classes of the welded joints, whereby only the most favourable notch class (butt joint of special quality) and the least favourable (longitudinal stiffener or gusset plate) are represented from DS 804. The classification of welded joints used in crane construction into the notch classes K0 to K4 in accordance with DIN 15018 is represented in Tables 6-10. As can be seen from Fig. 148, with the exception of the position of the upper curve limit (statically permissible stresses) there are nearly no differences between St 37 and St 52-3 as regards welded joints. The only difference is between the curves W0 and WI of the unnotched base materials St 37 and St 52-3. Somewhat higher stresses are permissible in the compressive range (mean stress negative) than in the tensile range (mean stress positive). The WI curve from DS 804 for the unnotched St 37 is considerably lower than the corresponding W0 curve from DIN 15018.

The permissible stress amplitude $2\sigma_{aper}$ can be calculated from the permissible (upper) stress σ_{per}:

$$2\sigma_{aper} = (1 - R)\sigma_{per} \tag{38}$$

In accordance with the IIW guideline[143] the permissible stress amplitude will be independent of the mean stress in future (justified by the high residual welding stresses in the structure), and therefore also $\sigma_{p\,per} = 2\sigma_{A0\,per}$.

The permissible shear stress τ_{per} is obtained from the permissible tensile stress σ_{per} in DIN 15018 and DS 804, for unnotched base material in accordance with the von Mises distortion energy hypothesis:

$$\tau_{per} = \frac{\sigma_{per}}{\sqrt{3}} \tag{39}$$

For welded joints τ_{per} is obtained with a modified formula, which corresponds approximately to the ISO formula for static loading:

$$\tau_{per} = \frac{\sigma_{per}}{\sqrt{2}} \tag{40}$$

The permissibility of multiple loading is established as follows in accordance with DIN 15018:

$$\left(\frac{\sigma_\perp}{\sigma_{\perp per}}\right)^2 + \left(\frac{\sigma_\|}{\sigma_{\|per}}\right)^2 - \left(\frac{\sigma_\perp \sigma_\|}{\sigma_{\perp per} \sigma_{\|per}}\right) + \left(\frac{\tau}{\tau_{per}}\right)^2 \leq 1.1 \tag{41}$$

The value 1.1 instead of 1.0 on the right side of the equation leads to a contradiction, because, if only one of the stress components $\sigma_\perp$, $\sigma_\|$, or τ is introduced, the corresponding permissible stresses are increased by the factor $\sqrt{1.1}$. However, according to code, if there are several possibilities of assessment, the most conservative result is relevant.

Assessment of service fatigue strength as in DIN 15018 includes not only the fatigue strength for infinite life (endurance limit) considered above, but also the fatigue strength for finite life and the service fatigue strength with load spectra. The assessment of service fatigue strength is obligatory for $N \geq 2 \times 10^4$ cycles. The permissible stresses can be increased considerably with respect to the endurance limit in the range $2 \times 10^4 \leq N \leq 2 \times 10^6$ (but not higher than the statically permissible stresses). This is based on the higher level of fatigue strength for finite life and, for load spectra, on the fact that not all the load cycles reach the maximum amplitude which must remain below the permissible stress. The possibility of increasing the permissible stresses is mainly used in the range of small (positive or negative) limit stress ratios, i.e. in the left sections of the curves in Fig.148. The extent of the increase of the permissible stresses depends on the loading class to which the individual structure may be allocated (includes number of cycles and shape of the spectrum). The factor $\sqrt{2} = 1.41$

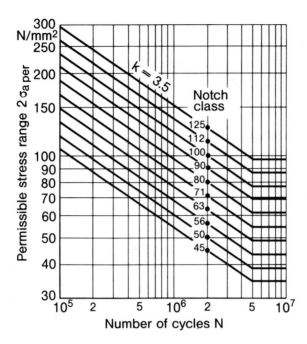

149 Permissible ranges of nominal stress (P_f = 0.023), gradient exponent k, structural steel, notch classes according to IIW recommendation.[143]

for welded joints or $\sqrt{1.41}$ = 1.19 for the unnotched base material corresponds to one loading class step as in Table 4. The factor for the welded joints is derived from equidistantly parallelised service life curves. Thus the permissible stress for welded joints can be increased by transition from the heaviest to the lightest loading class in the most favourable situation by the factor 1.41^5 = 5.65, provided that this does not cause the statically permissible stresses to be exceeded. More sophisticated safety concepts on the basis of advanced damage accumulation rules are considered in Ref.186, 187.

An up to date alternative, as regards method, for determining the permissible stresses is established in the IIW design recommendation 693-81[143] for welded structures made of steel and subjected to fatigue loading, Fig.149. Standardised S-N curves are given for the failure probability P_f = 0.023. The endurable (double) stress amplitude, $2\sigma_A$, is plotted over the number of cycles, N, for different notch classes. The notch class curve 125 designates a ground flush and fully radiographically inspected transverse butt weld, the notch class curve 50 designates a longitudinal stiffener, gusset plate or cover plate end. The curves in between cover the remaining welded joints. The notch class curve 45 applies to an under-dimensioned cruciform joint with failure in the fillet section propagating from the weld root, whereby the amplitude of the nominal stress in the weld seam is meant here instead of the nominal stress in the base material. It is noteworthy that the endurance limit is not reached until N = 5 × 10^6.

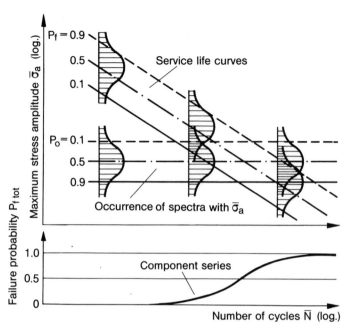

150 Failure probability for a component series, scatter band for the service life curves and scatter range of the load spectra; P_f and $P_{f\,tot}$ failure probability, P_o occurrence probability, after Haibach.[205]

An approach, which goes beyond the concept of permissible stresses as regards method, is the following simple form of a statistical safety evaluation for mass produced welded joints. Using the scatter band for the fatigue strength values of the joints and the scatter band for the service load spectra as a basis, the resulting failure probability can be defined by superimposing the two statistical distributions (simplified as Gaussian normal distributions or as Weibull distributions), Fig.150. The failure or survival probability, *i.e.* the risk of failure for the individual joint is thus represented dependent on the number of cycles endured. The approach is further expanded in Ref.206.

A general theory of reliability for structural systems is being considered nowadays with regard to the safety requirements *e.g.* for nuclear power stations, natural gas pipelines or railway bridges, to harmonise the safety concepts in the codes on this basis.[150,151,195-197] The theory is mostly simplified along the lines of exclusively (actual or approximated) Gaussian normal distributions for the random variables, which determine the failure situation, and by linearising the limit state equations. The determination of the safety index, β, as a measure for the failure probability is illustrated for the two base variables of structural theory, one load parameter and one resistance parameter, in a three dimensional representation of the probability density with relevant failure limit surface. Depending on the importance of the structure and the extent of the consequences of damage ('safety class') the aim is to achieve failure probabilities less than 10^{-5}-10^{-7}.

7.6.2 *Tank, boiler and pipeline construction*

Assessment of fatigue strength of tanks, boilers and pipelines is necessary if the number of cycles relevant for fatigue strength — primarily the number of start-up and shut-down operations — is unusually high. Static dimensioning with reference to the yield limit is adequate for $N \leq 10^3\text{-}10^5$ cycles during the whole life, the lower value applying to severely notched welded joints, which are disadvantageous as regards fatigue strength, the higher value to relatively smooth welded joints, which are advantageous as regards fatigue strength. Service load amplitudes which are reduced in elation to the design loads produce an increased total life corresponding to the normalised S-N curve (see TRB AD/Sl[160]).

The assessment of fatigue strength in tank, boiler and pipeline construction (TRB AD/S2[161] in particular is binding) is established on the basis of nominal stresses, which are determined as membrane stresses (bending stresses are superimposed only in special cases). The local increase of membrane stress, which occurs on nozzles, block flanges and other attachments, is included. This is a first step in the direction of the assessment of structural stress. The range of stress, $\Delta\sigma$, the double stress amplitude, $2\sigma_a$, from start-up and shut-down operations, including the temperature stresses in the wall, is to be assessed with respect to permissible values dependent on mean stress. The procedure is carried out as follows:

— stress amplitude $\sigma_{eq\ max}$ determined as maximum value of the three differences of the principal stresses (shear stress hypothesis according to Tresca), presupposing invariable principal directions, more complicated procedure if there is a variable principal direction;

— increase of mean stress and stress amplitude by the stress concentration factor, f_k, which is dependent on the notch class (K1 to K3, $f_k = 1.85\text{-}2.85$ for low tensile steel and endurance limit), on the number of cycles to be endured (f_k increasing with N), and on the static strength of the steel (f_k increasing with σ_U), and alternatively by the surface factor f_{k0}, which is dependent on roughness if the welded joint is ground flush.

— stress and strain correction in accordance with Neuber's macrostructural support effect formula if the yield limit is exceeded at the notch root (fatigue strength for finite life);

— fictitious increase of the stress amplitude corresponding to the reduction of the endurable stresses in the presence of increased mean stress (parabola in Haigh's fatigue strength diagram, factor $1/f_m$);

— fictitious increase of the stress amplitude corresponding to the reduction of the endurable stresses in the presence of increased operating temperatures (factor $1/f_T$, $f_T = 1.0\text{-}0.4$ for $T = 100\text{-}600°C$).

The range of the fictitious total stress $2\sigma_a^*$ is therefore given by

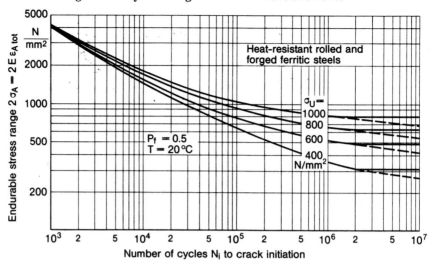

151 Endurable stress ranges, heat resistant ferritic structural steels, as in TRB-AD-S2.[161]

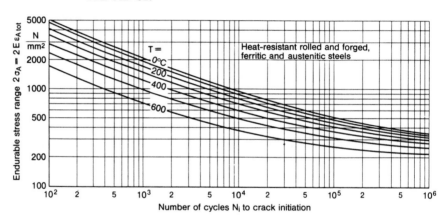

152 Endurable stress ranges, heat resistant ferritic and austenitic structural steels, as in TRD-301[162]

$$2\sigma_a^* = \frac{\sigma_{eq\,max}\,f_k}{f_m\,f_T} \qquad (42)$$

This stress range is to be compared with endurable values which represent elastically converted elastic-plastic total strains up to crack initiation (and which are higher than the actual fatigue strength values for finite life because of this conversion into fictitious values) Fig.151. Four S-N curves are given for four groups of static strength of the steel. The influence of static strength is opposite to the statements in section 3.1, the same low cycle fatigue strength with increasing endurance limits. S = 1.5 or S* = 10 is to be set as the safety factor.

The cyclic damage according to Miner, using the S-N curves above as a basis, and (for increased temperatures) the creep damage in accordance with an

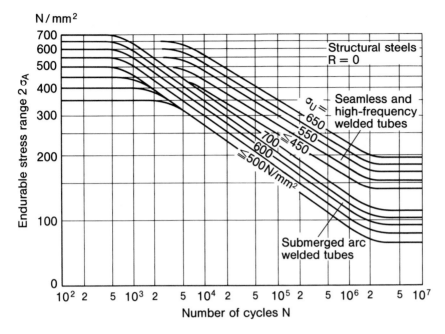

153 Endurable stress ranges for steel tubes as in DIN 2413.[163]

approach analogous to Miner, using the creep strength curves as a basis, are to be assessed in the case of service load spectra (see section 3.3.4).

A similar method is used in the older TRD 301[162] for boiler construction, stress concentration factors for outlet pipes f_k = 2.6-3.2, additional lack of roundness factor, always a surface factor (dependent on static strength of the steel) and, if required, a factor for the unmachined butt weld root, fillet welds with root slit not permissible, safety factor for number of cycles, S^* = 5, endurable stresses dependent on temperature specified for ferritic and austenitic steels, Fig.152.

The assessment of fatigue strength for seamless and high frequency resistance welded or submerged-arc welded tubes refers to endurable stresses shown in Fig.153, safety factor S = 1.5 or S^* = 5. Once again the grading of the curves dependent on static strength of the steel is worth noting. Long distance pipelines for crude oil and natural gas are usually designed on the basis of assessment of strength for static loading. Sensational cases of failure in the recent past renders this procedure questionable.

The nominal structural stress concentration factors below are introduced for boilers and pressure vessels in accordance with the ASME code: K_s = 1.0 for capped butt welds, K_s = 2.0 for butt welds which are fused on to a backing plate and K_s = 4.0 for fillet welds. S = 2.0 or S^* = 20 is to be introduced as the safety factor. The S-N curves for the permissible fictitious-elastic stresses for ferritic and austenitic steels according to the ASME code,[164] which also forms a part of the German KTA code[165] are shown in Fig.154. The ASME curve

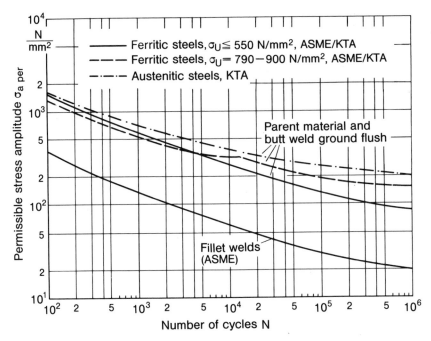

154 Permissible stresses for boilers and pressure vessels, derived from
endurable strains via modulus of elasticity and safety factor (ferritic
steels: E = 2.07 × 10⁵ N/mm², austenitic steels E = 1.79 ×
10⁵ N/mm²) in accordance with ASME code[164] and KTA code.[165]

for unstabilised) austenitic steels is identical to the ASME curve for low tensile
ferritic steels, the KTA curve for (stabilised) austenitic steels was higher, as
shown. A modification[199] is being discussed with regard to the British pressure
vessel standard.[198]

7.6.3   Aluminium structures

Permissible stresses, σ_{per}, and the relevant S-N curves as in Fig.155 are being
discussed as the basis for Italian standards.[75] Fracture propagation from the
weld root or from internal defects is excluded. Test results for alloys of the type
AlMg were evaluated and extended to include AlSi and AlZn in plate
thicknesses between 4 and 12mm. In the same way as for steel $\sigma_{P\,per}$ is set equal
to $(4/3)\sigma_{A0\,per}$.

The proposal[75] has been established on the basis of a normalised and
statistical evaluation of extensive test series from different laboratories. The
following criteria are fundamental for the evaluation:[75,76]

—   the S-N curve is represented by two intersecting straight lines within
     double logarithmic scales;
—   within a single test series the endured number of cycles is represented by a
     Gaussian normal distribution for a preset stress amplitude;

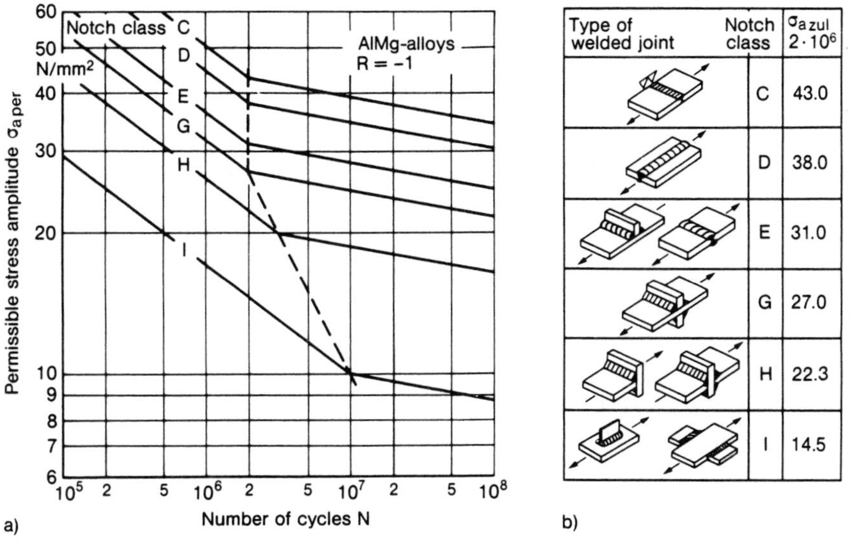
Notch class C
D
E
G
H
I
AlMg-alloys
R = -1
60
50
N/mm2
40
30
20
10
9
8
7
6
Permissible stress amplitude σaper
10^5 2 5 10^6 2 5 10^7 2 5 10^8
Number of cycles N
a)
Type of welded joint
Notch class
σa zul 2·10^6
C 43.0
D 38.0
E 31.0
G 27.0
H 22.3
I 14.5
b)

155 Permissible stresses, aluminium structures, notch classes for welded joints, after Atzori and Dattoma.[75]

- the 0.5 survival probability S-N curves from comparable test series at different laboratories are also represented by a Gaussian normal distribution;
- the results from test series with welded joints of varying joint geometry, type of loading and limit stress ratio, but made from the same base material may be represented in each case by a scatter band with a constant width and a uniform angle of inclination. The location of the bend in the S-N curve can be dependent on the sharpness of the notch;
- the scatter band for S-N curves from comparable test series at different laboratories is independent of the geometry of the welded joint, type of loading and limit stress ratio;
- the relative fatigue strength (for finite and infinite life) of welded joints is independent of the geometry of welded joints, type of loading and limit stress ratio (normalised S-N curve).

The values with 0.977 survival probability (two standard deviations from the 0.5 value) in the total scatter representation (test series and laboratories) are specified as permissible stresses. With reference to the values used for structural steels with 0.9 survival probability the following relation applies

$$\sigma_{a0.977} = \frac{3}{4}\sigma_{a0.9} \tag{43}$$

7.7   Assessment of structural stress

7.7.1   General ideas

Local stress which occurs immediately in front of the weld toe or weld end notch (or on the inside of the plate in front of the weld spot) is designated structural stress, σ_s. Another name is 'geometric stress' referring to the considered influence of the global geometric parameters. Its maximum value is designated 'hot spot stress' referring to the local heating effect of cyclic loading. This structural stress is largely identical to the surface stress, which can be calculated at the hot spot in accordance with structural theories used in engineering (without taking into account the notch effect).

The assessment of structural stress, which is beginning to be introduced for structures which are subjected to fatigue loading in shipbuilding, drilling platform, tank and vehicle construction,[200] to supplement the assessment of nominal stress, arose from the requirements and experience produced by practical conditions. Structural stress analyses, previously performed mainly by measurement and now mainly by finite element analysis, can be included in an assessment of structural stress in conformity with design codes as knowledge and experience is growing. The following stages are to be distinguished:

- removing structural stress peaks: even if the operational loading of the structure is inadequately known, the reduction of the structural stress peaks in the basic loading cases (*e.g.* axial loading, bending, torsion, internal pressure) is an extremely effective means of increasing the service fatigue strength of the structure. The pioneer of this procedure is W Kloth;[83]
- structural stress comparison, new design versus tried and tested design: in practice care is taken that the maximum structural stresses in the new design do not exceed those in the previous successful design;
- assessment of structural stress: the highest structural stresses determined must be shown to be smaller than the structural stresses, which lead to visible crack formation. The latter are evaluated from fatigue tests on welded structural components where the structural stress amplitude was locally recorded. They are specified as endurable structural stresses dependent on number of cycles. A safety margin from them is to be preserved. The pioneer of this assessment is E Haibach (see section 8.3.1).

7.7.2   Some difficulties

The first difficulty with assessment of structural stress is based on the fact that it is not the structural stresses but rather the notch stresses or even the crack stresses which determine fatigue strength. The hot spot stress can only be a useful reference value if the weld has a relatively uniform notch effect (in practice usually the toe of butt or fillet welds subjected to mainly transverse loading, excluding crack propagation from the weld root) or because the former is usually not uniform, even with the above-mentioned restrictions, if the increase in structural stress is predominant (*e.g.* an order of magnitude of 10 or 20 is possible for tubular joints). Examples are to be covered separately where the fatigue fracture is initiated at the weld root or in the case of fractures at the weld toe deviating from normal conditions in a positive manner (*e.g.* by grinding or dressing of the weld toe) or negatively (*e.g.* by undercuts). Therefore assessment of structural stress needs to be supplemented by an investigation of the notch stress if possible in the form of an assessment of seam weld or spot weld notch stress. However, assessment of the structural stress can also be interpreted as an attempt to transfer the fatigue strength values determined from simple butt, fillet or spot welded specimens subjected to axial loading via the local structural stress to welded joints of complex shape in the sense of a rough estimate. Thereby the notch effect does not appear explicitly.

The second difficulty in the assessment of structural stress lies in the necessity of separating the structural stresses from the notch stresses. When structural stress analysis first began, both by measurement and by calculation, this separation was unnecessary because it was not possible to determine the notch stresses in practical conditions. Bonded or mounted strain gauges could in fact be brought close to the weld toe but could scarcely include the notch effect as they had a measurement basis of ≥ 2mm. On the other hand the plate and shell models of the finite element method could not pick up any notch effects at all and even in solid element models of thick walled components the proportion of notch stress was small because the finite element mesh had to remain relatively coarse for economic reasons. With constant improvements in measuring and calculating techniques (*e.g.* miniaturised bonded strain gauges with a very small effective length or finer meshes and higher order elements), however, the question of separation required an answer. A scientifically 'clean' separation is not possible. The procedures below are used in practice in conformity with codes.

One procedure consists of a linear extrapolation of structural stresses on the surface of the component at adequate distances, perpendicularly in front of the weld on to the locus of the weld notch ('hot spot'), Fig.156. A distance is considered adequate at which the notch has hardly any remaining effect. For tubular joints, a first measuring point is selected immediately in front of the weld and a second point at a greater distance, dependent on wall thickness and tube diameter, as the base points for the extrapolation, Fig.157, Table 15. This special selection of measuring points made it possible to present the fatigue strength values of tubular joints (large versions for drilling platforms) with a relatively narrow uniform scatter band. The procedure is particularly suited to structural stress analysis by measurement on tubular joints. The procedure should fail in the presence of heavy transverse shear in the notch cross section because the notch stresses generated by transverse shear appear independently of the tensile and bending stresses in front of the notch.[201]

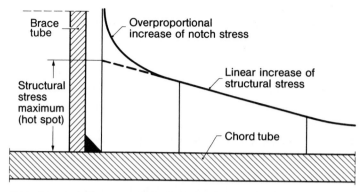

156 Structural stress maximum at the weld toe, elimination of the non-linear rise of notch stress, after Ref.159.

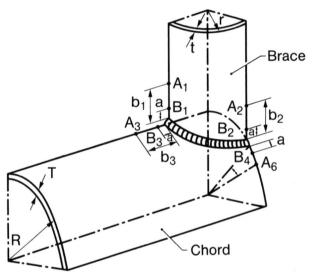

157 Measuring points used as the basis for extrapolation on to the structural stress maxima, after Ref.159.

Table 15. Positioning parameters for determining structural stress according to various authors[200]

Author	a	b_1, b_2	b_3
Dijkstra, de Back	$0.2\sqrt{rt}$	$0.65\sqrt{rt}$	$0.5\sqrt{RT}$
UK Guideline	$0.2\sqrt{rt}$	$0.65\sqrt{rt}$	$0.4\sqrt[4]{rtRT}$
Gurney, van Delft	$0.4\,t$	$0.65\sqrt{rt}$	$0.4\sqrt[4]{rtRT}$
API-Recommendation	$0.1\sqrt{rt}$	—	—

The other procedure consists of a linearisation (under constant resulting force and moment) of the stress distribution, which is non-linear to a high degree, particularly in the notch area, in sections perpendicular to the surface of the component, *e.g.* over the wall thickness, and using the surface stress obtained in this way as structural stress. If the stress analysis (as is often the case in practice) is performed according to one of the structural theories used in engineering with partly linearised stress distribution (*e.g.* Bernoulli's beam bending theory, Kirchhoff's plate bending theory) the surface stress determined in this way at the critical point becomes the maximum structural stress (within the framework of the model accuracy). Therefore the method of internal linearisation is particularly suited to the analysis of structural stress by calculation. It has been included in the ASME Code.[164] This method too should fail in the presence of heavy transverse shear in the notch cross section because the transverse shear notch stresses occur independently of the internal tensile and bending stress patterns under consideration.

The third difficulty with assessment of structural stress is because the elastic structural stresses no longer define the fatigue strength when approaching low cycle fatigue but rather the elastic-plastic structural strains. The latter are just as impossible to divide clearly into structural and notch effects as are the stresses. The procedures described for the stresses of surface extrapolation or internal linearisation can be used in principle but have not been tested. In the calculation methods the yield hypothesis, yield law and hardening law are to be taken into account. The steady state cyclic stress-strain curve is to be used as the basis for both measurement and calculation.

Two further points are to be taken into account in assessment of structural stress, which are not always considered adequately. The complete biaxial structural stress state on the surface of the component is to be determined and assessed by means of an equivalent stress hypothesis (important *e.g.* for the comparison of welded joints in tensile specimens with those in tank walls). The most unfavourable geometrical conditions caused by manufacture are to be considered in the calculation model (*e.g.* linear and angular misalignment) because they have an essential effect on the structural stress values.

Despite the difficulties, which have been explained, the results of fatigue tests on welded joints have occasionally allowed a relatively narrow scatter band to be reproduced in graphic representations evaluating the structural stress, *e.g.* for tubular joints, shipbuilding components, tank nozzles, particularly also the influence of a lack of roundness, of linear or angular misalignment, or of the different welding processes which cause them. The assessment of structural stress in connection with a detailed regulation of the determination of the structural stress concentration is well established in practice within limited sets of component shapes and loading cases. However, care is advised when using the method outside the groups which have been outlined.

7.7.3   Formalistic aspects

The following details refer to the formalistic aspects of structural stress assessment in technical literature.

The increase of structural stress, σ_s, (an equivalent stress if necessary) with reference to the nominal stress, σ_n, is expressed by the structural stress concentration factor, K_s:

$$\sigma_s = K_s \sigma_n \qquad (44)$$

The increase of structural stress is also stated as subdivided according to the factor, K_{s0} and the factor, K_{cw} as a result of an uneven distribution of bending stresses in flanges or cover plates (*e.g.* I section girder or box girder, subjected to transverse force bending), see equation (32). In a different source, uneven stress distribution in the joint is taken into account separately by means of the factor K_{ch}, see equation (33).

The structural stress concentration factor and the nominal stress are to be introduced separately for the (possibly superimposed) basic load cases: axial, bending and shear. Insofar as the increase of stress occurs at the same point and in the same direction for all three load cases, the total structural stress, σ_s, is obtained from:

$$\sigma_s = K_{sa} \sigma_{na} + K_{sb} \sigma_{nb} + K_{s\tau} \tau_n \qquad (45)$$

The following, less accurate, simplified formula is also used in practice:

$$\sigma_s = K_s(\sigma_{na} + \sigma_{nb} + \tau_n) \qquad (46)$$

7.7.4   Endurable and permissible stresses

The endurable structural stresses are reached when a clearly visible crack is initiated. The S-N curves given in the technical literature and in the codes generally limit the scatter band of the test results from below. When comparing the stress limit curves from the different codes, it is to be noted that the maximum structural stress value can be defined differently. In the case of a multi-axial structural stress state the von Mises distortion energy hypothesis or the Tresca shear stress hypothesis should be used. If there are loading spectra, the partial damages according to Miner on the basis of the above-mentioned S-N limiting curves have to be added.

The S-N limit curves for tubular joints and drilling platforms according to the American Welding Society[157] (AWS-XI and AWS-X2, X1 for improved

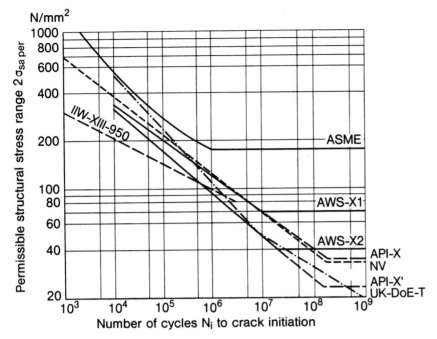

158 Permissible structural stresses for tube joints, according to various codes[156-159] compared with the ASME Code curve[164] for boilers and pressure vessels, after Wardenier[86] and Iida.[200]

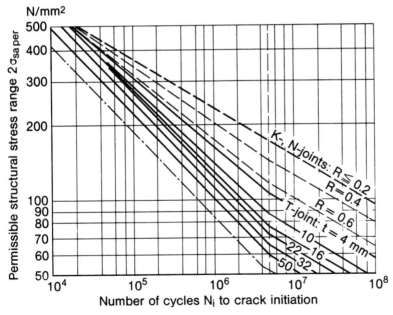

159 Permissible structural stresses for tubular joints with thickness and limit stress ratio in accordance with Eurocode 3[150], after Wardenier.[86]

weld toe), according to the American Petroleum Institute[156] (API-X and API-X′, X for improved weld toe), according to Det Norske Veritas[158] (NV) and according to the United Kingdom Department of Energy[159] (UK-DoE-T for unprotected node joints in seawater) are compared in Fig.158. In addition the limit curve in accordance with ASME for pressure vessels and pipelines is recorded, which permits relatively high stresses and is considered as unsafe. On the other hand the (tensile) limit curve proposed to the Japanese shipbuilding industry[90,91] (evaluation of the fatigue tests on the type of construction in accordance with section 5.2.1 to 5.2.3 with R = 0) is somewhat low in the fatigue strength for finite life range and in particular has a questionably small gradient (*i.e.* large gradient exponent k = 6.3 compared to k ≈ 3.5).

Further detailed information on axially loaded node joints of hollow cross sections is contained in the Eurocode,[150] Fig.159.

The continuous curves are applicable to any node configuration and mean load. They are stated for varying chord thicknesses, t, (reduction to factor 0.5 for large scale versions). The dotted curves apply to lightweight K and N type node joints subject to certain restrictions. The dot dash curve serves as a check on the nominal stress amplitude, $2\sigma_a$ in the chord. The endurable stresses, which are specified, are to be halved in the presence of corrosion.

7.7.5   Superimposition of notch stresses and crack stress intensities

Considering the notch stress concentration factor, K_k, at the point where the structural stress concentration, $K_s = \sigma_{smax}/\sigma_n$, (44) has been determined, Fig.160:

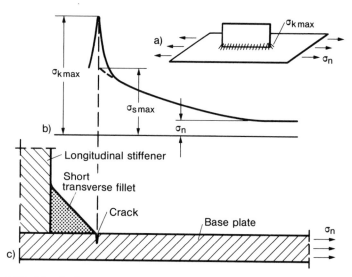

160 Nominal stress, σ_n, structural stress, σ_s, notch stress, σ_k, and crack stress intensity, K_I, superimposed; welded joint with longitudinal stiffener as an example.

$$K_k = \frac{\sigma_{kmax}}{\sigma_{smax}} \tag{47}$$

and then a small crack at the notch root (stress intensity factor, K_l, crack length, l, adjustment factor, $\varkappa$):

$$K_l = \sigma_{kmax}\sqrt{\pi l}\varkappa \tag{48}$$

the following superimposition formula holds true:

$$K_l = K_s K_k \sigma_n \sqrt{\pi l}\varkappa \tag{49}$$

With continuing crack propagation, at first the influence of K_k, and later the influence of K_s, is diminished.

These relations between local and global stresses conclude the fatigue strength considerations on the basis of global approaches in the first part of the book.

8

Notch stress approach for assessment of fatigue strength of seam welded joints

8.1   General fundamentals of the method

8.1.1   *Prerequisites and reference to practice*

Crack initiation in welded joints subjected to fatigue loading can be described quantitatively on the basis of the notch root stress or strain by expanding the aforementioned methods of nominal and structural stress assessment. The notch root stress or strain is also decisive, at least initially, for subsequent crack propagation. To predict the strength of the components, the notch root stress or strain is compared with the characteristic strength values of the material at the notch root which can be derived from tests with equivalent unnotched specimens. Thus the local notch stress approach can be used instead of the global nominal stress approach. Consideration of structural stress is an approach between the two.

As the notch root stress or strain is the dominant influencing parameter for the fatigue strength of welded structures and test specimens, the notch stress or notch strain approach is a key to evaluating strength in a way which is close to reality, to the transference of test results to structure and to measures for improving the strength of structures. A disadvantage of the approach is the much greater amount of measuring, testing and calculation required compared with the nominal stress approach. The knowledge gained from application experience, which is necessary for substantiating any new assessment method in practice, is also still in its infancy.

Prerequisites for predicting component strength successfully in accordance with the notch stress or notch strain approach are:

—   mastery of the theory of notch stress including the statements in it of general character as well as the approximation formulae;
—   availability of numerical and measuring methods for determining the notch root stress and strain in the individual case, *e.g.* the finite element and boundary element method,[210-213] the strain gauge measuring technique and photoelasticity;[6]

— knowledge of the geometrical, dimensional and shape parameters of the
   structure and the joint including the deviations from the design values
   caused by manufacture;
— knowledge of the local cyclic stress-strain curves, as well as of the
   characteristic material values (parent material and weld material) with
   regard to fatigue including the extent to which they are affected by
   microstructural transformation (*e.g.* increase of hardness), impurity (*e.g.*
   microcracks), and surface condition (*e.g.* roughness, corrosion);
— knowledge on damage accumulation in the material when subjected to
   varying operational loading cycles;
— availability of testing equipment for determining missing characteristic
   values for the materials, particularly a servo-hydraulic equipment for
   small scale specimens;
— knowledge of the residual stress state at the notch root, especially of the
   welding residual stresses.

These prerequisites can only be fulfilled completely in special cases.
However, there are simplifications of the notch stress or notch strain approach
possible which enable more general practical application of the method to take
place. In particular, this includes the comparison of the notch stresses in new
designs with those in older designs, which have an adequate service fatigue
strength or between differing new designs to find out that with the highest
fatigue strength. Notch stress analysis offers essential suggestions where cases
of damage are being investigated. Design optimisation with respect to service
fatigue strength is not possible without considering notch stresses. But even the
complete notch stress or notch strain approach will be performed more often
now as the level of knowledge increases in this matter.

8.1.2. *Assessment of notch stress*

The fatigue process is divided into the crack initiation phase (or crack front
orientation when there are pre-existing cracks) and the stable crack
propagation phase, which leads to unstable final fracture. The crack initiation
phase in engineering terms is considered to be finished at a crack depth of about
0.25mm, corresponding to the technical limitations of crack detection (mostly
via strain gauges at the point where the crack initiates). On the other hand, the
crack propagation phase, during which the structure is still serviceable, is
limited to a crack depth approximately equal to half the wall thickness,
presupposing adequate fracture toughness. The proportion of the crack
initiation and crack propagation phases in the technically utilisable total
lifetime can vary considerably. The crack initiation phase proportion increases
in the direction of high cycle fatigue strength.

The notch stress is decisive for crack initiation and for the initial stage of
crack propagation in the high cycle fatigue strength range although not to its full
extent, but rather modified corresponding to the elastic and elastic-plastic

support effect. The sharp notches of the weld profile have a strong strength-reducing effect, but not quite as strong as the increase of notch stress, characterised by the elastic stress concentration factor, K_k, would indicate. Instead of the stress concentration factor, K_k, the smaller fatigue notch factor, K_f, is introduced, which takes into account the material- and loading-dependent elastic support effect, using K_k as a basis (conventional definition with regard to fatigue strength for infinite life), but which principally can also represent the elastic-plastic support effect (with regard to fatigue strength for finite life) at least in the high cycle fatigue strength range. The stress concentration factor, K_k, can be determined numerically (finite element and boundary element methods), by measurement (strain gauge or photoelasticity) or on the basis of functional analysis solutions.[207,209] Formulae and methods from various authors[27] are available for the conversion of K_k to K_f. If the elastic-plastic support effect is additionally to be taken into account, this can take place in accordance with a simple formula derived by Neuber.[214] The fatigue effective notch stress, $\bar{\sigma}_k$, which is determined via K_f is compared with endurable or permissible notch stresses, σ_{ken} or σ_{kper} which are determined from tests on unnotched specimens of the material (S-N curve, fatigue strength diagram and strength hypothesis with regard to multi-axial loading). Instead of the mixed material at the weld toe and weld root, the base material or the weld material is used in the fatigue test in a hardened or softened state which corresponds to the state at the root of the notch (the fatigue strength can also be obtained from approximation formulae on the basis of σ_U). A possible difference in the surface roughness between the material specimen and the welded joint can be taken into account via the surface factor x^* ($x^* \leq 1$).

The stress concentration factor, K_k, designates the ratio of the maximum elastic notch stress, σ_{kmax}, to the nominal stress σ_n (stress amplitudes are considered in the following):

$$K_k = \frac{\sigma_{kmax}}{\sigma_n} \tag{50}$$

According to Ref.1 the fatigue notch factor, K_f, designates the ratio of the (amplitude) fatigue strength of the unnotched specimen, σ_A, to the (amplitude) fatigue strength of the notched specimen, σ_{Ak}.

$$K_f = \frac{\sigma_A}{\sigma_{Ak}} \tag{51}$$

The maximum fatigue effective notch stress (amplitude), $\bar{\sigma}_{k(max)}$, is derived from K_f:

$$\bar{\sigma}_k = K_f \sigma_n \tag{52}$$

The surface factor, x^*, designates the ratio of the (amplitude) fatigue strength of the unnotched specimens with rough (σ_A^*) and polished (σ_A) surface:

$$x^* = \frac{\sigma_A^*}{\sigma_A} \tag{53}$$

As a formula, the assessment of notch stress for high cycle fatigue strength or endurance limit is represented as follows ($\sigma_{k\,en}$ for N or S*N cycles, safety factor S or S*, all σ values stress amplitudes):

$$\bar{\sigma}_k \leqslant \sigma_{k\,per} \tag{54}$$

$$\sigma_{kper} = \frac{x^* \sigma_{ken}}{S} \quad (S = 1.2\text{-}1.5) \tag{55}$$

$$\sigma_{kper} = x^* \sigma_{ken} \quad \text{for } S^* \times N \, (S^* = 2.0\text{-}5.0) \tag{56}$$

The elastic and elastic-plastic support effect is taken into account in $\bar{\sigma}_k$ or K_k when using equation (52). Alternatively, a procedure is possible, in accordance with which only the elastic K_k defines those (fictitious) maximum notch stresses, which are compared with a fatigue strength diagram, modified corresponding to the respective elastic-plastic support effect (upper boundary formed by the 'shape-dependent strain limit', $K_{0.2}^*$, instead of by the yield limit, σ_Y or $\sigma_{0.2}$, see Ref.25, 214.

The fatigue strength values determined on unnotched specimens are to be introduced as endurable stress amplitudes, σ_{ken}. In steels, the alternating fatigue strength is $\sigma_{A0} \approx 0.4\text{-}0.6\sigma_U$, in aluminium alloys it is $\sigma_{A0} \approx 0.3\text{-}0.5\sigma_U$. The fatigue strength diagram for the unnotched specimen, amplitude strength, σ_A, dependent on mean stress, σ_m, can be drawn approximately on the basis of σ_{A0} and σ_U (e.g. in accordance with Fig.12a), using one of the following functional dependences (according to Goodman, Berber and Smith respectively, higher order functions according to Troost and El-Magd[215]):

$$\sigma_A = \sigma_{A0}\left(1 - \frac{\sigma_m}{\sigma_U}\right) \tag{57}$$

$$\sigma_A = \sigma_{A0}\left[1 - \left(\frac{\sigma_m}{\sigma_U}\right)^2\right] \tag{58}$$

$$\sigma_A = \sigma_{A0}\frac{1 - \dfrac{\sigma_m}{\sigma_U}}{1 + \dfrac{\sigma_m}{\sigma_U}} \tag{59}$$

Instead of the ultimate tensile strength, σ_U, based on the initial cross sectional area, the rupture strength, σ_R, based on the reduced cross sectional area at rupture, is sometimes inserted in the equations above which makes them more suitable for transference to the notch root (for structural steel $\sigma_R \approx 1.33\,\sigma_U$).

The S-N curves for the unnotched specimen, *e.g.* for alternating and pulsating loading can be drawn using σ_U and σ_{A0} or σ_p as the basis (*e.g.* in accordance with Fig.12b). More accurate 'synthetic' S-N curves can be found in Ref.217.

For more general practical purposes, only the notch stress assessment with regard to fatigue strength for infinite life solely taking account of the elastic support effect, is recommended. On the basis of the fatigue strength for infinite life of the welded joint, ensured by means of assessment of notch stress, either 'safe' S-N curves for nominal stress can be drawn (*e.g.* in accordance with Fig.14) or the well established normalised S-N curve (see Fig.40 and 74) is used to ensure fatigue strength for finite life.

The assessment of notch stress has proved itself successful[216,218,219] as regards fatigue strength in unwelded machine components (*e.g.* gear wheels, steering knuckles, crankshafts, connecting rods). It therefore suggests that this assessment could be used on welded joints too, even though additional complications are produced by the inhomogeneity of the material and the microdefects in the weld notch area, by the undefined profile of unmachined welds and by the residual welding stresses, which are inadequately known in most cases. However, comparative investigations with trend predictions should not produce complications*.

8.1.3   *Assessment of notch strain*

The notch strain is considered decisive for crack initiation and initial crack propagation in the low cycle fatigue range rather than the notch stress. This notch strain concept is actually a mixed concept because, in addition to plastic strain, elastic strain and therefore elastic stress are considered to be decisive for local strength. The notch strain assessment is extended to high cycle fatigue so that in the latter range we meet both with assessment of notch strain and of notch stress. Finally, the notch strain assessment is used in dealing with questions involving service fatigue strength. The assessment of notch strain is frequently not recognisable as such in the codes, because the total strains there are converted into fictitious elastic stresses.

The elastic-plastic stress and strain state at the notch root, if subjected to a constant load amplitude, is usually determined in accordance with Neuber's macrostructural support formula, equation (105), taking the cyclic stress-strain curve together with the elastic stress concentration or fatigue notch factor as a basis (the latter takes into account the elastic support effect). The residual stresses at the notch root should be included as far as they are known. The total strain at the notch root can also be measured with strain gauges. The separation of the total strain into the elastic and plastic proportion is achieved, for example, by subjecting an unnotched 'companion specimen' to the strain sequence measured at the notch root and determining the resulting

---

* This was the state of the art in 1984 when the German text book was prepared. Meanwhile substantial progess has been achieved, see section 8.2.5.

load sequence at the specimen from which the elastic stresses and strains follow directly. Endurable or permissible total strain amplitudes are stated, dependent on number of cycles (and independent of the mean stress), which consist of an elastic proportion (dominant for high cycle fatigue strength) and a plastic proportion (dominant for low cycle fatigue). These strains are determined by testing unnotched or slightly notched specimens, in which the highly stressed volume should correspond to the crack initiation volume in the notched component because of the size effect. They are represented by the strain S-N curve or by the corresponding equation (108) after Morrow and Manson. The procedure, which was only described verbally above (the 'simple' notch root approach as in section 8.3.2) was verified for welded joints by Mattos and Lawrence.[234,235]

As regards formulae, the assessment of notch strain for low and high cycle fatigue strength is represented as follows (by analogy with equations (54) to (56), x^* is omitted because of the receding influence of the surface and all ϵ values are strain amplitudes:

$$\epsilon_k \leqq \epsilon_{kper} \tag{60}$$

$$\epsilon_k = \epsilon_{kel} + \epsilon_{kpl} \tag{61}$$

$$\epsilon_{kper} = \frac{\epsilon_{ken}}{S} \, (S = 1.2\text{-}1.5) \tag{62}$$

$$\epsilon_{kper} = \epsilon_{ken} \text{ for } S^* \times N \, (S^* = 2.0\text{-}0.5) \tag{63}$$

With regard to the endurable strain amplitudes, see for example Fig.13, in which once again no distinction is made between parent material, weld material and mixed material as a first approximation.

If the load amplitudes are variable, simple damage accumulation calculations are possible by analogy with Miner's hypothesis using the strain spectrum and the strain S-N curve as a basis, provided that there is little change in the mean strain. The procedure becomes considerably more complicated if the mean strain is variable. In the latter case, the whole elastic-plastic notch stress and strain state, which is variable with time, must be determined by calculation or measurement and finally must be simulated on an unnotched 'companion specimen'; the purpose of the latter is to determine the number of cycles to fracture for the specimen, which is set equal to the number of cycles to crack initiation for the notched welded joint. A damage hypothesis can also be used instead of the fracture test. The procedure, which was only described verbally above, (the 'extended' notch root approach as in section 8.3.3) was verified for welded joints by Heuler and Seeger.[260]

Apart from in design comparisons, assessment of notch strain in the design phase is appropriate in cases where conventional simple fatigue or service fatigue tests cannot always be carried out because of limitations on acceptable expenditure (*e.g.* large scale components in nuclear reactor engineering) or because of limitations on the time available for development (*e.g.* automotive engineering). In the American automotive industry in particular, fatigue relevant materials data were extensively compiled and pertinent computer programs developed.[236] The extended notch root approach is also particularly suitable for investigating the effect of differing load spectra and load sequences on the same design. The assessment of notch strain, subsequent to design phase, the strain gauge measurement at the notch root of the manufactured structure, is suitable for acceptance tests and in service monitoring.

8.1.4   *Relevant knowledge from 'notch stresses at cutouts and inclusions'*

The following knowledge relevant to the assessment of notch stress or strain in welded joints is available in the text book 'Notch stresses at cutouts and inclusions':[209]

— theory and state of the art survey of notch stresses at cutouts and inclusions;
— stress concentration factor diagrams for a wide range of variations in shape, loading and dimensional parameters;
— methods for transferring the solutions to more complex situations, relevant to practical cases;
— examples of calculations for welded structures.

The notch problems treated in Ref.209 are to be designated structural stress problems in the sense of the present book's definitions because, although the solutions can be applied to cutouts, nozzles, reinforcements and other parts of welded structures, they disregard the actual weld notches. A solution involving functional analysis is suggested for these structural stress problems, which can replace or supplement the more expensive finite element or boundary element methods. Although the analysis of structural stress is a prerequisite for the assessment of notch stress or strain at the welded joint (see section 8.2.1), it is not possible to consider functional analysis, finite element and boundary element methods in detail within the framework of the present book. These methods of assessing structural stress and notch stress are not specific to welded structures and Ref.209 can be assumed to be available to the reader. Only the special method of multiplicative superimposition of structural stress and notch stress concentration factors is shown in section 8.2.11 in an improved form.

8.2.   Elastic notch effect of welded joints with regard to fatigue strength for infinite life

8.2.1   *Global and local structure, macro- and micronotch effect*

Determination of the elastic notch effect of welded joints serves directly in the assessment of the notch stress for the fatigue strength of structures for infinite life (endurance limit) and indirectly as the basis of the assessment of notch strain for the fatigue strength of structures for finite life. The statements here refer to seam welded joints, see Chapter 10 with regard to spot welded joints.

A distinction is made between determining notch stresses by measurement and by calculation. Considering measurement, the process involves using brittle lacquer coatings, calibrated with respect to cracking sensitivity, or photoelastic surface layers to determine the areas of the structure, which are subjected to the highest stresses. The increase of (biaxial) stress at the notch positions in these areas is subsequently measured, previously using mounted mechanical strain gauges, nowadays with bonded wire strain gauges. In welded structures the notch positions in front of, or on which the measurement takes place are the weld toes and the weld ends. Measurements on photoelastic models can support the procedure. The value of notch stress measurements of this type is limited. Only the surface and the external notches (in so far as they are accessible) can be subjected to measurement. The measured stresses are only averaged values of the measuring length. The measurement is often too late for the basic design decisions as a prototype has to be built before any measurements can be made. Neither can measurements, which are not based on theory, provide any suggestions on design improvements, which could be made. Thus, for the above reasons, the stress measurement is only used as a final check nowadays in the design development.

In contrast, numerical stress analysis is a development procedure of major practical importance, which does not have the above-mentioned limitations. A distinction is made between the global and the local structure in the numerical stress analysis of welded structures. Global structure refers to the load bearing structure composed from plates and shells (including plate and shell type bars), which is interconnected in the joint lines. It is usually analysed nowadays by the finite element method, using the technical theories on strength of structures as a basis. Local structure refers to the geometrical details of the welded joint lines,

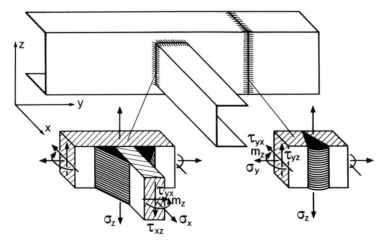

161 Global and local structure of welded joints.

the weld seam and the immediately adjacent joint components. It forms a solid structure, but which can be analysed within a two dimensional approximation, particularly advantageously by the boundary element method, or evaluated on the basis of general results of the notch stress theory. The mechanical behaviour of the global structure is reduced to internal forces transmitted in the welded joint lines, which act as external forces on the local structure to determine the notch effect of the total structure, Fig.161. Micronotches (if present) in the mixed material of the toe and root areas are not dealt with directly as a notch effect but rather indirectly via a possible reduction in the fatigue strength of the parent material at the notch root. On the other hand, micronotches in the form of microcracks can be dealt with on the basis of fracture mechanics.

8.2.2   Notch shape parameters of the local structure

Whereas the shape parameters of the global structure are set by the design drawings and the tolerance limits,[220] in the case of the local structure, only the thickness of the weld is prescribed by the designer. The representation of a weld by its graphical 'nominal shape', *e.g.* as in Fig.161 (see the weld seam profiles), is inadequate for investigations into the notch effect. The actual shape deviates considerably from the nominal shape. In addition, the sharp corners at the weld toe and the sharp slits ends at the weld root of the nominal shape make the investigation difficult because, assuming an elastic model, infinitely high notch stresses occur at these points. Therefore, if the actual notch stresses at the weld are to be determined, the actual shape of the weld is to be used as the basis of analysis.

The description of the actual shape of the weld is problematical in that the conventional procedure in machine construction and steel engineering cannot be used in a straightforward manner, namely, to consider the actual shape as a superimposing of the nominal shape and the shape deviations, whilst the surface inaccuracies, which are smaller by several orders of magnitude, are dealt with separately as 'roughness'. In welded joints, shape deviations and surface roughness are of the same order of magnitude, especially at the toe and root of the weld. The distinction between actual shape and surface roughness can only be made with any degree of success by introducing a non-geometrical criterion for demarcation, which is then used more as an estimate than for an exact procedure. The substitute microstructural length introduced in section 8.2.3 serves as the demarcation criterion, relevant to fatigue strength. Geometrical deviations less than about one tenth of the smallest possible substitute microstructural length ρ^* are considered as surface roughness, *i.e.* $0.1\rho^*_{min} \approx 0.01mm$, anything above this value is considered to be notch-effective.

The weld contour is composed section by section from straight lines and arcs using the described demarcation criterion, for establishing an averaged contour line, Fig.162. The most important geometrical parameters for the notch effect of the weld are the radius of curvature of the notch, ρ, and the notch slope angle, θ, at the toe and root of the weld, Fig.163.

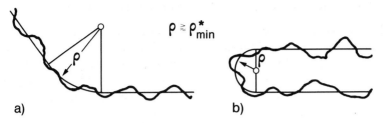

162 Separation of surface roughness and contour shape at weld toe (a) and weld root (b).

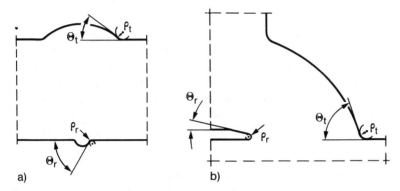

163 Idealised contour shape of butt weld (a) and fillet weld (b).

A measurement[221] of butt and fillet welds made of structural steel St 37 with varying plate thickness (5, 10, 20mm), welding process (manual arc, gas metal arc and submerged-arc welding), welding position (gravity position, vertically upward or downward, overhead) and electrode coating (titanium dioxide and low hydrogen) has the following result. The averaged radii lay between 0.01 and 3mm. Particularly small radii (0.01-0.1mm) occurred at the root of fillet welds and welded-through butt welds, and at the toe of welds produced by means of fully automatic submerged-arc welding. Particularly large radii (1-3mm) occurred at fillet weld toes produced in the gravity position. A large number of butt and fillet weld toes had radii of about 0.1mm. The plate thickness did not have a great influence on the radius. There was no difference between the cover pass and the root capping pass in the butt weld. The difference between hot and cold running electrodes was not significant. The averaged angle lay between 15-80°. Butt welds had 15-50°, fillet welds 25-80°, welded through V weld roots 65-70°, fillet welds, welded vertically downwards or overhead, 45-70°.

There are no research results available on the obvious question concerning the reasons for the different radii and angles. If the notch effect of an actual welded joint is to be investigated satisfactorily from a quantitative point of view in practical situations, the weld profile concerned is always to be determined first of all from sample welds carried out under the actual manufacturing conditions.[222]

8.2.3   Microstructural support effect and fatigue notch factor

The microstructural support effect with reference to the fatigue strength for infinite life can be obtained by averaging the notch stresses in planes perpendicular to the notch root over a small, material dependent substitute microstructural length, ρ^*, and comparing them with macrostrength values which remain the same. The basic idea of Neuber's formulation[214] of the microstructural support effect, preferred below, which comprises the geometrical size effect (further formulations by Siebel in the VDI-Guideline,[25] by Thum, Rühl, Bollenrath/Troost and Peterson, see survey on procedures in Ref.27) not to assess the local increase of notch strength directly by subsequent averaging of the calculated elastic notch stresses (mean stress hypothesis) nor to assess it indirectly by additional calculation of the stress gradient,[25] both in planes perpendicular to the notch root, but rather to calculate a (fictitious) maximum notch stress value $\tilde{\sigma}_{kmax}$ straight away, which reflects the actual strength reduction of the notch including the support effect. The support effect is covered by a fictitious increase of the notch root radius. In this procedure, according to Ref.214, the actual radius of notch curvature, ρ, is to be increased by the substitute microstructural length, ρ^*, multiplied by the support factor, s, to obtain the fictitious radius of notch curvature, ρ_f, in the model:

$$\rho_f = \rho + s\rho^*. \tag{64}$$

The support factor, s, is given in Table 16 for various strength hypotheses and loading types. The substitute microstructural length, ρ^*, determined from alternating fatigue tests on notched bars, is shown in Fig.164 for various groups of metallic materials dependent on their static strength. Conservative

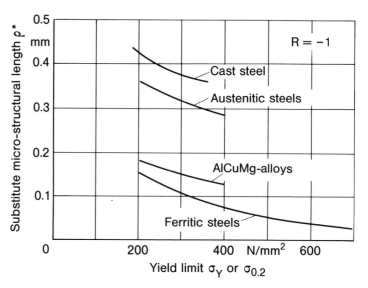

164 Substitute microstructural length, ρ^*, for metallic materials dependent on static strength.

Table 16. Support factor, s, of the microstructural support effect for notched bars (with Poisson's ratio, v), after Neuber[214]

	Tensile and bending load		Shear and torsional load
	Flat bar	Round bar	
Normal stress hypothesis	2	2	1
Shear stress hypothesis	2	$\dfrac{2-v}{1-v}$	1
Octahedral shear hypothesis Distortion energy hypothesis	$\dfrac{5}{2}$	$\dfrac{5-2v+2v^2}{2-2v+2v^2}$	1
Strain hypothesis	$2+v$	$\dfrac{2-v}{1-v}$	1
Strain energy hypothesis	$2+v$	$\dfrac{2+v}{1-v}$	1

estimates, $\rho^* = 0.4\text{mm}$, $s = 2.5$ and $\rho = 0\text{mm}$ (*i.e.* $\rho_f = 1.0\text{mm}$) have proved to be realistic for welded joints in structural steels with crack initiation in the cast material structure of the weld toe or root.

In accordance with the latter, the welded joints to be investigated with regard to the notch effect preferably in the form of transverse contour models subjected to in-plane loading, are to be rounded with radius $\rho_f = 1\text{mm}$ at the weld toe and at the weld root, Fig.165. The procedure is generally applicable. Difficulties only arise on the inner slits with smaller plate thicknesses ($t \leq 5\text{mm}$) because the load bearing cross section is severely weakened as a consequence of the relatively large radius. A correcting procedure for such cases can be

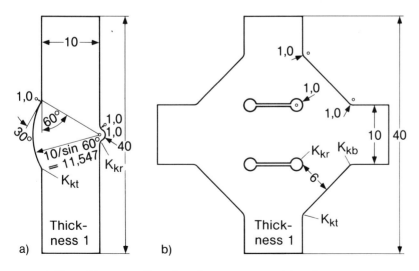

165 Fictitious rounding of notches on account of the microstructural support effect for butt joint (a) and cruciform joint (b).

found in section 10.5. In addition the notch effect of slits in the direction of the main loading, (e.g. transverse stiffeners or strapped joints with transverse fillet weld) is reflected as too severe. Thus the straight slit in the direction of tension has no notch effect at all (theoretically), whilst the circular hole, subjected to the same type of loading has a stress concentration factor around 3.0. In this case the circular hole not only simulates the support effect, but also simultaneously transverse cracks at the end of the slit with crack length $2\rho_f = 2mm$. The current slits on welded joints do in fact as gaps have a transverse extension, but only of 0.1mm with runaway values up to 0.5mm.[221] Moreover, it is to be noted that the local loading direction at the end of the slit, in the examples of welded joints above, deviate considerably from the direction of the slit. Finally, it is to be taken into account when defining the model that, in certain load cases, pressure is transferred across the slit surfaces, which can be simulated by appropriate boundary conditions at the slits. Taking the microstructural support effect into account in plane models for calculation or measurement of welded joints by rounding the toe and root has been pursued consistently only by the author.[223-226] In investigations into notch stress concentration with a different origin, the curvature radii of the notches in the (e.g. photoelastic) test specimens depend on the manufacturing process or are contained indirectly in tests or calculations with sharp toe or root notches in the averaging evaluations. Whereas the author determines the fatigue notch factor, K_f, immediately with some justification, in the latter case it is a stress concentration factor, K_k, which, at most, may serve for predicting trends of fatigue strength. The central idea of the author's approach, the fictitious rounding of weld toe and weld root, here put into effect with $\rho_f = 1mm$, was first applied in 1969 within a photoelastic notch stress analysis,[244] later on, in 1975 within a finite element notch stress analysis,[225] and at last, since 1983 exclusively within the boundary element notch stress analysis.

The fatigue notch factor, K_f, the ratio of the endured nominal stresses in the smooth and the notched specimen in accordance with equation (51) here results from relating the maximum notch stress, $\bar{\sigma}_{kmax}$, at the fictitiously rounded notch to the tensile nominal stress, σ_n, determined for the cross sectional area, A, subjected to the tensile force, F:

$$K_f = \frac{\bar{\sigma}_{kmax}}{\sigma_n} \tag{65}$$

$$\sigma_n = \frac{F}{A} \tag{66}$$

The relation to the reduction factor, γ, from equation (1) or (2) is expressed for structural steel by

$$\gamma = \frac{1}{0.89K_f} \tag{67}$$

The factor 0.89 takes into account the difference in strength between the reference specimens 'as-rolled bar' (for γ) and 'polished bar' (for K_f) (pulsating fatigue strengths for St 37 in accordance with Ref.3).

The fatigue notch factors, K_f, with the relevant reduction factors, γ, for different welded joints and loading types represent the strength relationships for loading the different joints with the same force amplitude (or nominal stress amplitude). The strength relationships for loading with the same displacement amplitude are expressed by corresponding factors, $\overline{K}_f$ and $\overline{\gamma}$, which are determined for the same displacement, Δl, of the load bearing end cross section converted to a comparison nominal stress, $\overline{\sigma}_n$, instead of the same nominal stress, σ_n, (bar length, l, elastic modulus, E), Fig.166:

$$\overline{K}_f = \frac{\overline{\sigma}_{kmax}}{\overline{\sigma}_n} \tag{68}$$

$$\overline{\sigma}_n = E\frac{\Delta l}{l} \tag{69}$$

$$\overline{\gamma} = \frac{1}{0.89\overline{K}_f} \tag{70}$$

The calculation results based on values ρ^* from Fig.164 apply to alternating loading; for pulsating loading $\gamma_P \approx \gamma_A + 0.1$ (see equation (13)).

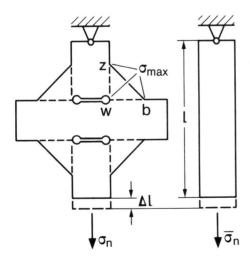

166 Parameters for determining the fatigue notch factors, K_f and $\overline{K}_f$.

8.2.4   *Notch effect of the welded joint subjected to transverse loading*

The notch effect will be described below for uniaxial transverse loading of the joint (equivalent to arbitrary in-plane loading of the contour model). The notch effect for biaxial oblique loading will be covered later on the basis of the above. Systematic investigations into basic types of transversely loaded welded joints were carried out by the author[226,227] using the boundary element method and by Rainer[230] using the finite element method. The accuracy of the calculated stress concentration factors is high in both investigations. Whereas in Ref.226 the fatigue notch factors are determined directly using Neuber's fictitious notch radius, in Ref.230 the stress concentration factors were converted into fatigue notch factors in accordance with formulae derived by Bollenrath and Troost. Both investigations prove that the calculated fatigue notch factors correspond acceptably with the results from fatigue tests. According to Mattos and Lawrence[234,235] stress concentration factors for individual welded joints, calculated by the finite element method, are converted into fatigue notch factors using Peterson's formula.

The joint models investigated in Ref.226 are shown in Fig.167. A distinction is to be made between butt welds with and without a root capping pass, double or single sided fillet welds, single bevel and double bevel butt welds with a root gap. The selected loading cases are cross tension, transverse bending with support on both sides (only possible in cruciform and T joints), and transverse bending with support on one side (possible in cruciform, T and corner joints, typical of the latter). The bending stresses usually determine the strength for transverse bending, as the following rough calculation illustrates. The tensile and bending nominal stresses for the superficially similar notch cases of the cruciform joint subjected to cross tension, the cruciform or T joint subjected to transverse bending and the corner joint subjected to transverse bending (with a double sided fillet weld in each case) show the ratio 1.0:1.8:3.7 despite the very small bending length assumed. Correspondingly different fatigue notch factors (referred to σ_n) are to be expected.

The tensile loading was applied as parallel displacement of the end of the tension member, as a consequence of which the secondary bending moments indicated in Fig.167 occur in asymmetrical joint shapes. They cause the particularly high fatigue notch factors in the corner joint. The support of the bending member was assumed to be bending free (*i.e.* a hinge) to generate bending stresses in the joint, which are independent of the joint type.

The fatigue notch factors, K_f, calculated using the boundary element method are included in Fig.167. They are subscripted according to notch position, tension member weld toe (t), bending member weld toe (b) and root of the weld (r). In addition, the fatigue notch factors $K_{fb}{}^*$ and $K_{fr}{}^*$ are also stated, which result from referring the maximum notch stresses (on the transverse member side) on to the nominal bending stress in the transverse member at the point of the toe or root notch. A change in the bending length only changes $K_{fb}{}^*$ by a very small amount, so that $K_{fb}{}^*$ is more generally applicable than K_{fb}. The reduction

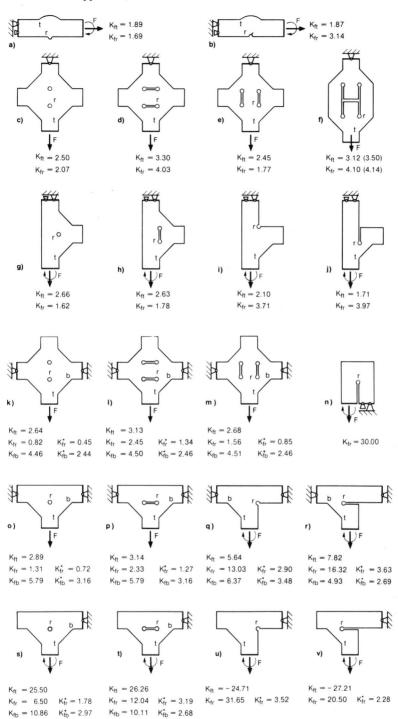

167 Contour models of welded joints, variations of shape, loading and
support conditions, relevant fatigue notch factors, after Radaj and
Möhrmann.[226]

factors γ and $\bar{\gamma}$, calculated from the largest K_f or $\overline{K}_f$ value of the welded joint concerned, are summarised in Tables 17 and 18. Thus, the maximum notch stresses were evaluated both for the same force and the same displacement at the end cross section of the different joints.

In Table 19, the calculated reduction factors, γ_p, (for $R = 0$) are compared with the relevant values known from fatigue tests as in Table 1 (for $R = 0$). Values from tests are only available for a few of the joint types which have been analysed by calculation. The comparison results in a good degree of correspondence, especially as the calculation refers to the least favourable sharp notch. The values from Table 1 for $P_f = 0.1$ are attributable to relatively sharp notches, too. Thus, the calculation method is suitable for comparative investigations and for optimising local shape as well as for dimensioning.

A further comparison was performed by Petershagen[229] after presentation of the author's IIW document[228] on the notch stress assessment of welded joints just described. The question posed was whether the proposed procedure would give results in conformity with the IIW design recommendation.[143] The comparison included transverse butt welds with root cap pass, two sided transverse stiffeners, cruciform joints with double bevel butt weld and cruciform joints with double fillet weld, the latter with root failure (joint types a, e, c, d in Fig.167). The appertaining K_f values for $R = -1$ and $P_f = 0.1$ were converted to stress range values $\Delta\sigma$ for $P_f = 0.023$ with an additional reduction for residual stresses. The correspondence with the IIW recommendation is satisfactory.

Principally, any design curve shows lower values than those from the described notch stress assessment because the former includes structural components with wall thicknesses and residual stresses higher than occurring in the specimens and because it represents a failure probability lower than $P_f = 0.1$. There is no necessity to differentiate between crack initiation and final fracture in the fatigue strength for infinite life range because there should be no fracture at all.

The author proposes the following procedure of comparison which is exemplified for the butt joint with root cap pass in as-welded condition considering low strength structural steel St 37. On the basis of $\sigma_{A0} = 145$ N/mm^2 and $\gamma_A = 0.6$, the result is $\Delta\sigma = 174$ N/mm^2 for $P_f = 0.1$, $\Delta\sigma = 130$ N/mm^2 for $P_f = 0.023$ (see (43) which includes the scattering of results from different laboratories) and $\Delta\sigma = 104$ N/mm^2 with factor 0.8 for residual stress and thickness higher than 10mm (the latter effect can principally be included in the notch stress analysis also). On the basis of $\sigma_p = 240$ N/mm^2 and $\gamma_p = \gamma_A + 0.08 = 0.68$ for pulsating load, on the other hand, the result is $\Delta\sigma = 163$ N/mm^2 for $P_f = 0.1$, $\Delta\sigma = 122$ N/mm^2 for $P_f = 0.023$ and $\Delta\sigma = 98$ N/mm^2 with factor 0.8. The $\Delta\sigma$ values 98 and 104 N/mm^2 correspond well to $\Delta\sigma = 100$ N/mm^2 for the considered joint in the IIW recommendation[143] or $\Delta\sigma = 90$ N/mm^2 in the Eurocode[150] or the ECCS recommendation.[151]

The difference between the fatigue notch factors, K_{ft} and K_{fr}, is frequently large, so that measures to reduce the larger K_f value are appropriate in the sense

Table 17. Reduction factors, γ_A, derived from
fatigue notch factors, K_f, for the welded joint
contour models in Fig.167 subjected to same force

a	0.60			b	0.36		
c	0.45	d	0.28	e	0.46	f	0.27
g	0.43	h	0.43	i	0.30	j	0.28
k	0.25	l	0.25	m	0.25	n	0.03
o	0.19	p	0.19	q	0.09	r	0.07
s	0.04	t	0.04	u	0.03	v	0.04

Table 18. Reduction factors, $\bar{\gamma}_A$, derived from
fatigue notch factor, $\bar{K}_f$, for the welded joint
contour models in Fig.167 subjected to same
displacement

a	0.58			b	0.36		
c	0.36	d	0.33	e	0.37	f	0.29
g	0.38	h	0.39	i	0.29	j	0.29
k	0.33	l	0.35	m	0.33	n	0.31
o	0.34	p	0.34	q	0.45	r	0.36
s	0.69	t	0.72	u	1.00	v	1.79

Table 19. Reduction factors, $\gamma_p = \gamma_A + 0.08$, calculation results from Fig.167 compared with
test results from Table 1

	Notch stress analysis	Fatigue test
Butt joint with capping pass	0.68	0.6-0.8
Cruciform joint with double bevel butt weld	0.53	0.4-0.6
Cruciform joint with fillet weld	0.35	0.3-0.5
Transverse stiffener with fillet weld	0.54	0.4-0.8

of a shape optimisation (*e.g.* by reduction of the slit length, by altering the
thickness of the weld, by a large radius at the weld toe).

As expected, the reduction factors, γ, are relatively large for the T joints and
extremely large for the corner joints (and dependent especially on the bending
length). However, surprisingly, the reduction factors, $\bar{\gamma}$, behave largely in the
opposite direction to the above statement. T and corner joints subjected to
loading by the same displacement can be assessed as being relatively
favourable (in part even an improvement compared with the smooth tension
member is possible: $\bar{\gamma} > 1.0$).

The following specific statements are applicable to joints subjected to
loading by the same force, Fig.167:

— butt joints subjected to cross tension, models a and b: open weld root
without capping pass particularly disadvantageous;

— cruciform and edge joints subjected to cross tension and double sided transverse stiffeners, models c to f: root gap acceptable for double bevel butt weld, fillet weld root in cruciform joint particularly at risk of failure, longitudinal slit in transverse stiffener joint, less at risk of failure, edge joints and cruciform joints approximately at the same risk;

— T joints subjected to cross tension and single sided transverse stiffeners, models g to j: double sided welds favourable as regards the weld root, single sided welds disadvantageous as regards the weld root;

— cruciform joints subjected to transverse bending, models k to m: toe notch on tension member equally effective, toe notch on bending member reducing strength considerably as a consequence of high bending stresses (but $K_{fb}^* \approx K_{ft}$), about the same strength reduction with and without slits as a consequence of lower stress level at the root notch;

— T joints subjected to transverse bending, models o to r: notch effect on bending member increased as a consequence of lack of stiffening on one side, single sided welds extremely disadvantageous as regards the weld root (but not in the displacement related presentation), transverse bending compared with through tension increases stress on toe and decreases stress on root, single bevel butt weld not much more advantageous than single sided fillet weld;

— corner joints, subjected to transverse bending, models s to v; notch effects once more increased considerably (but not in displacement related presentation), single sided welds predominantly subjected to bending stress (with compression at the weld toe);

— edge joints subjected to transverse shear, model n: exceedingly high notch and bending effect at the weld root.

In the case of highly stressed welded joints after having selected the type of joint and weld, it is necessary to optimise shape and dimensions with regard to fatigue strength. The dimension 'thickness of weld' can be varied within certain limits for the fillet welded joint (not the butt welded joint), including the depth of penetration or fusion. For example, in a cruciform joint subjected to cross tension as in model d, on the one hand the slit length is varied with the weld thickness remaining the same, on the other hand the weld thickness is varied with the slit length remaining the same. The calculated fatigue notch factors, K_{ft} and K_{fr}, are shown in Fig.168 dependent on the ratio slit length, s, to plate thickness, b, or weld thickness, a, to plate thickness, b, (with the other dimensions remaining the same). The point of intersection of the K_{ft} and K_{fr} curves, in which the fracture initiation should shift from toe to root or vice versa, is of particular importance with regard to realisation of a minimum weld thickness in design and with regard to verification in fatigue tests. However, no intersection occurs in the diagram on the right and in the left hand diagram the curves intersect with a small angle. This leads to the conclusion that the toe and

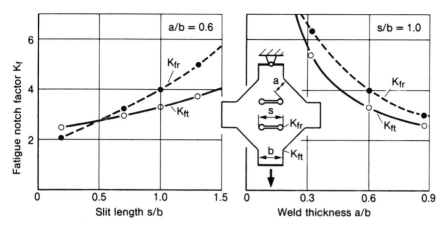

168 Fatigue notch factors, K_{ft} and K_{fr}, dependent on slit length and weld thickness, after Radaj and Möhrmann.[226]

root notch effects are influenced by approximately the same amount if the slit length or weld thickness is changed.

Figure 169 shows the welded joints investigated in Ref.230, $\rho = 1.0$mm is assumed at the weld toe, $\rho = 0.5$mm at the weld root. The geometrical parameters plate thickness, weld thickness and slit length are varied systematically. Tension, bending and torsion are applied as loading. The following general conclusions have been drawn from the results.

Small weld reinforcements and toe angles are advantageous for the butt joint. In the case of the cruciform joint, the reduction of the slit length is particularly advantageous for the notch effect at the toe and root. A small weld thickness is advantageous for the transverse stiffener, whereas the slit length has little influence. The notch effect is smaller in circumferential welds than in straight welds with the same contours. In contrast the conclusions drawn in Ref.230 with regard to the weld root of butt welds (high degree of weld reinforcement said to be favourable) and to crack initiation shifting from the root to the toe in the double fillet weld (said to occur for weld thicknesses larger than 1.2 times the plate thickness) are not generally acceptable.

The older notch stress investigations, mostly using photoelasticity, into welded joints subjected to transverse tensile or bending loading, which are not dealt with in detail here, are mainly concerned with the effect of geometrical parameters on notch stress, *i.e.* of the following parameters amongst others: toe angle of the fillet weld (flat toe angle advantageous), ratio of weld thickness to plate thickness (greater weld thickness advantageous for the cruciform joint, disadvantageous for the transverse stiffener), ratio of penetration depth to weld thickness (greater penetration depth advantageous) global dimensions of the joint (*e.g.* narrow transverse stiffener advantageous), plate thickness in the lapped joint (large thickness disadvantageous). The relevant literature is summarised in Ref.223-225.

No.	Welded joint	Joint type	Weld type	Loading
1			V-weld	M, F
2		Butt joint	X-weld	M, F
3			V-weld	M_t, M, F
4		Cruciform joint	K-weld	M, F
5			Fillet weld	M, F
6		Transverse stiffener	Fillet weld	M, F
7		T-joint	K-weld	M, F
8			Fillet weld	M, F
9		Cruciform joint	Fillet weld	M_t, M, F
10		Transverse stiffener	Fillet weld	M_t, M, F
11			Fillet weld	M_t, M, F
12		Cover plate joint	Fillet weld	M, F
13			Fillet weld	M, F
14		Strap joint	Fillet weld	M, F
15			Fillet weld	M, F

169 Welded joints, considered in the notch stress analysis by Rainer.[230]

8.2.5   Notch stress assessment suitable for codes

The above notch stress approach for welded joints is based mainly on theoretical considerations and evaluation of the relevant literature. A more comprehensive substantiation of the approach from specifically designed fatigue tests combined with numerical notch stress analysis including a procedure of notch stress assessment suitable for codes has recently been achieved by Olivier, Köttgen and Seeger.[231]

The welded joints made from low and higher tensile structural steels (*e.g.* St 37 to St 52) are evaluated uniformly. They are analysed as fictitiously rounded with ρ_f = 1mm at the weld toe and weld root. The determined global fatigue strength values (endurable nominal stresses) which are different for each joint type and load case are converted to local fatigue strength values (endurable notch stresses) which are the same for all joint types and load cases via the notch stress analysis using ρ_f = 1mm. So the local fatigue strength, $\bar{\sigma}_F$, is evaluated from fatigue tests on welded joints and not from fatigue tests on the base material but the result comes close to the endurance limit, σ_F achieved from the latter tests. The conclusion is that ρ_f = 1mm is a realistic choice. The central demand of the assessment procedure is the following (with safety factor, S):

$$\bar{\sigma}_k \leqq S\bar{\sigma}_F \,(\rho_f = 1\text{mm}) \tag{71}$$

The following fatigue strength values (endurance limits), $\bar{\sigma}_F$, have been determined for welded joints made of structural steel St 52-3:[231]

$$\bar{\sigma}_{A0} \approx 220\,\text{N/mm}^2 \quad (R = -1, P_f = 0.5), \tag{72}$$

$$\bar{\sigma}_p \approx 320\,\text{N/mm}^2 \quad (R = 0, P_f = 0.5). \tag{73}$$

The uniform notch radius, ρ_f, and the uniform local fatigue strength, $\bar{\sigma}_F$, for different structural steels corresponds to the rather uniform strength behaviour of the welded joints made of these steels as can be seen from the S-N curves in the catalogues[37] or from the normalised S-N curve for welded joints. The minor differences in the notch sensitivity of the considered steels in the welded condition are neglected.

The investigation[231] considers T and Y joints derived from crane construction and subjected to diagonal tension (equivalent to counter-bending) and through-tension, Fig.170. These joints cannot be assigned to definite notch classes of the code so that their use within conventionally designed crane structures is severely restricted. The specimens were made from structural steel St 52-3, plate thickness 8, 15 and 40mm. The fatigue tests were conducted with alternating load, tensile and compressive pulsating load, with and without welding residual stresses. Both plate and weld fractures were initiated at the weld toe and at the weld root. The fatigue strength values determined from tests on three stress levels were extrapolated via the normalised S-N curve for welded joints to fatigue strength for $N = 2 \times 10^6$ cycles. The 'true' endurance limit in the range $10^6 \leqq N \leqq 10^7$ was subsequently ascertained by further individual tests within this range. The local fatigue strength values stated above have finally been determined from about 350 test specimens totally on the basis of the fatigue notch factors calculated for ρ_f = 1mm.

Besides the stated $\bar{\sigma}_{A0}$ and $\bar{\sigma}_p$ values, the influence of mean stress, welding residual stresses and plate thickness on the local fatigue strength has been evaluated. The mean stress influence is expressed by the fact that the endurable stress amplitude for pulsating load is about 20-30% lower than that for

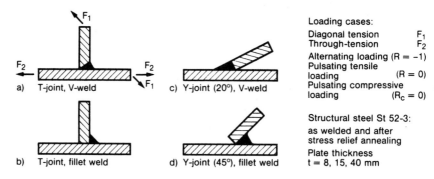

170 Welded joints of crane constructions investigated by Olivier, Köttgen and Seeger[231] applying the notch stress approach.

alternating load. The influence of the welding residual stresses on the local fatigue strength can be neglected in many cases, but in some cases the fatigue strength is decreased or increased by up to 30%. The decrease of strength is caused by tensile residual stresses. The increase of strength occurs with fractures initiated at the weld root and is caused by compressive residual stresses at the weld root.

The decrease of the global fatigue strength with plate thickness found from tests with stress relieved specimens was proportional to approximately the cubic root of the thickness ratio as were the appertaining notch stress maxima, Fig.171. For comparison, the investigations by Bell, Vosikovski and Bain[232] on the fatigue strength of T joints with double bevel butt welds produced with

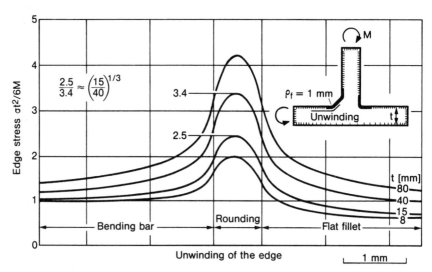

171 Boundary stress curves for different plate thicknesses, notch stress increasing with plate thickness according to a cubic root function, after Olivier, Köttgen and Seeger.[231]

undercut defects and subjected to transverse bending characterise the influence of plate thickness by approximately the 3.5th root and Eurocode[150] and ECCS recommendation[151] prescribe a reduction of the permissible stresses according to the fourth root for plate thicknesses larger than 25mm (see Ref.233 also).

The test results[231] have been evaluated also according to the local strain approach proposed by Haibach (see section 8.3.1). The conversion of the strain measured in front of the fusion line of the weld toe into the maximum notch stress of the weld toe on the basis of numerical notch stress analysis results in $\tilde{\sigma}_{A0} \approx 235$ N/mm². Finally, the investigated joint types and load cases were assigned to the notch classes of the relevant codes on the basis of the test results, whereby different safety concepts behind these codes (see Fig.147) had to be taken into account.

In continuation of the research project,[231] the basic types of conventional weld joints (butt joint, transverse stiffener, cruciform joint and lap joint) are investigated according to the above notch stress approach. A first evaluation of the S-N curve compilations[37] combined with calculated fatigue notch factors (see Fig.167) had the result shown in Fig.172. The local pulsating fatigue strength, $\tilde{\sigma}_p = 320$ N/mm² (for $P_f = 0.5$), stated above just as the reference fatigue strength, $\sigma_p = 240$ N/mm² (for $P_f = 0.1$), stated in section 2.1.1 are confirmed. Furthermore, the influence of plate thickness and welding residual stresses is considered in more detail within the continuing project.

Welded joint (structural steels)	Fatigue notch factor β_k	Global fatigue strength (R = 0) σ_A [N/mm²], $P_f =$			Local fatigue strength $\tilde{\sigma}_p = 2\beta_k \sigma_A$
	Fracture initiation	0.1	0.5	0.9	
Butt joint	1.89 Weld toe	61	78	99	294
Transverse stiffener	2.45 Weld toe	52	69	91	338
K-weld Cruciform joint	2.50 Weld toe	54	67	83	336
Strap joint	3.12 Weld toe	47	55	65	344
Fillet weld Cruciform joint	4.03 Weld root	32	43	57	346

172 Local pulsating fatigue strength, $\tilde{\sigma}_p$, (endurance limit) of typical welded joints fictitiously rounded at weld toe and root with $\rho_f = 1$mm, conversion of compiled data for t = 10mm, $P_f = 0.1, 0.5$ and 0.9 on the basis of calculated fatigue notch factors, after Olivier, Köttgen and Seeger.[231]

8.2.6   *Shape optimisation at thickness transitions*

The issue concerning the most advantageous transition as regards stress (and therefore as regards fatigue strength) between a smaller and a larger plate thickness is of fundamental and practical importance; this is relevant to butt joints between plates of differing thicknesses and cruciform or T joints at the transition from tension plate to transverse bending plate. The transition, optimised with regard to shape, can be preformed alongside the weld (forged, rolled or cast) or dressed at the weld toe (milled or ground). The problem of shape optimisation which occurs in this connection has been addressed repeatedly both theoretically and experimentally (photoelasticity, fatigue tests).

The theoretical numerical investigations were originally based on the (partial) analogy of flow field (plane potential field) and stress field (plane bipotential field with potential behaviour near boundaries). The jet of liquid emerging freely downwards from the circular opening in the bottom plate of a cylindrical tank contracts, with a constant flow speed at the free boundaries. The boundary shape of the jet transferred to the notch problem proves to be particularly favourable as regards stress (after Baud[241]). Applied to the notch problem in a less direct way, the requirement can be made for the optimised transition to have a constant (not increased) boundary stress. The catenary curve proves to be an exact solution of this requirement for the flat bar of infinite length (after Neuber[240]). In more empirical investigations, ellipses, parabolas and Cassini curves turned out to be advantageous (but not optimum) curves for the transition (see Ref.209, *ibid*, page 119).

The theoretical optimisation problem was dealt with for practical purposes (by Schnack[238,239]) to the extent that a connecting curve was sought between two fixed points as the boundary of a given region which produces constant boundary stress either along the whole length or the greatest possible length between the fixed points. The optimisation of the transition curve for stepped flat bars with different width ratios in accordance with the above-mentioned criteria, produced the solutions shown in Fig.173 (from Ref.238) by means of a finite element model (altering the local curvature iteratively, which predominantly only influences the local notch stress). The curve according to

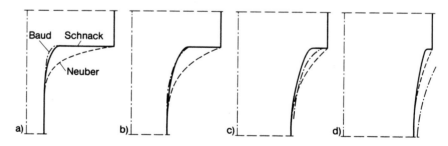

173 Optimised shape of thickness transition after Schnack,[238] compared with
solutions by Neuber[240] (dash line) and Baud[241] (dot-dash line), with
various shoulder heights (a) to (d), after Schnack.[240]

Schnack is compared with Baud's streamline solution, which is useful only for a marked shoulder step, and Neuber's solution, which is only exact for a bar of infinite length. Schnack's solution, which is closer to reality, reveals constant boundary stresses, right up to the shoulder. Neuber's 'more theoretical' solution reveals a decrease of stress to zero at the end point of the transition curve. Obviously the shape optimisation of the transition between the two widths is mainly a matter of increasing the width initially very slowly, whilst there are various shapes acceptable in the remaining part of the transition.

The shape optimisation[242] close to practical requirements for the weld toe of the cruciform joint, taking note of the above-mentioned more theoretical results leads to a cruciform joint of 'special quality' (the result can also be transferred to the butt joint). A double bevel butt weld with a 30° fillet weld covering it, the toe of which is ground to a relatively large radius, is recommended. In Fig.174 version (a) manufactured on the spot is compared

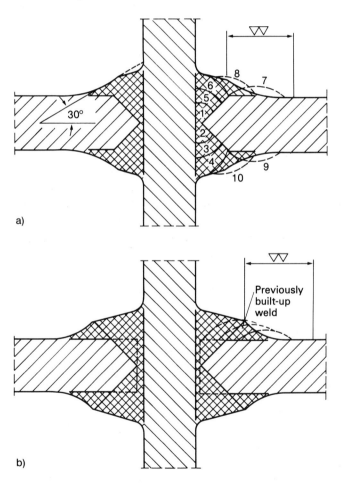

174 Optimised shape of thickness transition of double bevel butt weld with fillet weld in cruciform joint, after Ruge and Drescher.[242]

with version (b) with a premanufactured transition. Weld toe and weld root (if present) only show a very small remaining notch effect.

The result of a photoelastic shape optimisation of the nozzle in a spherical vessel, subjected to internal pressure, is shown in Fig.175. The internal bulge has not become accepted in practice. The reason for this is to be found in the deposition of corrosive remnants in the region shielded by the bulge.

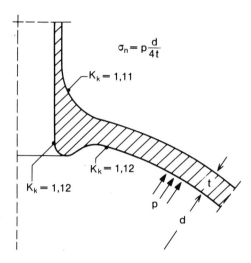

$$\sigma_n = p\frac{d}{4t}$$

$K_k = 1{,}11$

$K_k = 1{,}12$

$K_k = 1{,}12$

175 Optimised shape of transition for nozzle in spherical vessel, after Peterson.[208]

The issue of shape optimisation for reinforced cutouts has been investigated theoretically (Ref.209, *ibid*, page 109-113). So-called 'neutral holes' without any notch stress increase can be achieved either by a steady thickening towards the edge of the cutout or by reinforcing the edge only in combination with single hinged connecting rods within the cutout. A special edge shape which depends on the load case is necessary in both cases.

8.2.7   *Notch effect of the welded joint subjected to biaxial and oblique loading*

Basic definitions

On the basis of the stress concentration factor, $K_{k\perp}$, for transverse loading of the welded joint (or of the fatigue notch factor, $K_{f\perp}$, determined taking the microstructural support effect into account) the stress concentration factors K_{k1} and K_{k2}, referring to the first and second principal stress at the notch root subjected to biaxial oblique loading of the welded joint can be determined via a simplified procedure.[227,243] The biaxial and oblique basic loading of joints is the usual case in practice. The oblique position and size of the biaxial basic stresses result from the stress analysis of the global structure consisting of plate- and shell-like components (in most cases using the finite element method). In this model, the welded joints are reduced to joint lines. The basic stress at the joint line can also be determined by measurement, in which case the principal direction of the stress state is usually determined using brittle coatings.

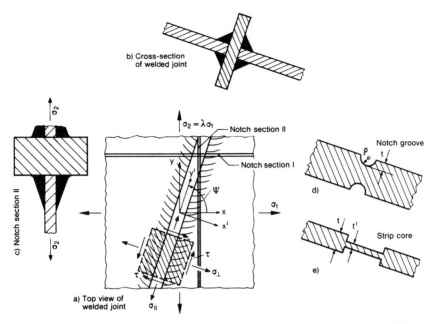

176 Biaxially and obliquely loaded welded joint (a, b) and models 'oblique notch in solid body' (a, d) 'oblique strip core in plate' (a, e) and 'oblique notch sections' (a, c).

The welded joint is considered in an oblique position with regard to the biaxial basic stress state, Fig.176a (horizontal projection) and Fig.176b (cross section). The cruciform joint, in the figure, represents any desired, approximately straight, welded joint with marked notches at the weld toe and root. The principal components of the basic stress state are σ_1 and $\sigma_2 = \lambda\sigma_1$. $\lambda = -1$ designates pure shear, $\lambda = 0$ uniaxial tension and $\lambda = +1$ biaxial tension identical with all-sided tension.

The welded joint is inclined with the notch angle, ψ, relative to the direction of the first principal stress. The joint, subjected to transverse loading, considered in previous sections, is characterised by $\lambda = 0$ and $\psi = \pi/2$; $K_{k\perp}$ is related to this case. The pure shear loading, not only of the base plate, but also of the welded joint notch, is characterised by $\lambda = -1$ and $\psi = \pi/4$. As an alternative to ψ, the 'oblique angle' $(\pi/2 - \psi)$ which complements ψ, is also used in the stress concentration factor diagrams below. The joint-oriented co-ordinate system x'-y' is rotated by the oblique angle relative to the principal stress oriented co-ordinate system, x-y.

The stress concentration factor, $K_{k\perp}$, of the welded joint subjected to transverse loading is assumed to be known (see section 8.2.4). The stress concentration factors, K_{k1}, K_{k2} and K_{keq} are to be determined as functions of $K_{k\perp}$, λ and ψ:

$$K_{k1} = \frac{\sigma_{1max}}{\sigma_1}, \tag{74}$$

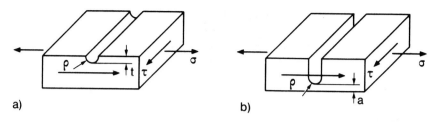

177 Single sided shallow (a) and deep (b) notch in plate subjected to tensile and (longitudinal) shear loading.

$$K_{k2} = \frac{\sigma_{2max}}{\sigma_1},$$ (75)

$$K_{keq} = \frac{\sigma_{eqmax}}{\sigma_1} = \sqrt{K_{k1}^2 + K_{k2}^2 - K_{k1}K_{k2}}$$ (76)

The stress concentration factor, K_{keq}, characterises the maximum equivalent stress in accordance with the distortion energy hypothesis after von Mises which is applicable to the fatigue processes. The functional dependences $K_{k1/2/eq} = f_{1/2/eq}(\alpha, \lambda, \psi)$ are determined in accordance with the model 'oblique notch' in solid body and 'oblique notch sections', whilst the model 'oblique strip core in plate' is considered outdated.

Model 'oblique notch in solid body'

The solutions, derived by Neuber[207] based on the theory of elasticity, for the geometrical limit cases of the 'shallow' and 'deep' double sided notch groove in a solid body subjected to tensile and longitudinal shear loading (the latter named 'twisting' in Ref.207), Fig.177, make an exact determination of the stress concentration factors, $K_{k1/2}$, possible (and therefore also of K_{keq}) of an equivalent notch corresponding to the welded joint, Fig.176d, subjected to the basic stress state, $\sigma_\perp, \sigma_\parallel, \tau$, Fig.176a, with

$$\sigma_\perp = c_\perp \sigma_1, \qquad c_\perp = \frac{1}{2}[\lambda + 1 + (\lambda - 1)\cos 2\psi]$$ (77)

$$\sigma_\parallel = c_\parallel \sigma_1, \qquad c_\parallel = \frac{1}{2}[\lambda + 1 - (\lambda - 1)\cos 2\psi]$$ (78)

$$\tau = c\sigma_1, \qquad c = \frac{1}{2}(\lambda - 1)\sin 2\psi$$ (79)

The stress concentration factor equations in accordance with Ref.207 (*ibid*, pages 42, 56, 147, 148) are as follows for the shallow and deep notch subjected to tensile or longitudinal shear loading (notch depth, t, of shallow notch, half width, a, of narrowest cross section of deep notch, radius of curvature, ρ, of notch root):

$$K_{k\perp} = \frac{\sigma_{\perp max}}{\sigma_{\perp}} = 1 + 2\sqrt{\frac{t}{\varrho}} \tag{80}$$

$$K_{k\tau} = \frac{\tau_{max}}{\tau} = 1 + \sqrt{\frac{t}{\varrho}} \tag{81}$$

$$K_{k\perp} = \frac{\sigma_{\perp max}}{\sigma_{\perp}} = \frac{2\left(\dfrac{a}{\varrho} + 1\right)\sqrt{\dfrac{a}{\varrho}}}{\left(\dfrac{a}{\varrho} + 1\right)\arctan\sqrt{\dfrac{a}{\varrho}} + \sqrt{\dfrac{a}{\varrho}}} \tag{82}$$

$$K_{k\tau} = \frac{\tau_{max}}{\tau} = \frac{\sqrt{\dfrac{a}{\varrho}}}{\arctan\sqrt{\dfrac{a}{\varrho}}} \tag{83}$$

A statement worthy of note, drawn from these equations, is that the notch effect of the shallow notch, after subtracting the basic stress state, is only half as large for shear loading as for tensile loading. Furthermore, numerical evaluations for shallow and deep notches, when compared, reveal that the ratio of tensile to shear notch effect, expressed by the corresponding stress concentration factors, $K_{k\perp}$ and $K_{k\tau}$, is identical for shallow and deep notches within accuracy limits, which are of practical interest, if only the same value of $K_{k\perp}$ is used as a basis (not the same value of t/ϱ and a/ϱ):

$$\frac{K_{k\perp}}{K_{k\tau}} = 1.0\text{-}2.0 \tag{84}$$

The values close to 2.0 in equation (84) occur when there is a strong notch effect, the value 2.0 itself refers to the crack tip. The values close to 1.0 occur when there is a weak notch effect, the value 1.0 itself refers to the smooth plate. The above-mentioned identity makes it possible to use the simpler formula for the shallow notch both for shallow and deep notches in the derivations below. Moreover, it also indicates that the equivalent shallow notch covers the notch effect of the welded joint, which obviously is to be classified between the effect of shallow and deep notches, with a sufficient degree of accuracy. Finally, it is to be suspected that the above-mentioned identity is also applicable in the less frequent bending load cases as well as for the single sided notch effect, if certain boundary conditions are observed.

Therefore on the basis of equations (80) and (81):

$$K_{k\tau} = \frac{1}{2}(K_{k\perp} + 1) \tag{85}$$

The longitudinal notch stresses, $\sigma_{\parallel max}$, result from the basic stress component, $\sigma_{\parallel}$, increased by the stress, $\nu (\sigma_{\perp max} - \sigma_{\perp})$. The basic stress component, $\sigma_{\perp}$, is to be subtracted because it is connected with a free transverse contraction. Therefore

$$\sigma_{\parallel max} = \sigma_{\parallel} + \nu (\sigma_{\perp max} - \sigma_{\perp}) = \sigma_{\parallel} + \nu \sigma_{\perp} (K_{k\perp} - 1) \qquad (86)$$

With the known equation for the principal stresses

$$\sigma_{1/2max} = \frac{1}{2} (\sigma_{\perp max} + \sigma_{\parallel max}) \pm \sqrt{\frac{1}{4} (\sigma_{\perp max} - \sigma_{\parallel max})^2 + \tau_{max}^2} \qquad (87)$$

it follows that the stress concentration factors are

$$K_{k1/2} = \frac{1}{2} [K_{k\perp} c_{\perp} + c_{\parallel} + \nu c_{\perp} (K_{k\perp} - 1)] +$$

$$+ \frac{1}{2} \sqrt{[K_{k\perp} c_{\perp} - c_{\parallel} - \nu c_{\perp} (K_{k\perp} - 1)]^2 + [(K_{k\perp} + 1)c]^2} \qquad (88)$$

This equation applies exactly, provided that the introduced relation between $K_{k\perp}$ and $K_{k\tau}$ is applicable.

Model 'oblique strip core in plate'

A plate in which the oblique notch is replaced by a strip core of lesser thickness, Fig.176e, is designated a strip core model (a limit case of the inclusion or core of higher or lower rigidity, simulated by increased or reduced plate thickness). The increase of stress is caused exclusively by the reduction in thickness, the notch effect of the stepped transition between plate and core, remaining unconsidered. The thickness ratio, δ, is equated with the stress concentration factor, $K_{k\perp}$:

$$\delta = \frac{t}{t'} = K_{k\perp} \qquad (89)$$

The model is suitable for correctly simulating the stress and strain boundary conditions corresponding to the oblique notch position. Moreover, the notch effects for transverse loading ($K_{k1} = K_{k\perp}$) and for longitudinal loading ($K_{k1} = 1$) are simulated precisely. However, the increase of notch stress for shear loading turns out to be too large, namely equally large as for tensile loading:

$$K_{k\tau} = \delta = K_{k\perp} \qquad (90)$$

The stress transformation equations (77) to (79), equation (86) for $\sigma_{\parallel max}$ and the principal stress formula (87) remain valid. The result is:

$$K_{k1/2} = \frac{1}{2}[K_{k\perp} c_\perp + c_\parallel + \nu c_\perp (K_{k\perp} - 1)] +$$

$$+ \frac{1}{2}\sqrt{[K_{k\perp} c_\perp - c_\parallel - \nu c_\perp (K_{k\perp} - 1)]^2 + [(2K_{k\perp} c]^2} \quad (91)$$

This equation is only applicable with the stated restriction as regards the shear component. The model is to be considered outdated.

Model 'oblique notch sections'

The third model is based on oblique notch sections in the two principal directions of the basic stress state, Fig.176a and c. It is founded on the observation that the principal direction of the stress, made visible for practical purposes using brittle coatings, passes through the notch root of welded joints largely without changing direction.[243] The constancy of direction is actually more global than local. The principal direction and the corresponding brittle coating crack make a slight swerve at the root of sharp notches in an oblique position. The deviation in direction, which occurs when entering the notch root, is cancelled on exit. The fact that this model is not quite accurate also becomes apparent from the differences in direction between $\sigma_\perp$ and $\sigma_{\perp max}$, which can be calculated exactly for the 'shallow' or 'deep' oblique notch.[244] Despite the local deviations, the plane section and the plane stress or strain state, which can be assigned to it (the transverse contraction at the notch root is more or less completely suppressed), represents a good, general approximation. However, the global constancy of the principal direction, which is a precondition of the model cannot be fulfilled for multi-plate joints in certain load cases *e.g.* 'counter shear loading' as in Fig.178. However, in this case too constancy of the principal direction up to the centreline of the joint is a well founded approximation assumption.

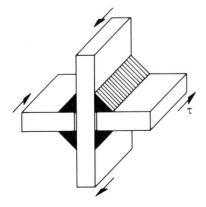

178 Cruciform joint subjected to counter shear loading.

The second starting point for the oblique notch section model is the fact that the notch, when cut obliquely in the principal direction, reveals a larger radius of curvature, a smaller slope angle and therefore a correspondingly weaker notch effect. As a trend, this corresponds to the weaker notch effect in the presence of shear loading compared with tensile loading, since as the oblique angle of the notch increases, the shear component, τ, arising from σ_1 or σ_2 also increases relative to the tension component, $\sigma_\perp$. When calculating the stress concentration factors, the oblique notch sections in the principal directions are not actually analysed (with the determination of the relevant stress concentration factors, *e.g.* for the section II in Fig.176c and for the section I perpendicular to it) but are only used for the derivation of the transformation equation relating, $K_{k\perp}$ to K_{k1} and K_{k2}.

The radii of curvature of the comparison notch, which is assumed as 'shallow' in the sense of Ref.207, increase with $1/\sin^2\psi$ or $1/\cos^2\psi$ (factors derived from the apex curvature of the section ellipses) for the notch section models in the direction of the first and second principal stress whereas notch depth, t, remains constant. It follows that

$$K_{k1} = 1 + 2\sqrt{\frac{t\sin^2\psi}{\varrho}} \tag{92}$$

$$K_{k2} = \lambda\left(1 + 2\sqrt{\frac{t\cos^2\psi}{\varrho}}\right) \tag{93}$$

The additional notch stress proportion, arising as a consequence of the restraint on transverse contraction (actually not complete) connected with the other principal notch stress is to be superimposed. Only the excess notch stress $(\sigma_{kmax} - \sigma_1)$ or $(\sigma_{kmax} - \sigma_2)$ is restrained as regards transverse contraction (with Poisson's ratio, ν):

$$K_{k1} = 1 + 2\sqrt{\frac{t}{\varrho}}(\sin\psi + \nu\lambda\cos\psi) \tag{94}$$

$$K_{k2} = \lambda + 2\sqrt{\frac{t}{\varrho}}(\lambda\cos\psi + \nu\sin\psi) \tag{95}$$

With $2\sqrt{t/\varrho} = (K_{k\perp} - 1)$ in accordance with equation (80) it follows that:

$$K_{k1} = 1 + (K_{k\perp} - 1)(\sin\psi + \nu\lambda\cos\psi) \tag{96}$$

$$K_{k2} = \lambda + (K_{k\perp} - 1)(\lambda \cos \psi + v \sin \psi) \tag{97}$$

Equations (96) and (97) are valid as approximations, but, as is shown below, with a high approximation quality for K_{k1} and K_{k2}. They are considerably less complex than the exact equation (88).

Result of a comparison calculation

The result of a comparison calculation for K_{k1}, K_{k2} and K_{keq} in accordance with the models described above is given in Fig.179 and 180. The oblique notch section model results in relatively accurate approximation values for almost all the stress concentration factors and loading cases: only in the case of K_{k2}, $\lambda = 1$, $0 < \psi < \pi/2$ do relatively large deviations occur, which, however, disappear in K_{keq}, the parameter decisive for the strength, caused by the peculiarity of the strength hypothesis.

The oblique strip core model results, as has already been stated, in relatively large deviations from the exact values in large areas. In the case of all-sided tension ($\lambda = 1$) the exact values K_{k1}, K_{k2} and K_{keq} are reproduced for any ψ values.

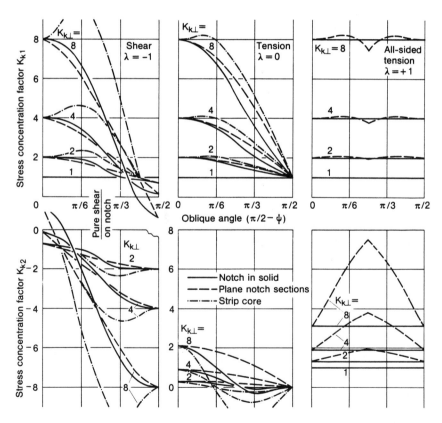

179 Stress concentration factors, K_{k1} and K_{k2}, of the oblique notch dependent on oblique angle and notch severity for the basic load cases 'shear', 'uniaxial tension' and 'all-sided tension', after Radaj.[243]

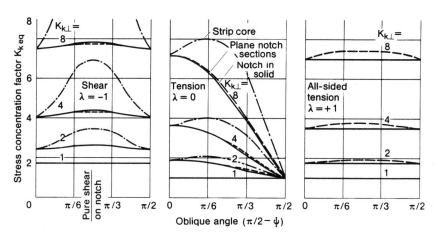

180 Stress concentration factor, K_{keq}, dependent on oblique angle and notch severity for the basic load cases 'shear', 'uniaxial tension', and 'all-sided tension', after Radaj.[243]

The dependence itself of the stress concentration factors (in accordance with the oblique notch in solid body model) is also worthy of comment: with regard to K_{k1} and K_{k2} strong dependence on ψ for $\lambda = -1$ and $\lambda = 0$, no dependence on ψ for $\lambda = +1$ (because of all-sided tension); with regard to K_{keq} almost no dependence on ψ for $\lambda = -1$ (remarkable) and $\lambda = +1$ (in accordance with expectations), strong dependence on ψ only for $\lambda = 0$. Therefore, the oblique position of the weld only has a practically important influence on the fatigue strength from the point of view of notch stress analysis if there is dominant uniaxial basic loading (further aspects are of importance from the point of view of crack propagation and, for light metal alloys from the point of view of heat softening). The advantage of the oblique weld position has been proved repeatedly in fatigue tests on spiral welded tubes and tanks.

It should be noted with regard to the K_{keq} diagrams that the K_{keq} values for $\lambda = -1$ are reduced by the factor $1/\sqrt{3}$ if σ_{eqmax} is referred to σ_{eq} (the equivalent stress of the basic loading) instead of σ_1. The more advantageous notch behaviour when subjected to shear loading, which has been mentioned, then becomes apparent.

8.2.8   Notch effect at the weld ends

It is possible to assess the fatigue strength of welded joints and welded structures reliably by the procedure described above, with the exception of the weld ends. Disregarding the weld ends is not adequate for industrial practice because weld ends occur relatively frequently and are a favourite area for fatigue crack initiation. Therefore, the issue of the notch effect of the weld ends must be solved.

As the solution of the three dimensional notch stress problem at the weld end is too expensive, the oblique notch section method, described above, is a possibility here too as an approximation. First of all, this procedure can be used unchanged on the weld cross section contour at the weld end, although the assumption of considering a continuous weld is not fulfilled here. In addition, the notch effect of the weld longitudinal section contour at the weld end is to be checked. It can be assumed as a simplification that the cross section and longitudinal section contours describe the shape of the weld end adequately, in the case of machined weld ends by the intersection of the two orthogonal contour cylinders, in the case of unmachined welds as limiting shapes, between which the intermediate sections can be interpolated. The fictitiously rounded longitudinal contour model is to be subjected to the same internal forces in the principal direction to which the transverse contour model has already been subjected in the investigation. The contour transition towards the weld seam is to be simulated realistically. A rigid support is provided on the side away from the transition, Fig.181.

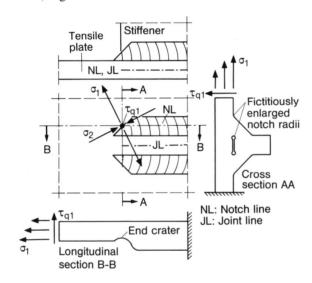

181 Longitudinal and transverse contour model for determining the notch effect of weld ends (notch lines assumed coinciding with weld seam centrelines).

When determining the internal forces in the principal directions at the ends of the weld lines of the global structure it is to be noted that a stress singularity can occur here, which is blurred in the finite element solution in an uncontrolled manner. Suggestions are made in section 8.2.9 below for dealing with the stress singularity in a controlled manner.

A stress singularity in the global structure also occurs at butt welds which end at a corner (*e.g.* gusset plate joint, Fig.182a). The fictitious rounding of the notch can be provided here in the global structure and the notch stress can be calculated using the boundary element method. If, in contrast, the (unmachined) butt weld ends at a straight or curved smooth edge, Fig.182b, then the notch effect of the local indentation or projection of the weld end (determined on a tension bar with corresponding edge notching) is to be

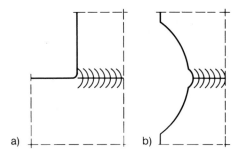

182 Fictitious rounding of end notch of butt weld at corner (a) and curved
edge (b).

superimposed multiplicatively on the notch effect of the global structure (in the
case of a curved edge).

8.2.9   *Practical performance of the notch effect calculation*

Problems and solutions

There are a series of difficulties involved in the practical realisation of the notch
effect calculations described above, which can, however, be overcome.[245,246]
The individual problem points and the measures for solving them are listed in
Table 20. Two main problems have to be solved:

—	the singularity of the internal forces in the corners of the global structure,
	Fig.183, represented in a 'blurred' manner in the finite element solution);
—	the partial or complete lack of constancy of the principal direction in the
	joint area.

This leads to the following remedial measures:[246]

—	evaluation of the internal forces in the notch line instead of the joint line,
	which, however, disturbs the equilibrium of the external forces at the
	contour model of the welded joint (here external forces), or, alternatively,
	evaluation of the internal forces in the joint line at the same distance from
	the corner as the notch line;
—	constancy of principal direction only up to centreline of the joint a well
	founded approximation, which causes an additional imbalance at the
	contour model of the welded joint.

To take up the imbalance forces the contour model of the welded joint must
be supported in a line approximately at the centre of the model. This support is
particularly influential with respect to the notch stress results, which can be
shown by changing the position of the support line. The optimum selection of
this line is therefore of great importance with regard to the accuracy of the
results. Point supports are completely unsuitable because of the disturbing
effect of local force concentration. An alternative, more expensive, possibility
consists in supporting the single joint end one after the other, applying the
external forces to the remaining ends, evaluating the notch stresses for the
individual support cases and finally calculating mean values from these
evaluations.

Table 20. Notch stress analysis for welded joints, problems and solutions

Problem	Solution
Re-entrant corners of the structure (with weld end), butt or T jointed, cause stress singularities in the global structure	Evaluation of internal forces in notch line (instead of joint line) of the global structure
Model 'oblique notch sections' excludes weld ends	Notch stress analysis for longitudinal contour model of weld end as well as for transverse contour model
Internal forces in joint line deviate from internal forces in notch line	Evaluation of principal internal forces in notch line
Principal internal forces in notch line transferred to transverse contour model are not in equilibrium	Line support of the contour model in centre of joint at some distance from the notches
Principal direction in multiple plate joint not necessarily constant (*e.g.* cruciform joint subjected to counter shear)	Constancy of principal direction up to centre of joint gives justification to transfer all principal internal forces to the same transverse contour model (produces non-equilibrium forces)
Principal directions of membrane and bending stresses may differ	The normally only slight differences are neglected. The maximum principal stresses, σ_1 or σ_2, on top and bottom side are transferred
Principal directions of membrane plus bending stress and of transverse shear stress may differ	Only the transverse shear stress in the section parallel to the weld is notch-effective and therefore transferred
Finite element analysis too inaccurate for global structure	Finer mesh, higher order elements, elements with singularity
Bending moment of the transverse shear force shifted from notch cross section to end cross section of contour model	Superimposition counter-rotating bending moment when shifting the transverse shear force

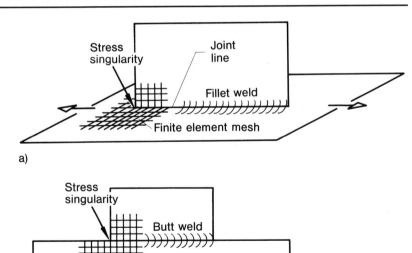

183 Stress singularity at corners of global structure: tensile plate with longitudinal stiffener (a) and gusset plate specimen (b).

The above-mentioned problems can only then be avoided completely if the three dimensional notch problem of the local structure can exceptionally be solved without the contour model approximation.

Principal internal force method

Calculation of the notch effect of the welded structure is performed in the following steps:

— finite element analysis for the global structure model, in which mesh lines or element centrelines are laid in the notch lines or on both sides of them (for subsequent interpolation of the internal forces);
— evaluation of the internal forces in the above-mentioned mesh lines or element centrelines: the absolutely larger principal stress, σ_1 or σ_2, on the top and bottom surface and the transverse shear stress, τ_{q1}, acting in the sectional face of σ_1 or σ_2;
— interpolation of the internal forces with respect to the notch line (unnecessary if a mesh or element centreline coincides with the notch line);
— transference of the internal forces to the transverse or longitudinal contour model with superimposition of the counter-rotating bending moment caused by the shifting transverse force from the notch section in the global structure to the end sections of the contour model;
— line support of the transverse contour model in the centre area of the model and of the longitudinal contour model in an equivalent position;
— notch stress analysis for the contour models with fictitious rounding of the weld notches;
— determination of the fatigue notch factor by referring the calculated maximum notch stress on to the nominal stress or on to a characteristic structural stress of the joint.

The 'internal force splitting method',[246] as an alternative to the above method, is more consistent as regards the rules of mechanics, but cannot be used in a straightforward manner at weld ends and joint corners because of the stress singularity there.

Analysis example 'frame corner joint'

As an example, calculation of the notch stresses for welded joints in accordance with the principal internal force method described above is carried out for the frame corner illustrated in Fig.161, consisting of side rail and cross member, subjected to torsion loading of the cross member. The selected simple joint design cannot be used for practical purposes because the torsion bending moment in the transverse member causes excessively high torsion bending stress at the edges of the flanges in the joint, which result in accordingly increased notch stresses, which then initiate fatigue processes very early. However, the marked increase of structural and notch stress makes this particular design especially suitable for demonstration purposes. The notch stresses are to be calculated (at the corners) of the cross member flange joint with fillet weld and at the edge of the side rail flange with butt weld.

The finite element model of the global structure is shown in Fig.184, including constant stress elements and mesh refinement in the joint area. The element stresses in the joint area are shown in Fig.185 with regard to magnitude, biaxiality and direction. High bending stresses are superimposed on membrane stresses. The flange ends of the transverse member are clearly visible in the cyclic pattern of the stress crosses on the web-plate of the side rail. This pattern corresponds to the torsion bending loading. Bending stresses dominate in the flange of the transverse member and cause the membrane torsion bending stresses to recede, which leads to the stress distribution in the web of the side rail which has just been mentioned (longitudinal compression

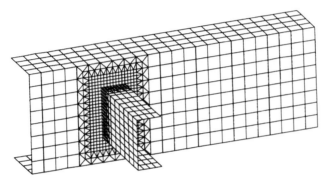

184 Finite element model for corner joint of frame.

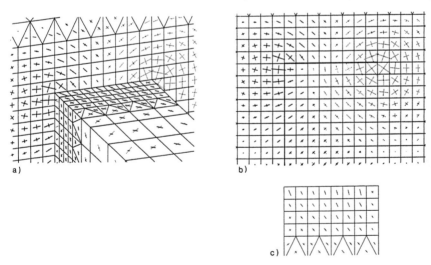

185 Principal stress crosses on a frame corner joint subjected to torsion loading of transverse member, oblique view (a), front view of side rail, inside of web (b), top view of cross member, outside flange (c), after Radaj.[245,246]

would be expected in the right flange edge). The linear interpolation of the internal forces in the centre points of the elements, inside (i) and outside (o) of the notch line, top side (t) and bottom side (b) of the elements, carried out with respect to the notch line (k), is illustrated in Fig.186. The stresses of the counter-rotating bending moment of the transverse force shifted to the end cross section (e) of the model are to be superimposed. The triangular shear stress distribution of the external transverse force corresponds to the restricted input facilities of the boundary element system, which was used.

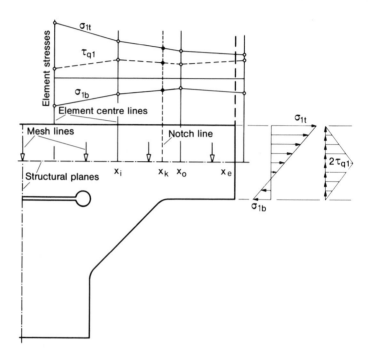

186 Interpolation of internal forces and loading of the end cross section, on the basis of finite element evaluation in the centre points of the elements.

The boundary element models of the local structure subjected to the internal forces of the global structure and the notch stress results are presented in part in Fig.187. A double fillet weld is compared with a single bevel butt weld with fillet weld and root gap at the transverse girder flange. The loading forces in the end cross sections and the support forces in the central cross section are visible in addition to the boundary stresses.

The calculated fatigue notch factors are summarised in Fig.188. The torsion bending stress, σ_ω, which would arise under warping-rigid support conditions according to the torsion bending theory,[247] was taken as the reference stress. In contrast, on the butt welded joint the bending stress σ_b at the outer surface of the flange according to the beam bending theory was assumed as the reference stress. The fatigue notch factors themselves are entered in boxes, the positioning of which corresponds to the positions of the weld notches in the

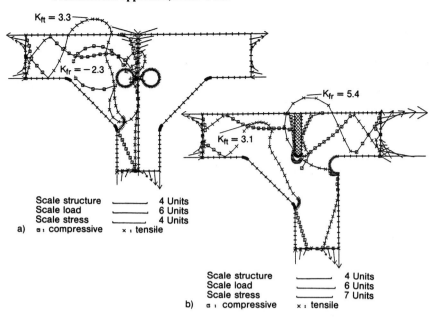

$K_{ft} = 3.3$

$K_{fr} = -2.3$

$K_{fr} = 5.4$

$K_{ft} = 3.1$

Scale structure ———— 4 Units
Scale load ———— 6 Units
Scale stress ———— 4 Units
a)   □ : compressive   × : tensile

Scale structure ———— 4 Units
Scale load ———— 6 Units
Scale stress ———— 7 Units
b)   □ : compressive   × : tensile

187 Notch stresses for double fillet welded joint (a) and single bevel welded joint (b), after Radaj.[245,246]

sections of the local structure above the boxes. The transverse and longitudinal sections which are considered are marked on the other hand in the global structure at the top of the figure as intersecting lines. The dimensions are also to be found at the top. The moment vector is reversed compared with finite element solution. The two weld types, double fillet weld and single bevel butt weld with fillet weld and root gap, are compared for section AA (however, the eccentricity of the single bevel butt-weld was not taken into account in the model of the global structure). The boxes for the weld toe notches are marked by a solid line, a dotted one is used for the weld root notches. The correspondence to the joint ends is indicated by Roman numerals. One positive and one negative value are entered for the fillet weld root corresponding to the multiple stress peaks at the slit tip.

Fatigue notch factors of a realistic magnitude were determined throughout. The torsion bending loading of the cross member enhances the risk of failure in the web of the side rail ($K_f = 2.7$-3.3 or 5.4) not in the flanges of the transverse member ($K_f = 0.3$-0.6). The double fillet weld, which runs around the end of the flange reveals approximately the same notch effect at this point in the three orthogonal sections ($K_f = 2.7, 3.3, 3.3$). A particularly severe notch effect occurs at the root gap of the single bevel butt weld ($K_f = 5.4$). The notch effect of the inner slit of the double fillet weld is smaller than that of the weld toes and equal to it in one case ($K_f = 1.3$-2.7). The butt weld is not critical despite a severe notch effect at the weld root ($K_f = 7.0$) because of the low reference stress. None of the determined values has the high degree of accuracy of a closed solution on the basis of the theory of elasticity because of the large number of introduced simplifications but they are to be considered as approximations for practical purposes.

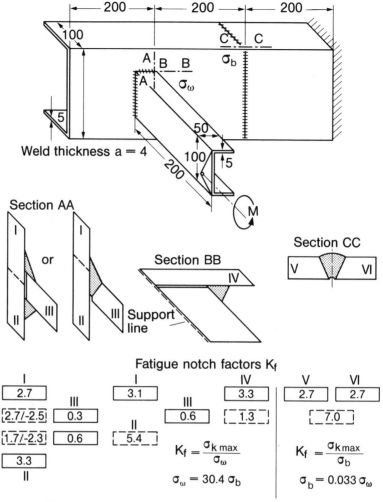

188 Fatigue notch factors, K_f, for frame corner joint, after Radaj.[245,246]

8.2.10 *Analysis examples from steel engineering and vessel design*

Web stiffener on I girder of a bridge

Transverse web stiffeners on I girders are required to maintain the cross sectional contour even under high loading, especially when introducing transverse forces (see section 5.1.5). The transverse weld connecting the stiffener with the tensile flange of the girder is at risk of fatigue failure because of the high notch stresses at its toe and root. In the following, results of notch stress analysis are summarised for the transverse stiffener with and without a cutout at the inner corner.[248] The investigation was carried out for a slender welded girder common in bridge construction, Fig.189, which was subjected to pure bending and transverse bending respectively. A double fillet weld was compared with a double bevel butt weld with root gap. The tension plate with single sided stiffener was included in the investigation.

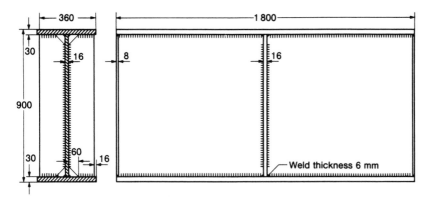

189 Welded I girder with transverse web stiffener.

The following conclusions were drawn from the finite element results for the global structure, Fig.190 and 191, which are the basis of the subsequent notch stress analysis. The changes in the direction of principal stresses in the tensile flange caused by the stiffener are negligibly small. The stiffener presses into the flange from above, with pure bending load to a lesser degree and transverse bending load to a geater degree. Thereby the structural stresses on the inside of the tensile flange are decreased and those on the outside increased. This seems to be advantageous because of the superimposed notch effect on the inside whereas there is no notch effect on the outside. The result for the tension plate with unloaded stiffener therefore is on the safe side. With transverse bending,

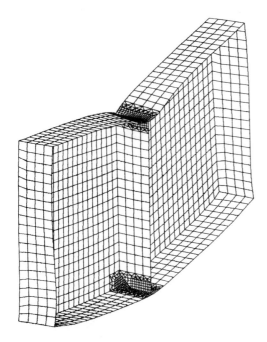

190 Finite element mesh and deformation of stiffened girder subjected to transverse bending.

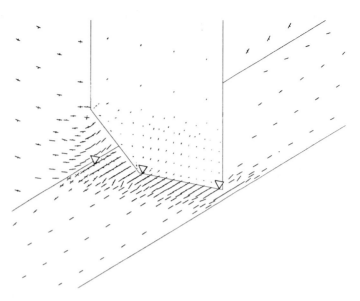

191 Structural stress distribution in area of weld joint between transverse
stiffener and tensile flange; subsequent notch stress analysis at points
marked by triangles.

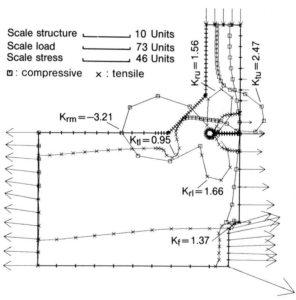

192 Boundary stresses for contour model of stiffener-to-flange joint with
fillet weld for I girder subjected to transverse bending, tensile forces in
flange and compressive forces in stiffener, stiffener with cutout, notch
position near cutout; K factors indexed: t, toe; r, root; f, flange; u, upper;
l, lower; m, median.

there is an uneven distribution of the tensile stresses in the flange over the flange width, increased stresses at the web, decreased stresses at the flange edge. Cutting out the inner corner of the stiffener is favourable in the case of transverse bending because the superimposed notch effect is avoided in the area of the cutout where the structural stresses are increased.

The results of the subsequent notch stress analysis, Fig.192, were converted via equation (67) and (13) into fatigue strength reduction factors, γ_P, for pulsating tensile load taking the maximum tensile notch stresses from the weld toe (for γ_t) and the weld root (for γ_r) into account, Table 21. The reduction factor, γ_f, designates the structural stress increase on the outside of the flange without any notch. All the reduction factors refer to the maximum nominal bending stress of the girder, *i.e.* the fatigue strength of the girder without stiffener.

The following general conclusions can be drawn for design and dimensioning from the lowest (tensile) reduction factor in each case of Table 21. The transverse stiffener is associated with a reduction $\gamma = 0.55$-0.73 for pulsating load (compared with $\gamma = 0.25$-0.75 for alternating load[248]). The high values can only be put into effect in the case of pure bending if measures are taken to improve the weld toe *e.g.* by grinding or dressing (in this case even $\gamma > 0.73$ seems possible for pure bending) or if the transverse forces are relatively high (with crack initiation on the outside of the flange). Deep penetration welding or double bevel butt welds are recommended to prevent crack initiation at the weld root in the case of transverse bending. Cutting out the inner stiffener corners improves the notch conditions considerably in the case of high transverse forces. Cutting out is also recommended because of the excessive gap and the limited accessibility at the inner corner. The tension plate with transverse stiffener can only be correlated to pure bending loading of the girder.

Table 21. Fatigue strength reduction factors, γ_P, (lower limit, evaluation of tensile notch stresses only) for I girder with transverse stiffener; design and load variants; results from calculation

				Reduction factors		
Component	Stiffening	Type of weld	Loading	γ_f	γ_t	γ_r
I girder	None	None	Pure bending	1.02	–	–
I girder	Transverse stiffener without cutout	Fillet weld	Pure bending	0.95	0.55	0.64
I girder	Transverse stiffener without cutout	Fillet weld	Transverse bending	0.74	0.71	0.57
I girder	Transverse stiffener with cutout	Fillet weld	Transverse bending	0.73	1.08	0.75
I girder	Transverse stiffener with cutout	Double bevel butt weld	Transverse bending	0.68	0.96	1.21
Tension plate	Transverse stiffener	Fillet weld	Tension	1.00	0.51	0.71
Tension plate	Transverse stiffener	Double bevel butt weld	Tension	1.00	0.52	0.78

The above reduction factors correspond with comparable minimum values in the German standard for crane design, DIN 15 018,[152] see Ref.248. In the IIW design recommendation,[141] the range of stress $\Delta\sigma = 80$ N/mm² with N = 2×10^6 (number of load cycles endured) is assigned to the transverse stiffener on girders as well as on tensile plates; the failure probability is $P_f = 0.023$, the reduction due to residual stresses has been taken into account. On the other hand, with the reduction factor $\gamma_P = 0.55$ obtained by the investigation described above, after conversion from $P_f = 0.10$ to $P_f = 0.023$, and after adjustment for residual stresses, the range of stress corresponds quite well to the IIW design recommendation, see Petershagen.[229]

Stress relief groove in welded pressure vessels

In the simplest case, welded pressure vessels are composed of a cylindrical shell and flat ends, Fig.193. The groove at the inner corner of the vessel is intended to facilitate welding and to relieve the notch stresses at the weld root which are caused by internal pressure loading of the vessel. The questionable relief effect of the groove was investigated by a notch stress analysis, see Ref.249. The predominant loading in the transition area between bottom and shell is meridional bending so that comparative investigations can be performed on the basis of this loading alone. There is no finite element analysis necessary in the considered case.

The transverse contour model for the notch stress analysis is shown in Fig.194. The groove dimensions radius, ρ_g, and depth, d_g, are variable, the radius, ρ_r, of the fictitious rounding of the weld root is constant, $\rho_r = 1$mm. The notch stress distribution for a typical parameter combination is shown in Fig.195. The fatigue notch factors, K_{fr}, and K_{fg}, are referenced to the nominal bending stress in the unnotched shell section. The two notch effects seem to be rather uncoupled. The higher notch effect occurs at the weld root notch. The result of a parametric investigation is shown in Fig.196. There is only a minor influence of ρ_g/t_{bo} on K_{fr} for $d_g = \rho_g$. A further small reduction of K_{fr} is possible

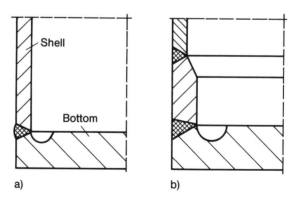

a)   b)

193 Welded joint between end and shell of vessels; stress relief groove in bottom plate; after Neumann.[3]

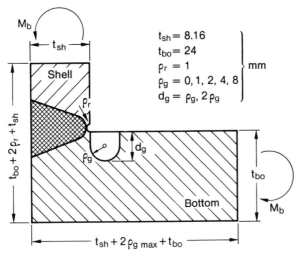

194 Transverse contour model of end-to-shell joint for parametric notch stress investigation.

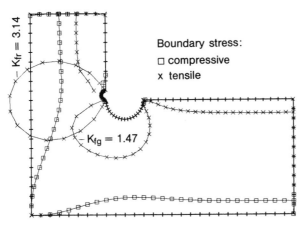

195 Boundary stresses for contour model with relief groove: model subjected to bending moment; dimensions t_{sh} = 16mm, t_{bo} = 24mm, ρ_r = 1mm, ρ_g = 4mm, d_g = 4mm.

by deepening the semi-circular groove into a U groove with d_g = $2\rho_g$. The maximum total improvement compared with the design version without a groove (K_{fr} = 3.40) is about 15%. The decrease of K_{fg} with ρ_g/t_{bo}, on the other hand, is of no practical value because K_{fr} is substantially higher.

The conclusion is, that the notch stress relief at the weld root achieved by the groove is low and presumably not worth the expenditure in manufacturing. A far better relief is possible by a (machined) concave fillet weld (with or without a root gap) in the inner corner of the vessel, K_f = 1.37, Fig.197 and 198. Even for an unmachined flat fillet weld, the stress relief is remarkable, $K_f \approx 2.50$, Fig.199.

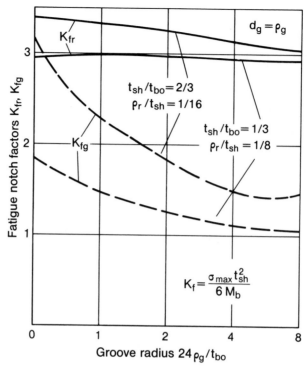

196 Fatigue notch factors for weld root and relief groove, K_{fr} and K_{fg}, as a function of groove radius, ρ_g, for two ratios, t_{sh}/t_{bo}.

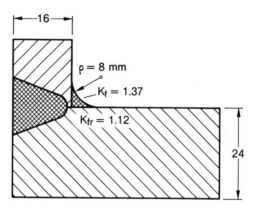

197 Alternative design for end-to-shell joint, fillet weld at inner corner of vessel.

The fatigue notch factors $K_f = 1.37\text{-}3.40$, which have been determined, can be compared with the weld correction factors, f_k, of the German design code for pressure vessels, AD-S1.[160] The factors stated here for low strength structural steels $(R_m = 400\,\text{N/mm}^2)$ and fatigue strength for infinite life $(N = 2 \times 10^6)$ are as follows: $f_{k0} = 1.4$ for the sheet with rolling skin (notch case K0), $f_{k1} = 1.8$ for a

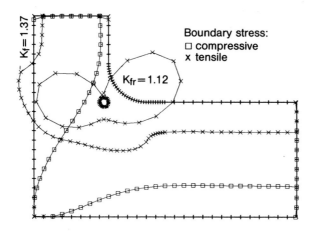

198 Boundary stresses for contour model with concave fillet weld with root gap; dimensions t_{sh} = 16mm, t_{bo} = 24mm, ρ_r = 1mm, gap length 2mm.

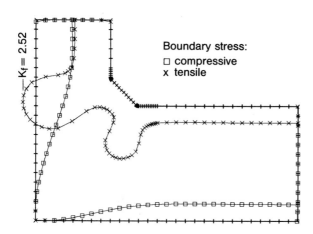

199 Boundary stresses for contour model with flat fillet weld without root gap; dimensions t_{sh} = 16mm, t_{bo} = 24mm, ρ_r = 1mm, weld thickness 4mm.

relatively weak notch effect (notch case K1), f_{k2} = 2.1 for a medium notch effect (notch case K2) and f_{k3} = 2.8 for a relatively strong notch effect (notch case K3).

The most favourable end-to-shell joint of those examined is thus in the strength range of the sheet material with rolling skin (f_{k0} = 1.4 for this case is set relatively high, though), the least favourable is clearly poorer than notch class K3. It should be noted with this allocation that the bending stress in the end-to-shell transition has to be determined as the reference stress. The K_f values are higher when the nominal membrane stress in the vessel shell is taken as the reference stress.

End-to-shell joint of a boiler

The notch stress approach for the assessment of the fatigue strength of welded structures has so far been represented as a procedure purely based on numerical analysis. A procedure based solely on numerical analysis provides a decisive advantage for practical application as the strength of the structure can be assessed at the design stage. Sometimes, however, a procedure based on measurement or on a combination of calculation and measurement is appropriate, especially if the approach based on measurement is cheaper, faster and more accurate.

The combination of structural stress measurement, notch stress calculation and fatigue strength assessment via notch classes was demonstrated for a three pass boiler end designed for heating of water.[250] The joint between flat end and cylindrical shell and the appertaining diagonal stay joint in the inside vertex area of the boiler which are considered to be at risk of crack initiation were analysed assuming internal pressure loading of the boiler.

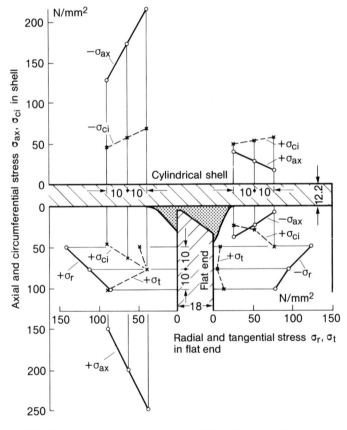

200 Structural stresses in end-to-shell joint between the upper diagonal stays, internal boiler pressure p = 10.2 bar; indexing: ci, circumferential; ax, axial; t, tangential; r, radial.

By means of strain gauges, the structural stresses were measured at three points in each case in front of the weld seam on the inside and outside of the end and the shell, Fig.200. Subsequently the complete internal force state was determined which is required for the notch stress analysis in the transverse contour model of the welded joint (tensile force and bending moment from the stresses directly, transverse force from the gradient of the stresses). It is a remarkable fact that the maximum structural stresses did not occur at the (well designed) ends of the diagonal stays, Fig.201.

On the basis of the fatigue notch factors, K_f = 2.3 for the end-to-shell joint and K_f = 2.9 and 3.9 for the diagonal stay-to-shell joint, the appropriate notch classes were determined within the framework of the (low cycle) fatigue strength assessment in accordance with the German codes TRB-AD S2[161] (Technical Specifications for Pressure Vessels) and TRD 301[162] (Technical Specifications for Steam Boilers). The permissible number of load cycles was indicated on that basis. Moreover, from the structural and notch stress analyses, advice concerning design optimisation was given.

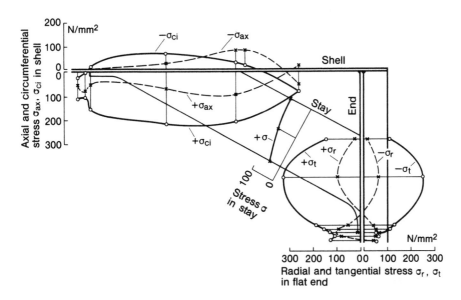

201 Structural stresses at diagonal stay, internal boiler pressure p = 10.2 bar; indexing: ci, circumferential; ax, axial; t, tangential; r, radial.

8.2.11 *Superimposition of structural and notch stress concentration factor for cover plates and edge reinforcement*

The issue of the notch effect of a welded joint in an oblique position with regard to a biaxial basic stress state occurs in practice, as mentioned, particularly in the analysis of plate and shell structures including thin walled bars, in which the actually extended welding joint is reduced to a joint line to simplify the model.

The internal forces in the joint line, of the bar, plate or shell structure form the basis of the subsequent consideration of the notch stress in the oblique notch.

Two specific classes of welded joints, which are important in practice, namely the cover plate joints and the edge-reinforced cutouts (welded joint between edge reinforcement and cutout edge) including nozzles and manholes, etc, can be dealt with advantageously via a multiplicative superimposition of the stress concentration factors (or of the fatigue notch factors when taking the microstructural support effect into account). The difference from the common plate and shell structural analysis is of a methodological and formal type. As regards method, these two types of welded joints can be considered as plane notch stress problems of the conventional type, the coverplate joint (plate thickness, t, total thickness in the cover plate region, t′) as an inclusion or core of increased rigidity (elastic modulus of the plate, E_p, elastic modulus of the core, E_c):

$$\frac{E_P}{E_c} = \frac{t}{t'} \tag{98}$$

The edge reinforced cutout on the other hand, can be considered as a plate cutout with a tension-stiff and possibly also bending-stiff edge (detailed representation in Ref.209, *ibid*, pages 91-109). The two above-mentioned notch stress problems can be formulated on the basis of functional analysis and solved numerically. This results in the stress concentration factors, K_{s1} and K_{s2}, (maximum stresses at edge of core or cutout, σ_{s1max} and σ_{s2max}, *i.e.* structural stresses in the sense of the superimposition, nominal plate stress, σ_n):

$$K_{s1} = \frac{\sigma_{s1max}}{\sigma_n} \tag{99}$$

$$K_{s2} = \frac{\sigma_{s2max}}{\sigma_n} \tag{100}$$

The structural stress concentration factors, K_{s1} and K_{s2}, are to be found in Ref.209 in diagrams or tables numerically evaluated for easy access in a large number of practically important shape and loading cases. The use of non-dimensional stress concentration factors instead of edge stresses with a dimension constitutes the above-mentioned formal difference.

At the edge point of K_{s1} and K_{s2}, the notch effect of the oblique welded seam in accordance with equations (96) and (97) is to be superimposed in which procedure the notch stress concentration factors K_{k1} and K_{k2} are used (total maximum notch stresses σ_{1max} and σ_{2max} at the cover plate or at the edge reinforcement):

$$K_{k1} = \frac{\sigma_{1max}}{\sigma_{s1max}} = 1 + (K_{k\perp} - 1)(\sin\psi + \nu\lambda\cos\psi) \tag{101}$$

$$K_{k2} = \frac{\sigma_{2max}}{\sigma_{s2max}} = \frac{1}{\lambda}\frac{\sigma_{2max}}{\sigma_{s1max}} = 1 + (K_{k\perp} - 1)(\cos\psi + \frac{\nu}{\lambda}\sin\psi) \tag{102}$$

It follows from the definition equations for K_{s1} and K_{k1} or K_{s2} and K_{k2} equations (99) to (102) for the total stress concentration factors K_1 and K_2:

$$K_1 = \frac{\sigma_{1max}}{\sigma_n} = K_{s1}K_{k1} \tag{103}$$

$$K_2 = \frac{\sigma_{2max}}{\sigma_n} = K_{s2}K_{k2} \tag{104}$$

The procedure of superimposing the stress concentration factors multiplicatively is illustrated in Fig.202 using the example of the tensile plate with square protuberances rounded at the corners and transitions (external shape of cover plates with concave fillet weld), on top of the component, on the left the elastic core, on the right the oblique notch, the cross section of which is taken from the component and the oblique position and loading parameters (including direction) of which follow from the stress analysis for the elastic core.

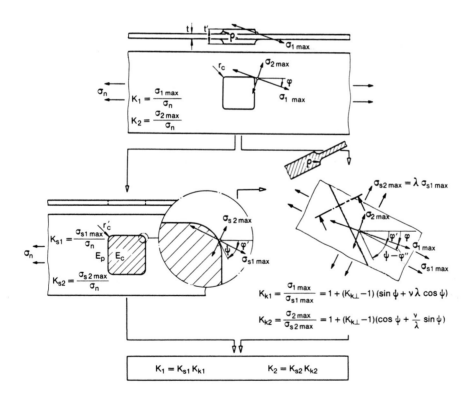

202 Multiplicative superimposition of notch effects for tensile plate with square protuberances (external shape of cover plates with concave fillet welds).

Here too, if the core or the reinforced cutout has sharp corners or transitions, this causes stress singularities. Stress evaluation at the toe notch line or alternatively expansion of the core or the reinforcement to this notch line with rounding of the corners leads to realistic stress values.

8.3   Elastic-plastic notch effect of welded joints with regard to fatigue strength for finite life and service fatigue strength

8.3.1   *Assessment of notch strain with regard to high cycle fatigue strength*

The assessment of notch strain for welded joints with regard to high cycle fatigue strength (including the endurance limit) on the basis of strains measured with a strain gauge in front of or directly on the weld toe is well proved. This approach is close to the assessment of structural stress but does not completely neglect the actual notch effect of the weld toe.

In the case of transversely loaded welded joints of various shapes subjected to different load cases, Haibach[53,251] has established a sufficiently safe averaged notch strain amplitude, which was measured immediately in front of or directly on the toe notch with a 3mm strain gauge at approximately 2×10^5 cycles to failure, Fig.203, and which was converted quasi-elastically via the normalised S-N curve to 2×10^6 cycles. The endurance limit in terms of the (elastic) strain amplitude of the strain gauge measured on welded joints, made of structural steel St 37 and St 52, is 525 µm/m (*i.e.* 0.05%), which corresponds to a uniaxial stress of 105 N/mm². The preconditions are as follows: plate thickness 10mm, alternating fatigue loading, joints relieved from residual stress and fracture initiation at the toe notch.

In a subsequent investigation[38] in addition to smaller and larger weld and plate thicknesses, pulsating fatigue strength and as-welded joints were also covered. The endurance limit for the strain amplitude lay between 300 and 700 µm/m. The thinner weld (combined with thinner plates) endured the higher

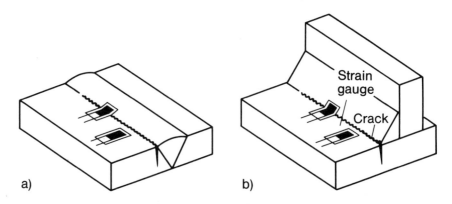

a)   b)

203 Strain gauges at weld toe of butt weld (a) and fillet weld (b), after Gassner and Haibach.[251]

strain amplitudes. The residual stresses had a strength reducing effect (reduction factor 0.7) particularly for alternating loading. In the case of pulsating loading the residual stresses were partly relieved by plastic deformation. The influence of mean strain on fatigue strength is characterised by a ratio of alternating to pulsating strain amplitude of 1.13-1.60.

An endurance limit $(N = 2 \times 10^6$ cycles) of the alternating strain amplitude of 1300 μm/m was determined for the hardenable aluminium alloy AlZnMg1, see Ref.252.

The endurable strain amplitudes stated above for $N = 2 \times 10^6$ cycles can be converted (quasi-elastically) to strain amplitudes endurable in the high cycle fatigue range via the normalised S-N curve (see Fig.40 or 74 or equation (16)).

The assessment of notch strain by measurement presupposes that the welded joint to be considered is already manufactured. However, the endurable strain amplitudes can also be used in connection with results from finite element analysis.

The graphic separation into primary and secondary notch effect, as in Fig.204 is characteristic of the explanatory considerations on notch effect of welded joints in Ref.38, 251. The 'deep' (in Neuber's definition) primary notch is said to have a strain concentration factor of up to 2.0, the 'shallow' secondary notch according to strain measurements should have similar values. The total stress concentration factor follows via multiplicative superimposition of the two stress concentration factors just mentioned. This consideration can only be used for a qualitative explanation of trends but not for a quantitative prognosis.

The correlation between the locally measured averaged strain and the maximum notch stress has been established by Olivier, Köttgen and Seeger.[231] The strain gauge with a gauge length of 3mm was applied with its centre 2-2.5mm in front of the fusion line of the weld toe. Strain gauges of shorter gauge length turned out to be less suitable. The local fatigue strength

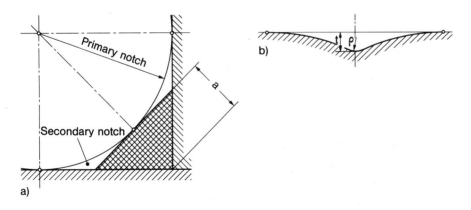

204 Primary and secondary notch of fillet weld (a), unwinding of secondary notch (b), after Gassner and Haibach.[251]

(endurance limit) of the structural steel St 52-3, stated as endured strain amplitude $\bar{\varepsilon}_A$ = 525 μm/m for N = 2×10^6 cycles,[53,251] was converted to the maximum notch stress for ρ_f = 1.0mm on the basis of numerical notch stress analysis. The conversion factor relating the measured strain, $\bar{\varepsilon}$, to the maximum notch stress, $\bar{\sigma}_k$, depends on the plate thickness (and perhaps on the type of loading), Fig.205. The uniform result in Ref.231 was $\bar{\sigma}_{A0} \approx$ 235 N/mm² for P_f = 0.5.

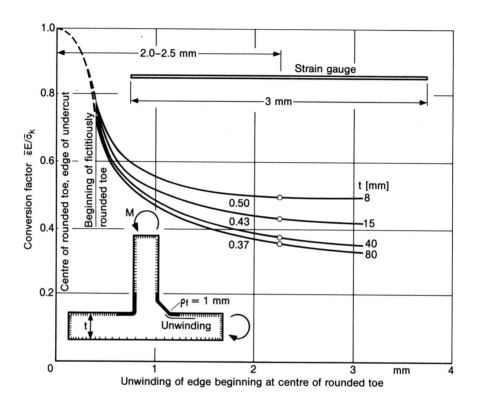

205 Conversion of strain, $\bar{\varepsilon}$, measured in front of weld into maximum notch stress, $\bar{\sigma}_k$, at the fictitiously rounded weld toe after Olivier, Köttgen and Seeger.[231]

8.3.2   Assessment of notch strain with regard to low cycle fatigue strength, 'simple' notch root approach

The assessment of notch strain with regard to low cycle fatigue strength (also to be carried out at increased temperatures in particular) is characterised by the determination of the complete elastic-plastic notch stress and strain state. The strains are combined into the total strain and compared with endurable values determined from unnotched specimens dependent on number of cycles and mean stress but independent of the mean strain. The high cycle fatigue strength is also consistently covered by this approach. The assumption that to begin with, in the presence of constant amplitude loading, the material at the notch root behaves locally just as it does globally in a small unnotched tensile specimen is termed the 'simple' notch root approach.[254-260]

The total strain at the notch root can be determined either by numerical analysis or by measurement. According to Neuber[214] once the yield limit, σ_Y, has been exceeded locally, the macrostructural support formula

$$K_k^2 = K_\sigma K_\varepsilon \tag{105}$$

is applicable with notch stress concentration factor in the elastic state, K_k, stress concentration factor in the elastic-plastic state, K_σ, and notch strain concentration factor in the elastic-plastic state, K_ε.

In corresponding equations by other authors (*e.g.* Seeger, Dietmann and Saal; Stowell, Hardrath and Ohmann), which are more precise under certain conditions, the ultimate load stress concentration factor, K_U, is introduced in addition to K_k, see Ref.255.

To determine stress and strain at the notch root the respective relation between stress and strain must also be known. In the simplest case (as regards the formula) of hardening being completely disregarded, ($\sigma_{kmax} = \sigma_Y$) the maximum notch strain, ε_{kmax}, is directly derived from equation (105):

$$\varepsilon_{kmax} = \frac{\sigma_n \, \sigma_n}{E \, \sigma_Y} K_k^2 \tag{106}$$

However, in the low cycle fatigue range in particular, hardening or softening caused by cyclic loading are to be taken into account, *i.e.* the hysteresis loops according to Fig.206a (in the small tensile specimen just as at the notch root), via the relevant cyclic stress-strain curve, Fig.206b which can deviate considerably from the monotonic curve, Fig.206c. The cyclic stress-strain curve is described by

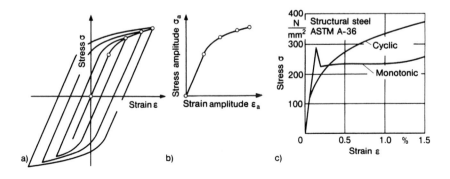

206 Cyclic loading: hysteresis loop (a), cyclic stress-strain curve (b) and comparison with monotonic stress-strain curve (c).

$$\frac{\Delta\epsilon}{2} = \frac{\Delta\epsilon_{el}}{2} + \frac{\Delta\epsilon_{pl}}{2} = \frac{\Delta\sigma}{2E} + \left(\frac{\Delta\sigma}{2K'}\right)^{1/n'} \tag{107}$$

with total strain range, $\Delta\epsilon$, elastic strain range, $\Delta\epsilon_{el}$, plastic strain range, $\Delta\epsilon_{pl}$, stress range, $\Delta\sigma$, elastic modulus, E, cyclic hardening modulus, K', and cyclic hardening exponent, n'.

The cyclic stress-strain curve is determined using unnotched tensile specimens in alternating fatigue tests with strain control. It represents the connecting line of the maxima and minima of the stabilised hysteresis loops. As the determination of the cyclic stress-strain curve is rather expensive when performed in a series of constant amplitude tests, other loading sequences are applied, which determine the stabilised curve in a shortened procedure. The most widespread method is the incremental step test, Fig.207.

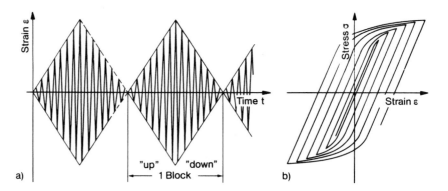

207 Incremental step test, strain cycles (a) and hysteresis loops (b).

The (temperature-dependent) cyclic stress-strain curve is considered to be largely independent of mean stress and mean strain. However, cyclic relaxation of mean stress does occur when strain is controlled as does cyclic creep of mean strain when stress is controlled, Fig.208.

The elastic-plastic notch strain, ϵ_{kmax}, causes crack initiation to the extent expressed by the endured strain range dependent on number of cycles, $\Delta\epsilon = 2\epsilon_A$, determined using unnotched tensile specimens in strain-controlled fatigue tests. The strain occurring in the tensile specimen as at the notch root can be expressed for $N \geqslant 10$, subdivided into elastic and plastic proportions, by the 'four parameter equation' according to Morrow and Manson (the strain S-N curve), Fig.209.

$$\frac{\Delta\epsilon}{2} = \frac{\Delta\epsilon_{el}}{2} + \frac{\Delta\epsilon_{pl}}{2} = \frac{\sigma'_U - \sigma_m}{E}(2N)^b + \epsilon'_U(2N)^c \tag{108}$$

The four parameters in the equation are fatigue strength coefficient, σ'_U; cyclic ductility coefficient, ϵ'_U; fatigue strength exponent, b; and cyclic ductility exponent, c. There is the additional influence of the mean stress, σ_m, and N is the number of cycles to crack initiation.

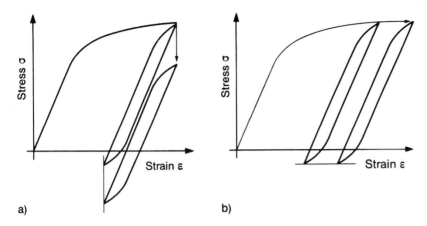

208 Cyclic mean stress relaxation (a) and cyclic creep (b).

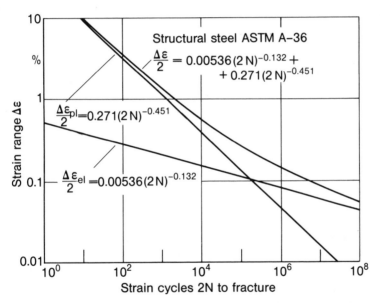

209 Strain S-N curve with elastic and plastic proportion, after Mattos and
Lawrence.[234]

The first part of equation (108) corresponds to Basquin's equation for the
S-N curve, the second part to Manson, Coffin and Tavernelli's low cycle fatigue
law (with $c \approx -0.5$). Equating σ'_U and ϵ'_U with the ultimate values σ_U and ϵ_U of
the static tensile test ($N = 0.5$) is incorrect. Whilst the mean stress, σ_m, is a preset
quantity at the tensile specimen, it is dependent on the loading history at the
notch root, therefore in the constant amplitude test, particularly on the loading
build-up to full amplitude.

The representation of the S-N curve after Basquin is identical to equation
(16). The representation of the plastic proportion after Manson,

Coffin and Tavernelli is known in a simplified form as the low cycle fatigue law both for the notch root and for the unnotched tensile specimen:

$$N\Delta\epsilon^2 = \ln\frac{1}{1-\psi} \tag{109}$$

There is a lack of uniformity with regard to the meaning of $\Delta\epsilon$. It is defined partially as the plastic strain range and partly as the total strain range. The reduction in cross sectional area at rupture, ψ, is introduced in place of the (true) strain at failure, ϵ_U, which can be converted via the condition of constant volume in the contracting region (specimen length, l, cross sectional area, A):

$$\epsilon_U = \ln\frac{l_U}{l_0} \tag{110}$$

$$\psi = 1 - \frac{A_U}{A_0} \tag{111}$$

$$l_U A_U = l_0 A_0 \tag{112}$$

The relationship originally suggested by Manson contains the ultimate values of stress and strain in the static tensile test explicitly:

$$\Delta\epsilon = 3.5\frac{\sigma_U}{E}N^{-0.12} + \epsilon_U^{0.6}N^{-0.6} \tag{113}$$

The material characteristic values of the equations (107) and (108) are compiled for some parent materials used in welded construction in Table 22. The presented parameters σ_U', b, ϵ_U', c of the 'four' parameter equation are useless as individual values because they scatter greatly as a consequence of the margin for adjustment, and are therefore not material characteristic values in the strict sense. Further (more incomplete) data sets are to be found in papers by Heuler[257] and Beste[255] (StE 690, X10 CrNiTi 18.9, 49 MnVS 3, AlMg 4.5 Mn) and by Klee[253] and Hanschmann[254] (AlCuMg 2, AlSi 7 Mg). Comprehensive compilations of data on cyclic material behaviour have been published by Boller and Seeger[236] and Boyer.[237] Databanks of this type have been available to the American automobile industry for many years, based particularly on contributions by Landgraf.[236] The characteristic values of the heat affected zone and the filler material may deviate from the values for the parent material dependent particularly on the hardening state.

The results of a (high cycle) fatigue strength calculation up to crack initiation using the parent material characteristic values is shown in Fig.210 for the butt weld, reinforced on one side, (plate thickness 25.4mm, groove angle 90°, slope angle of reinforcement 60°, toe radius 1.3mm). The proportion of crack initiation phase compared with total life was established on the basis of a supplementary fracture-mechanical analysis of the crack propagation process up to final fracture, Fig.211. The results of the calculation up to crack initiation

Table 22. Material characteristic values for the 'simple' notch root approach (σ'_v, b, ϵ'_U, c from strain controlled tests, $\sigma'_{0.2}$ refers to the cyclic $\sigma\epsilon$-curve)

Material		Source	E 10^3 N/mm²	$\sigma_{0.2}$ N/mm²	σ_U N/mm²	$\sigma'_{0.2}$ N/mm²	K' N/mm²	n'	σ'_U N/mm²	b	ϵ'_U	c
Structural steel	St 37	Ref.253	214	295	435	273	988	0.207	873	−0.100	0.557	−0.518
Structural steel	St 52		210	400	597	389	1228	0.185	1105	−0.098	0.621	−0.541
Structural steel	BHW 25	Ref.260	209	460	614	424	1220	0.170	966	−0.083	0.253	−0.484
Structural steel	StE 320		209	375	558	368	1033	0.166	813	−0.078	0.237	−0.466
Structural steel (high tensile)	StE 690	Ref.256	214	736	798		1163	0.127	1096	−0.078	0.626	−0.615
Casting steel (high ductile)	13 MnNi 63		203	312	501		896	0.148	654	−0.078	0.118	−0.529
Titanium alloy	Ti 6 Al 4 V		120.4	1006	1034		1898	0.146	2201	−0.126	2.819	−0.861
Aluminium alloy	AlCuMg 2		74.5	378	486		625	0.051	973	−0.107	0.286	−0.610
Aluminium alloy	AlZnMgCu 1.5		71.5	382	462		1162	0.194	989	−0.140	0.433	−0.724
Aluminium alloy	AlMg 4.5 Mn		72	293	363		546	0.074	707	−0.106	0.731	−0.839
Aluminium alloy	AlZnMg 1		70	280	360		689	0.127	646	−0.093	0.607	−0.733
Structural steel (high tensile)	HY 130	Ref.234					1515	0.100	1488	−0.060	0.90	−0.64
Structural steel (high tensile)	HY 80						1309	0.146	1350	−0.096	0.89	−0.62
Structural steel (low tensile)	A 36		190	224	413	232	1089	0.249	1013	−0.132	0.27	−0.45
Aluminium alloy	7075-T6						730	0.049	1847	−0.172	1.65	−1.29
Structural steel (high tensile)	A 514	Ref.235	209	890	938	604	1090	0.091	1304	−0.079	0.975	−0.699
Aluminium alloy	5083		71	131	300	292	594	0.114	725	−0.122	0.405	−0.692

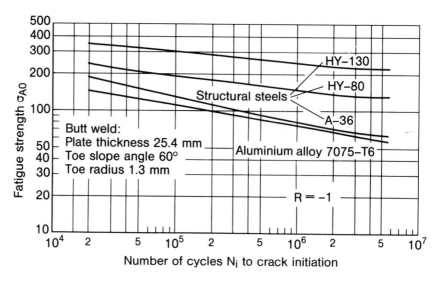

210 Calculated S-N curves for crack initiation of butt welds of identical shape, various structural steels and one aluminium alloy, after Mattos and Lawrence.[234]

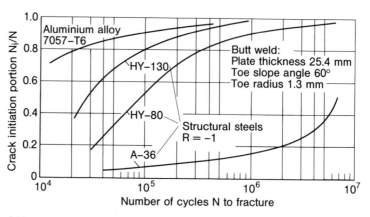

211 Calculated ratio of crack initiation cycles to total life cycles for butt welds of identical shape, various structural steels and one aluminium alloy, after Mattos and Lawrence.[234]

were checked by fatigue testing of welded specimens made of structural steel A 36. The crack depth of 0.25mm, which can be measured reliably using a strain gauge, was evaluated as the end of the crack initiation phase. It leads to complete fracture in the small unnotched tensile specimen very quickly, whereas there is a relatively long crack propagation phase in the larger welded specimen with the notches. The relatively long crack initiation phase for high tensile steels is in contradiction to the statements at the end of section 4.1.

8.3.3   Assessment of notch damage with regard to service fatigue strength, 'extended' notch root approach

In questions of service fatigue strength (*i.e.* load amplitudes and mean load are variable) of notched structural components tracking the notch root stresses and strains as illustrated in Fig.212 and determining the relevant local damaging from these is essential for a realistic prediction of the life of the component up to crack initiation. The nominal stress approach, combined with the linear damage accumulation hypothesis, fails in this case and even improved hypotheses including load sequence effects or the relative Miner rule only solve the problem in individual cases.

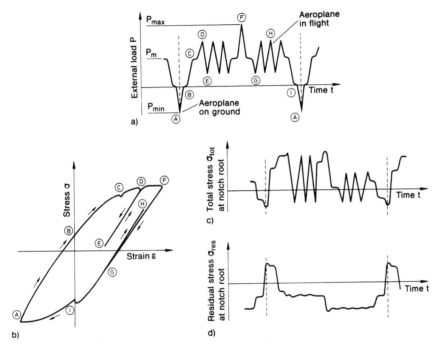

212 Stress sequence at notch root connected with strong mean stress variations (underside of aircraft wing subjected to service load cycles); load sequence (a), stress-strain hysteresis loop (b), notch stress sequence (c), residual notch stress sequence (d), after Haibach, Schütz and Svenson.

The assumption that for any load sequence the material behaves locally at the notch root just as it does globally in a small tensile specimen, is termed the 'extended' notch root approach.[254-260] The basic pattern of the procedure is shown in Fig.213. In extension of the simple notch root approach the damage behaviour is to be considered quantitatively. The cyclic stress-strain curve must be evaluated in a more differentiated form. Moreover the issue of transferability to the structural component is to be clarified.

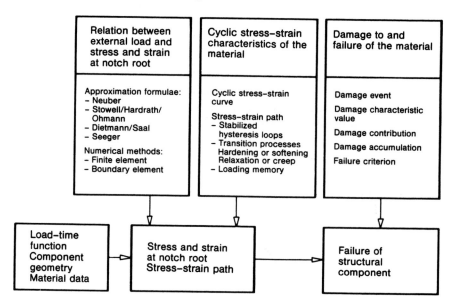

213 Basic pattern of the calculation of life up to crack initiation for notched specimens and components in accordance with the 'extended' notch root approach, after Beste.[255]

The purely experimental procedure in accordance with Fig.214 is called the 'companion specimen method'. The total strain sequence measured uniaxially at the notch root for the loading sequence of the notched specimen or component is transferred to the considerably smaller unnotched companion specimen up to crack initiation. Thus the companion specimen simulates the damage effects at the notch root and accumulates them. The notch stresses, which cannot be measured directly at the notch root, result as a by-product in this procedure. A process computer generates the load sequence on the notched specimen and monitors the control processes.

The combined numerical-experimental procedure in accordance with Fig.215 is termed the 'Neuber control method' in its simpler form or 'notch strain simulation' in its more refined form. The stresses and strains at the notch root are calculated in accordance with Neuber's formula, equation (105) including the 'loading memory', on the basis of the load sequence so that the notched specimen with its testing machinery can be omitted. The small tensile specimen simulates and accumulates the notch root damage, so that the most problematic part of the purely numerical analysis below is avoided.

The purely numerical procedure follows the basic pattern shown in Fig.213. The stresses and strains at the notch root are calculated with sufficient accuracy using Neuber's formula or another approximation formula, if necessary also using finite element methods. The cyclic stress-strain curve needed for this, equation (107), is determined for the increasing and decreasing load amplitude sequence, Fig.207, of the incremental step test which corresponds to the mixture of load amplitudes in reality. The hysteresis loops are represented as regards their shape by an approximation in the form of the cyclic tension or

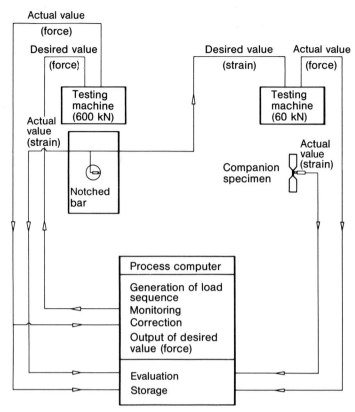

214 Companion specimen method for determining the life up to crack initiation of notched specimens and components, after Heuler.[257]

compression curve with doubled amplitudes. Further approximation equations are necessary for cyclic relaxation and cyclic creep, which provide the cyclically produced change in mean stress or mean strain dependent on amplitude value and corresponding number of cycles. The loading memory consists in the fact that when a smaller hysteresis loop has been passed through and a return is then made to the larger loop, the latter is continued and not the former, Fig.216.

The damage and failure behaviour (surface cracking at the notch root up to a width of 0.5mm can be considered as failure), which is described completely for constant amplitude loading by the four parameter equation (108) of the strain S-N curve, needs to be supplemented here to take into account the influence of mean stress (cycle by cycle). The damage is considered to be defined by σ_a, ϵ_a and σ_m (*i.e.* without ϵ_m). The most frequently used damage parameter, P, after Smith, Watson and Topper states:

$$P_{SWT} = \sqrt{(\sigma_a + \sigma_m)\epsilon_a E} \tag{114}$$

Another suggestion, derived by Haibach and Lehrke with the assumption of crack-like defects (designations in accordance with Fig.217) states:

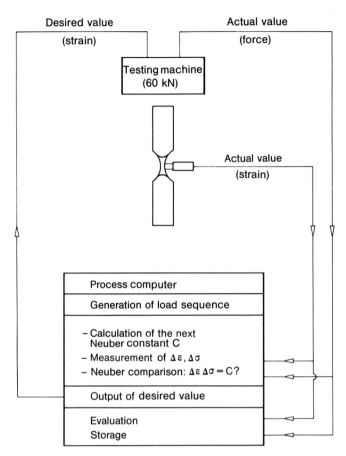

215 Neuber control method for determining the life up to crack initiation of notched specimens and components, after Heuler.[257]

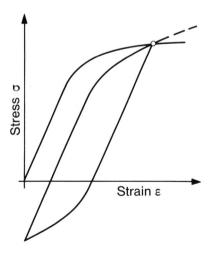

216 Loading memory of the material.

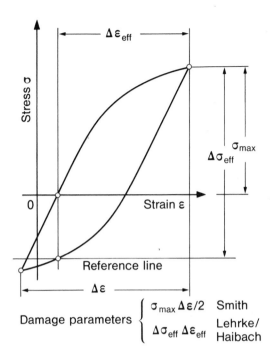

217 Damage parameters derived from hysteresis loop.

$$P_{HL} = \Delta\sigma_{eff}\,\Delta\epsilon_{eff} \tag{115}$$

Correcting terms may be inserted in the above equations to take account of the sequence effect, which already occurs for the unnotched specimen. By means of the damage parameter it is possible to reduce numbers of cycles with any mean stress to numbers of cycles with the same damage based on zero mean stress. It is possible to derive damage S-N curves from the strain S-N curves (endurable damage parameters dependent on the number of cycles) and to compare the damage produced by the more general loading sequence with them. If statistical methods are used for counting the cyclic stress amplitudes, the variable mean stress is also to be evaluated. Therefore, range pair plus mean value counting and rain flow counting (see section 3.2.2) are suitable here. How precisely the desired number of closed hysteresis loops is covered by these counting methods, is not yet definitely clarified.

Finally, within the framework of the notch damage assessment the problem of transference is to be noted, *i.e.* the possible deviation of the behaviour of the notch root in the specimen or structural component from the behaviour of the companion specimen:

— caused by manufacture *i.e.* by different material production processes;
— caused by notch shape *i.e.* by differing stress gradients normal to the notch root;

—   caused by statistical microstructural defect distribution *i.e.* by differing material volumes;
—   caused by surface condition *i.e.* by differing hardening depths for example.

Having approximately the same material volumes at the notch root and in the tensile specimen is advantageous. In addition, a specimen with a shallow notch groove ($K_k \approx 1.5$) is recommended instead of the completely unnotched specimen.

Varying multi-axiality of the stress and strain state at the notch root has not yet been quantified in the assessment procedures. This represents a shortcoming with respect to application in practice. However, in cases of proportional (*i.e.* synchronous) multiaxial stresses it should be sufficient to work with the von Mises equivalent stress instead of the uniaxial stress and to proceed by analogy in the case of strain. The strength behaviour under non-proportional multiaxial stresses (caused by non-proportional multiple loading) is the object of actual research programmes.[258,259]

In the presence of increased temperatures, damage caused by creep is also to be taken into account.[57] A corresponding influence on service fatigue strength is observed from cyclic frequency, cyclic shape and operational breaks.

The extended notch root approach has been applied to assessment of life of welded joints, see Ref.260, 96, 97.

9

Fracture mechanics approach for assessment of fatigue strength of seam welded joints

9.1   Principles of the approach

The fracture processes in welded joints can be described (to a limited extent) from the point of view of fracture mechanics. The stress intensity factor (or corresponding energy parameters) at the crack or slit tip is used as the basis of this approach. The theoretical and practical principles of fracture mechanics are assumed to be known for the purposes of this chapter, see Ref.261-265. Reference 266 gives a summary of its application to welded joints. General information on fatigue crack growth is available from Ref.267-269.

The process of fatigue failure is subdivided into the stages of crack initiation, stable crack propagation and unstable final fracture. Crack initiation comprises the motion of dislocations in the slip planes with the formation of very small separations in the material on the slip bands in areas which are smaller than the grain size. The slip bands preferably occur where the stress is locally increased, *i.e.* at notches, defects, inclusions, cavities, and cracks (the latter require the realignment of the crack front perpendicular to the local principal stress direction). A microcrack capable of propagation is reached at the latest when the crack size more or less equals the grain size. Stable crack propagation comprises the cyclic growth of a microcrack to a macrocrack with dimensions in the order of magnitude of relevant component dimensions, *e.g.* the plate thickness. However, experience has shown that for the larger part of the service fatigue life the microcrack is relatively small. Finally, when the crack size has become large enough, instantaneous final fracture occurs.

The application of fracture mechanics to questions of fatigue strength is generally accepted where predictions of the behaviour of a detected crack-like defect, *e.g.* a surface crack with a depth of 0.25mm, is concerned. Fitness-for-purpose concepts can thus be substantiated as well as fail-safe concepts. Fracture mechanics is the essential basis for safety analyses and for back-tracing actual failures.

The fatigue life concept based on fracture mechanics, originally developed by Maddox[270-274] and taken up by Hobbacher,[293] Haibach[311] and Franke[296] is not generally accepted. According to this concept, the fatigue life of welded joints is exclusively propagation of very small cracks occurring in the mixed material zone of filler and parent materials at the toe and root of the weld seam, the propagation can be described on the basis of the crack intensity factor range, ΔK_I.

It can be demonstrated not only for the transition zones of welds but also for smooth specimens of base materials that approximately 90% of the crack initiation time to a crack depth of 1mm (which is an initial crack size relevant for the engineer) is spent on microcrack propagation.[275,276] However, the microcrack is embedded in a plastic zone so that the simple ΔK_I concept of crack propagation can no longer be applied. Crack propagation is described in this case more correctly by means of a modified and extended J integral which includes the cyclic elastic-plastic stress and strain and is called Z integral. By means of phenomenological investigations, however, it has been demonstrated that the microcrack propagation is highly influenced by microstructural inhomogeneities (*e.g.* hard inclusions, grain boundaries, surface roughness) so that the Z integral is not always valid.

In accordance with Ref.277, 278 as well, short cracks (crack length a = 0.25-1.00mm, plastic zone not small as compared with crack length) propagate in a way incompatible with the ΔK_I concept valid for large cracks. The deviations tend towards the unsafe side (da/dN is increased and ΔK_{th0} is reduced). Differing crack closing behaviour of short and long cracks is pointed out as being the reason for the deviations.

On the other hand, by means of the potential drop method with extremely high resolution, it has been shown[313-315] that crack propagation at the weld toe (from a crack depth of 0.17mm on and described by the ΔK_I concept) is only prevailing with high tensile mean stress.

The following can therefore be stated in conclusion with reference to the crack propagation concept of fatigue life:

—   the metal physicist's microcrack propagation is the main part of the engineer's macrocrack initiation. The engineer prefers S-N curves for macrocrack initiation from tests with constant or variable amplitude loading. For the engineer, the issue of microcrack propagation is thus a purely academic one;

—   the simple ΔK_I concept is not applicable to microcrack propagation or only if the microcrack is relatively large, and even the more sophisticated Z integral concept may be on the unsafe side. The interest of the engineer as regards microcracks is limited to questions of damage accumulation.

The fracture mechanics approach, independent of its degree of formalisation, complements the nominal stress, structural stress, and notch stress assessments, which are the only basis for dimensioning and design of welded joints, by an assessment of failure safety and service life on the basis of detected actual defects or undetected assumed crack-like defects.

9.2 Crack propagation equations

9.2.1 *Basic equations*

The typical curve of the crack propagation rate da/dN (limes of the ratio of the increase of crack length, Δa, per number of cycles, ΔN) for metallic materials, measured in specimens with macrocrack, dependent on the (constant) stress intensity factor range, ΔK, at the crack tip subjected to pulsating tensile load is shown in Fig.218. Here and in the following, ΔK is the stress intensity, ΔK_I in accordance with crack opening mode I, as fatigue cracks take such a direction and shape that mainly or exclusively ΔK_I occurs (in this case, the superimposed stress intensities ΔK_{II} and ΔK_{III} are negligible).

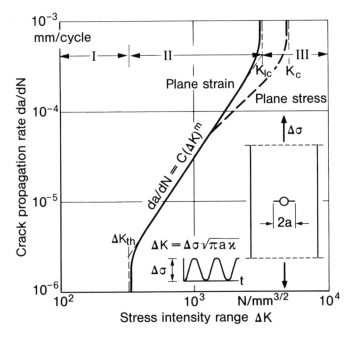

218 Crack propagation rate due to fatigue, dependence on the range of the crack tip stress intensity factor, numerical data typical of structural steel.

In the low stress intensity range (range I in Fig.218), the crack does not propagate at all:

$$\frac{da}{dN} = 0 \text{ for } \Delta K < \Delta K_{th} \tag{116}$$

The threshold stress intensity factor range, ΔK_{th} is the maximum range of the stress intensity factor which is endured for infinite life, *i.e.* the endurance limit of components with a crack. For structural steels without crack propagation $K_{th0} = 180 \text{ N/mm}^{3/2}$ for $R = 0$ can be generalised as the lower limit of scattering values. An approximation independent of materials is (with elastic modulus, E in N/mm^2):

$$\Delta K_{th0} = 0.5 - 1.5 \times 10^{-3} E \ N/mm^{3/2} \tag{117}$$

The influence of static mean stress on ΔK_{th} can be represented as follows (materials constant, m, from equation (119):

$$\Delta K_{th} = \Delta K_{th0} (1-R)^{1/m} \tag{118}$$

The limit stress intensity ratio, $R = K_1/K_u$, is the ratio of lower and upper stress intensity. ΔK_{th0} is the threshold stress intensity for pulsating loading ($R = 0$). With high (tensile) mean stress, ΔK_{th}, approximates zero according to equation (118), whereas with zero mean stress (alternating loading) it increases and reaches its maximum value with compressive mean stress.

In the medium stress intensity range (range II in Fig.218) the crack propagation rate, da/dN, is proportional to the stress intensity factor range, ΔK, with the material constant, m as exponent. This dependency is represented by a straight line in logarithmic scales, Fig.218; it is described by the equation of Paris and Erdogan:

$$\frac{da}{dN} = C(\Delta K)^m \ for \ \Delta K_{th} < \Delta K \ll K_{Ic} \tag{119}$$

In accordance with Hartman and Schijve, $(\Delta K)^m$-$(\Delta K_{th})^m$ instead of $(\Delta K)^m$ is introduced in equation (119). For structural steels, m = 2.0-3.6, for aluminium alloys, m $\approx$ 4.0. Typical values for da/dN in range II are 10^{-6}-10^{-3}mm per cycle.

For equation (119), the plane deformation state of sufficiently wide crack fronts is assumed for the total range up to final fracture. The transition to the plane stress state of thin plates is revealed in the crack propagation rate curve by a kink with approximation to K_c which is dependent on plate thickness (instead of K_{Ic} which is independent of plate thickness) in the further course of the curve. The total life results obtained using equation (119) are therefore on the safe side.

In the high stress intensity range (range III in Fig.218) crack growth is accelerated and finally an instantaneous fracture occurs as soon as the material's characteristic critical stress intensity factor, K_{Ic}, is reached. K_c is used instead of K_{Ic} for the plane stress state ahead of a through-crack in a thin plate.

The crack growth rate in range III (including range II) can be represented by a formula proposed by Forman, which also covers the influence of tensile mean stress deviating from the pulsating load case:

$$\frac{da}{dN} = \frac{C'(\Delta K)^m}{(1-R)K_{Ic} - \Delta K} \tag{120}$$

This equation can be simplified for microcracks introducing the condition $\Delta K \ll (1-R)K_{Ic}$, but the reservations with respect to the ΔK_1 concept stated in section 9.1 must be kept in mind:

$$\frac{da}{dN} = \frac{C(\Delta K)^m}{1-R} \tag{121}$$

With alternating and compressive pulsating load, crack closure increasingly takes place with corresponding crack growth retardation. With highly or frequently changing load amplitudes, strain hardening and residual stresses additionally occur at the crack tip, which may also have a retarding effect on crack growth. The above equations do not comprise this mutual influence of successive load amplitudes of differing height called 'interaction effect'. By means of appropriate more sophisticated hypotheses and concepts the effect can be simulated and calculated in advance, for example in accordance with Ref.284-286. In the case of constant amplitude loading, it is sufficient to introduce ΔK_{eff} instead of ΔK in the above equations. The effective stress intensity range simply disregards the negative proportion of the ΔK-values as related to crack closure. The crack propagation curve in Fig.218 is identical for all $R \neq 0$ values if ΔK is substituted by ΔK_{eff}.

If the crack propagates in a corrosive environment, this process is considerably accelerated due to electrochemical effects at the crack tip (stress corrosion and fatigue corrosion cracking). These phenomena are not included in the above equations. The mainly intergranular stress corrosion cracking occurs if the relatively low stress intensity factor, K_{Iscc} (scc = stress corrosion cracking), is exceeded, which happens quite frequently with welded joints due to the tensile residual stresses alone. The mainly transgranular fatigue corrosion cracking is characterised by the considerable acceleration of the fatigue crack growth, air as compared with a vacuum, humid air as compared with dry air, a saline solution as compared with pure water, hydrogen as compared with air. Nevertheless, at increased temperatures, crack propagation by fatigue and corrosion may be slower than crack propagation by creep (the fatigue strength for finite life of the component with a crack is then higher than its creep strength).

9.2.2   Cyclic range of stress intensity

The cyclic range of the stress intensity factor, ΔK, appearing in the above equations is dependent on the range of the basic stress, $\Delta \sigma$, and the crack length, a, (in the case of a through-crack) or the crack depth, a, (in the case of a surface or internal crack):

$$\Delta K = \Delta \sigma \sqrt{\pi a} \, \varkappa \tag{122}$$

The adjustment factors, $\varkappa$, for the details of geometry, loading and support in the crack environment are compiled in Ref.287-290. An approximation procedure to determine $\varkappa$ is provided in Ref.291. Surface cracks and internal cracks are of special significance in practical application; they are approximated as semi-elliptical or elliptical cracks for determining the stress intensity factors. In the case of the semi-elliptical surface crack at the root of the notch on a plate surface, $\varkappa$ is represented according to Ref.292 as follows (analogous formula for the elliptical internal crack):

$$\varkappa = \varkappa_s \varkappa_b \varkappa_f \varkappa_g K_k \tag{123}$$

The individual factors are assigned to the following influences on the crack tip stress intensity:[292]

x_s = 1.0-1.12 for a/c = 0-1.0, with crack width, 2c, at the crack mouth, adjustment for free surface of plate at the crack mouth;

x_b = 1.0-1.2 for a/t ⩽ 0.7 with plate thickness, t, adjustment for free backsurface opposite to the crack mouth;

x_f = 0.6-1.0 for a/c = 1.0-∞ with crack width, 2c, at the crack mouth, adjustment for curved crack front;

K_k = 1.0-10.0 notch stress concentration factor, normal range for welded joints (including the microstructural support effect);

x_g = 0.1-1.0 for K_k = 10.0-1.0, adjustment for stress gradient perpendicular to notch root from bending and notch effect.

In the case of elliptical cracks the designation M/φ was originally used instead of x, φ being an elliptical integral which depends on the dimensional ratio a/c. Instead of the notch stress concentration factor, K_k, the structural stress concentration factor, K_s, is also used (*e.g.* in the case of tube joints), which — as structural stress is defined as bending and membrane stress — results in more uniform x_g values related only to the effects of bending.

The typical curve of stress gradient adjustment times stress concentration versus crack length is shown in Fig.219 by the example of a tensile loaded notched bar. In the limit case of the microcrack at the notch root, provided it is small enough and the maximum notch stress is acting as the basic stress, x_g = 1.0. The growing microcrack spreads into an area of considerably decreasing

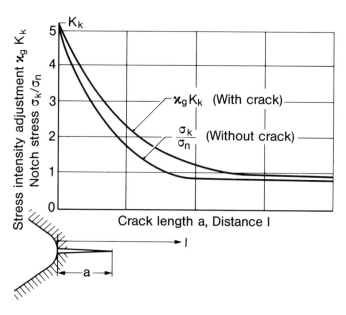

219 Adjustment for stress gradient times stress concentration versus crack length, illustrated for a notched bar, after Zettlemoyer and Fisher.[301]

notch stresses. The decrease of stress intensity is not quite as strong as the decrease of notch stress in the notched bar without a crack. It is of particular interest that in general the value of $x_g K_k$ does not drop below 1.0. The function $x_g K_k$ has to be determined by a relatively precise analysis of the notch stress perpendicular to the notch root.

The typical curve of stress intensity, $K_{(I)}$, versus the crack depth, a, for a semi-elliptical crack at the fillet weld toe is shown in Fig.220. The first peak of the curve results from notch stress, the second peak from the crack approaching the back side of the plate. The K curve changes for bending loading of the considered joint because the crack penetrates into the initial compressive bending stress zone only slowly. The crack propagates faster in the transverse direction than it does in the depth direction.

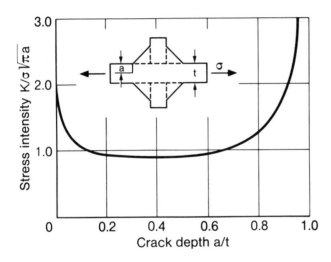

220 Stress intensity factor for semi-elliptical surface crack at fillet weld toe of transverse stiffener joint dependent on crack depth, after Albrecht and Yamada.[302]

In the case of a sharp notch combined with a small crack it is recommended to limit the notch radius to the dimensions given by the microstructural support effect, at least in the endurance limit range, to avoid unprofitable limit case considerations. In the latter range the fatigue notch factor, K_f, is to be inserted into equation (123) instead of the notch stress concentration factor, K_k, K_f being determined via fictitious rounding of the sharp notch.

9.2.3   Crack propagation life

Fatigue strength for finite life can be determined on the basis of the crack propagation relations as far as crack initiation can be neglected. Integrating equation (121) from the initial crack length, a_i, to the critical crack length, a_c, (critical with respect to final fracture) assuming $a_i \ll a_c$ and $\Delta K \ll (1 - R)K_{Ic}$ and neglecting the dependence $x = x(a)$:

$$N = \frac{2(1-R)(\Delta\sigma)^{-m}}{(m-2)C\pi^{m/2}x^m a_i^{(m-2)/2}}$$

(124)

Provided the assumption is correct that the fatigue strength or fatigue life of welded joints exclusively depends on crack propagation, equation (124) should correspond to equation (16) of the S-N curve for finite life. In fact, the exponents, m (for crack growth) and k (for S-N curve), deviate slightly from each other; for structural steels $m = 2.0\text{-}3.0$ and $k \approx 3.75$, for aluminium alloys $m \approx 4.0$ and $k \approx 4.3$ (see the normalised S-N curves for welded joints in Fig.40 and 74. By adjusting a_i, on the other hand, every C resulting from the crack propagation tests can be inserted into equation (124) in a consistent way.

The fatigue strength diagram can also be derived by means of the crack propagation relations. The dependence of the fatigue strength for infinite life on mean stress results from equations (118) and (9). It is expressed by the endurable upper stress $\sigma_{Up} = f(R)$:

$$\sigma_{Up} = \frac{\Delta\sigma_F}{1-R} = \Delta\sigma_{F0}(1-R)^{(1-m)/m}$$

(125)

This equation describes a hyperbola which begins at $0.63\sigma_{F0}$ for $R = -1$, and converges to infinity for $R = 1$ via σ_{F0} for $R = 0$. The upward deviation for $R \geq 0.5$ when compared with the actual finite fatigue strength curve can be explained by the fact that the introduced simplification, $\Delta K \ll (1-R)K_{Ic}$, does not apply in this range. The differing profiles of the fatigue strength curves for welded joints with different notch effects are obviously not represented.

The fatigue strength for infinite life (endurance limit) of the conventional type can be set in relation to the threshold stress intensity ΔK_{th} on the basis of equation (122):

$$\Delta\sigma_F = \frac{\Delta K_{th}}{\sqrt{\pi a_i x}}$$

(126)

However, an excessive fatigue strength for infinite life results from the assumed small crack lengths of fatigue strength calculations for finite life (e.g. $a_i \approx 0.05\text{mm}$). On the other hand the initial crack length which can be calculated from equation (126) is too large when the actual fatigue strength for infinite life is taken as a basis. For example, the result is $a_i \approx 0.2\text{mm}$ for $\Delta\sigma_{F0} = 220\,\text{N/mm}^2$ (butt weld machined flush), $\Delta K_{th0} = 180\,\text{N/mm}^{3/2}$ (structural steel) and $x \approx 1.0$ (wide surface crack).

The evaluation of the equations (124) and (126) (equation (124) with $R = 0$ and without assumption $a_i \ll a_c$), considering a semi-elliptical surface crack in the tensile loaded steel plate, shows clearly the influence of the fracture mechanical parameters on the fatigue strength of welded joints.[279-281] The size of the initial crack, a_i, and the stress range, $\Delta\sigma$, have a dominating effect on the

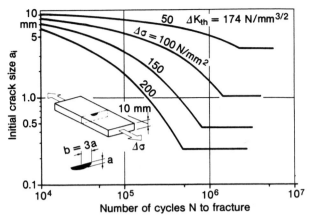

221 Effect of initial crack size, a_i, and range of stress, $\Delta\sigma$, on endured number of cycles, N, for structural steel, after Hirt.[281]

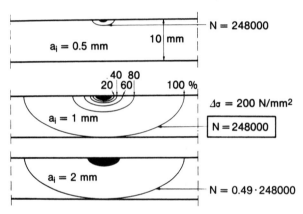

222 Majority of stress cycles endured with small crack size, after Hirt.[281]

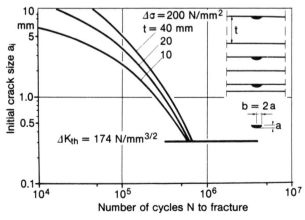

223 Effect of initial crack size, a_i, and plate thickness, t, on endured number of cycles, N, for structural steel, after Hirt.[281]

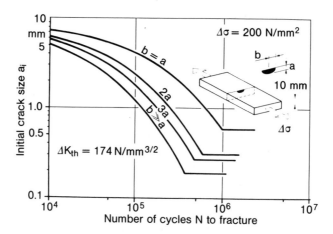

224 Effect of crack shape parameter, b/a, on endured number of cycles, N, for structural steel, after Hirt.[281]

number, N, of stress cycles endured, Fig.221. The majority of stress cycles is endured, when the crack is still small, Fig.222, which indicates the significance of short crack behaviour. The plate thickness, t, has little influence on the result, Fig.223. The crack shape parameter, b/a, has a stronger influence, Fig.224. From this, the conclusion can be drawn that it is particularly the initial crack size and the stress range which have to be introduced into the fracture mechanical calculation as close to reality as possible.

9.2.4   Dimensionless representation

The equations above can be transformed into a non-dimensional form which is more suitable for fundamental research purposes by means of reference to the state variables of the initial crack at the fatigue strength for infinite life.[293,294,296] Thereby independence from the uncertain initial crack size is achieved (the assumption $a_i \ll a_c$ is sustained):

$$\frac{d\bar{a}}{dN} = \frac{C_0}{1-R}(\Delta\bar{\sigma}\sqrt{\bar{a}}\,\bar{x})^m \text{ for } \Delta\bar{\sigma}\sqrt{\bar{a}}\,\bar{x} > (1-R)^{1/m} \tag{127}$$

$$\Delta\bar{\sigma} = \frac{\Delta\sigma}{\Delta\sigma_{F0}} \tag{128}$$

$$\bar{a} = \frac{a}{a_i} \tag{129}$$

$$\bar{x} = \frac{x(a)}{x(a_i)} \tag{130}$$

9.3   Input parameters of the fracture mechanical strength or life evaluation for welded joints

9.3.1   Initial crack size

Initial cracks and crack-like defects which are relevant in terms of fracture mechanics occur especially at the transition between parent plate and weld seam (including the toe and the root of the seam), where molten filler and parent materials mix with tiny slag and gas residues possibly resulting in cracks and defects. The highest notch stresses occur simultaneously at the weld toe and root. In the published calculations $a_i = 0.01$-0.05mm is assumed for aluminium alloys and $a_i = 0.1$-0.5mm for steels. The surface cracks which strongly affect fatigue characteristics are simulated as semi-elliptical at a ratio of crack depth to crack width, $a/c = 0.1$-0.5. The possibility to determine the change of crack shape with crack propagation more precisely by means of calculation is too complex and time consuming for practical application.

9.3.2   Material parameters

The materials-dependent crack propagation parameters, C and m, show relatively small deviations from the common mean value within a group of technical materials and microstructural states (*e.g.* parent material and heat affected zone) which is advantageous for practical application. However, less favourable parameter values have to be expected when a crack is propagating in the direction of plate thickness-than when it is propagating in the plane of the plate. A narrow scatter band of crack propagation rate values dependent on ΔK is shown in Ref.264 for several steels. On the other hand, larger discrepancies can also be found in the literature. For structural steels the following parameters were determined (da/dN in mm per cycle, ΔK in $N/mm^{3/2}$):

$$m = 2.0\text{-}3.6, C = 0.9\text{-}3.0 \times 10^{-13} \tag{131}$$

The following values are quoted in Ref.295 for the structural steels St 37 and St 52 $(R = 0)$:

$$m = 3.33, C = 0.137 \times 10^{-13} \tag{132}$$

$$m = 3.18, C = 0.339 \times 10^{-13} \tag{133}$$

$$m = 4.0$$

$$C = 7.4 \times 10^{-16} \text{ for } P_f = 0.975$$

$$C = 1.7 \times 10^{-15} \text{ for } P_f = 0.995 \tag{134}$$

In accordance with Ref.2, C is presented as a function of m:

$$C = 1.315 \times 10^{-4} \times 895.4^{-m} \tag{135}$$

Approximation equations, as well, are known for m as a function of ultimate tensile strength or Vickers hardness.

The possible discrepancy between the exponents, m, resulting from crack propagation measurement and k resulting from S-N curves has already been pointed out. Often improved (formal) suitability for practical application of the fracture mechanical calculation is attained as a result of adjusting m as equal to k (k taken from the normalised S-N curve).

The following values are to be inserted into the dimensionless representation of the fracture mechanical equations in accordance with Ref.293 for welded joints made of structural steel:

$$m = 3.5, C_0 = 4.22 \times 10^{-15} \tag{136}$$

The following dependence is indicated in Ref.2 as the mean value of ΔK_{th} for structural steels:

$$\Delta K_{th} = 240 + 173R \text{ N/mm}^{3/2} \tag{137}$$

In accordance with Ref.283 the relation is:

$$\Delta K_{th} = 190 + 144R \text{ N/mm}^{3/2} \tag{138}$$

In accordance with Ref.293 the following value is recommended as the lower limit of scatter:

$$\Delta K_{th0} = 180 \text{ N/mm}^{3/2} \tag{139}$$

The following values are given in Ref.295 for the structural steels St 37 and St 52 at R = 0:

$$\Delta K_{th0} = 236 \text{ and } 245 \text{ N/mm}^{3/2} \tag{140}$$

The material's characteristic value K_{lc} (or K_c) is higher than 600-3000 N/mm$^{3/2}$ for structural steels and aluminium alloys under service conditions.

In general, different materials' characteristic values, m, C and K_{th0}, are determined for the base material, the filler material and the different heat affected zones, respectively. The statistical analysis[282] of the characteristic parameters for a longitudinal crack in or in front of the weld seam showed in a particular case that there is no dependence on material (four structural steels), stress ratio, R, and (welding) energy input per unit of length.

9.3.3   Stress intensity factors

A sufficiently precise determination of the stress intensity range ΔK for the rather complex crack and joint geometries of welded structures requires greater expenditure. Generally, the superimposition of the known solutions for simpler limit cases, including the notch stress concentration without the crack, with microstructural support effect if necessary, gives only a first approximation. This is based on the compilations of notch stress concentration and crack stress intensity factors.[207-209,287-290] An additional simplification is offered in Ref.291 for macrocracks with a considerable bending stress proportion at the crack tip.

The more precise determination of ΔK or $x_g K_k$ requires a solution in accordance with elasticity theory for the weld joint with a crack, slit or gap

which is mostly approximated as a transverse contour model. This has been
carried out, up to now, mainly by means of the infinite element method, for
example, for the butt welded joint and the cruciform joint with double bevel
butt weld or double fillet weld,[266,297] for the transverse stiffener and the strap
joint with edge fillet weld,[301-302] each with an initial crack at the weld toe;
moreover, the double bevel butt weld and double fillet weld joints have been
considered with different slit or gap lengths equivalent to different penetration.
The stress intensity factor is generally determined from the rate of the elastic
deformation energy released per unit of crack length. An advantageous
alternative is to determine the notch stress distribution without the crack, in the
expected crack plane and compensate it at the crack faces by superimposing
counteracting forces there. The stress intensity then results from an
easy-to-solve integral on the notch stress distribution in the crack plane.[301] The
grid line spacing of the finite element mesh at the notch with initial crack should
be smaller than the crack length. This necessitates extremely fine meshes in this
area.

Less fine meshes are required for determination of the stress intensity of the
relatively larger slit or gap of a cruciform joint with double bevel butt weld or
double fillet weld. The stress intensity factors, K_I and K_{II}, for the slit or gap tip in
the cruciform joint subjected to tensile load are shown in Fig.225 and 226 in

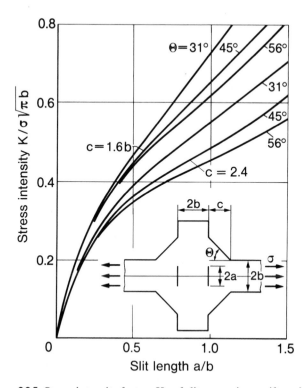

225 Stress intensity factor, K_I, of slit or gap in cruciform joint, dependent on
slit length, leg width and slope angle of weld, after Gurney.[297]

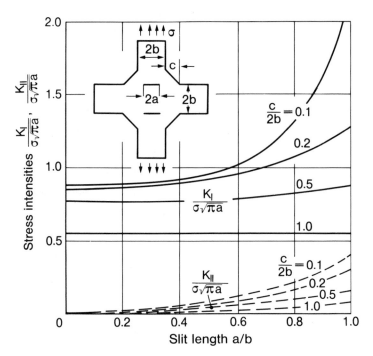

226 Stress intensity factors, K_I and K_{II}, of slit or gap in cruciform joint, dependent on slit length and weld leg width, after Usami and Kusumoto in Ref.266.

dependence of slit length, a, leg width, c, and slope angle, θ. A similar solution by Hijikata, Yoshioka and Inoue including tensile and bending load is recorded in Ref.290. There is also a solution for the slit in the lap joint in the above reference.

For T joints with single bevel butt weld and root backing plate subjected to transverse bending as in Fig.227a, the K_I and K_{II} values of the unfused gap at the root plate (gap depth considered as crack length, the crack subjected to mixed

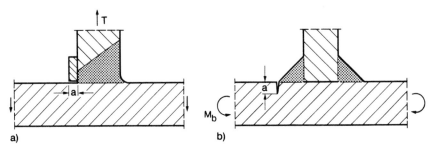

227 Welded joints analysed for stress intensity factors: T joint with single bevel butt weld and root backing plate (a) and tension bar with single sided transverse stiffener (b), after Sprung and Zilberstein[298] and Niu and Glinka.[299]

mode loading) are presented in dependence of the spacing between the supports of the bending bar.[298]

In another investigation[299] using a special weight function formulation (without finite elements) the stress intensity factor, K_I, is determined for a bending loaded specimen with single sided transverse stiffener connected via double fillet welds, see Fig.227b. The stress intensity is presented depending on fillet slope angle, θ, and fillet toe radius, ρ, (relative to plate thickness, t). The slit or gap in the joint is not taken into account. It is shown that K_I mainly depends on the ratio ρ/t for very short crack lengths and on slope angle, θ, for larger crack lengths.

The stress intensity factors for the (semi-elliptical) crack propagation at the fillet weld toe of a T joint (double bevel butt weld with fillet welds) subjected to transverse bending are recorded in Ref.300.

9.3.4   *Effect of welding residual stresses*

The effect of welding residual stresses must be taken into account when dealing with welded joints left in the as-welded state, as da/dN and ΔK_{th} depend via R on the mean stress. Tensile residual stresses accelerate crack propagation substantially, compressive residual stresses retard crack propagation. In tensile residual stress fields, crack propagation occurs even when subjected to compressive stress cycles. The stresses at the crack tip are displaced into the tensile range.[307] Whereas residual stresses acting in transverse direction to the weld take effect on crack propagation in the same way as mean stresses by external loads, special approaches are required in the case of residual stresses acting in longitudinal direction to the weld seam.[303-306] The normal distribution of longitudinal welding stresses can be described by a curve showing a tensile stress maximum, σ_0, in the centreline of the seam, a drop nearby with zero level crossing at $x = l_0$, a smaller compression stress maximum and convergence towards zero from below. It can be assumed that this residual stress distribution is independent of plate depth for plate thicknesses $t \leq 20mm$. The distribution is typical for cooling-down processes without microstructural transformation.

If a transverse crack crosses the residual stress field, the residual stress intensity factor reaches its maximum at $c/l_0 \approx 0.6$, and subsequently drops down to zero, Fig.228. The following approximation equation is proposed in Ref.303, 304:

$$K_I = \sigma_0 \sqrt{\pi c}\, e^{-0.42(c/l_0)^2} \left[1 - \frac{1}{\pi}\left(\frac{c}{l_0}\right)^2 \right] \tag{141}$$

The actual occurrence of the compressive stress intensity factor for $c/l_0 > \sqrt{\pi}$ depends on the course of the residual stress curve. It occurs if there is a marked compressive stress maximum with a rapid drop to zero nearby. In the compressive stress intensity range, crack propagation is expected to be considerably retarded. Related crack propagation investigations on the basis of ΔK_{eff} have been conducted numerically and experimentally.[305]

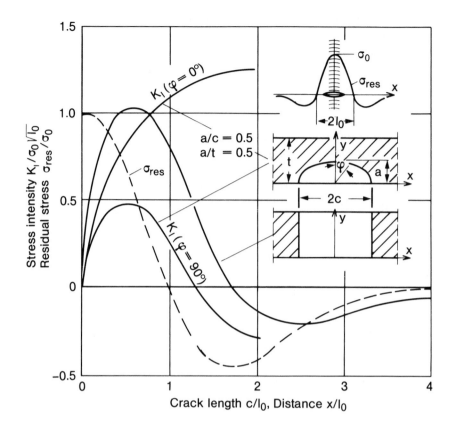

228 Residual stress intensity factor, K_I, for transverse crack in longitudinal weld through-crack and surface crack, residual welding stresses in longitudinal direction of weld, after Terada[303] and Wu.[306]

The semi-elliptical surface crack in the above residual stress field shows a decreased stress intensity factor in the surface points of the crack front and an increased stress intensity factor at the crack root, Fig.228, so that such an initial crack propagates at a slower rate on the surface of the plate and at a faster rate in the interior. As a result, the surface crack tends to change into a through-crack.

With plate thicknesses t ≥ 20mm (mostly jointed by multilayer welding) considerable differences in residual stress can occur between surface and interior areas (in many cases tensile stress on the surface, and compressive stress in the interior). As a result, crack growth is accelerated on the surface and retarded in the interior. Procedures for the calculation of the stress intensity factor of semi-elliptical surface cracks in such inhomogeneous stress fields — where crack propagation is correspondingly difficult to predict — are presented in Ref.306. It is known from a practically orientated investigation that — in the case of approximately elliptical internal cracks — it is above all the crack propagation in the direction of plate thickness (*i.e.* of unfavourable tensile residual stresses) which is decisive for the residual life of the component.

9.4   Application for assessment of fatigue strength for finite and infinite life

The results of formalised fracture mechanical calculations for welded joints with an assumed initial crack can be summarised in the following way (concerning residual life calculation by fracture mechanics combined with crack initiation analysis by the notch stress approach, see section 9.5).

The fatigue strength for finite and infinite life can be assessed on the basis of the fracture mechanical equations — but only after the assumed initial crack length and the material parameters have been adjusted. Within this procedure the varying notch stress concentration factor has a direct strength-reducing effect.

For the cruciform joint with double bevel butt weld or double fillet weld with slits (or internal gaps) of different initial lengths, the crack propagation equation has been integrated to determine endurable stress cycles.[271,309,310] There is an optimum throat thickness with regard to fatigue strength for finite and infinite life. Insufficient thickness leads to an early seam rupture initiated at the weld root, and excessive thickness to an early plate rupture initiated at the weld toe (the stress concentration factor, K_k, of the weld toe decreases with increasing throat thickness). The result of the fracture mechanical calculation[271] is shown in Fig.229 for crack propagation in the weld starting from the weld root, in comparison with crack propagation in the plate starting from the weld toe (limit curves, initial crack length 0.15mm with plate thickness 12.5mm). The calculation was adjusted to fit the fatigue test results obtained with the cruciform joint.

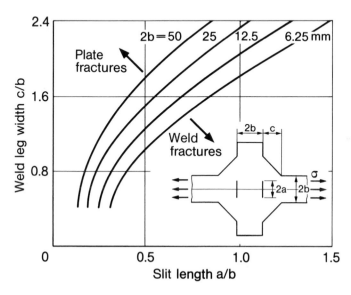

229 Weld and plate fractures dependent on plate thickness, weld leg width and slit length, after Maddox.[271]

In accordance with the ECCS recommendation,[151] welded components can be assigned to notch classes on the basis of a fracture mechanical crack propagation analysis. This, however, is objected to because the crack initiation stage is completely neglected, because the scatter bands of the basic parameters for the fracture mechanical calculation are rather wide and because the amount of calculations required is rather large.[312] Assignment to notch classes on the basis of structural stress or notch stress assessment seems to be the better way.

The basic tendency of the size effect on the fatigue strength of welded joints (larger welded joints, despite geometrical similarity, have a lower fatigue strength in terms of nominal stress) is correctly expressed by the fracture mechanical calculation,[311] provided that the initial crack length is introduced proportional to the other dimensions (the latter assumption requires verification in the individual case), Fig.230.

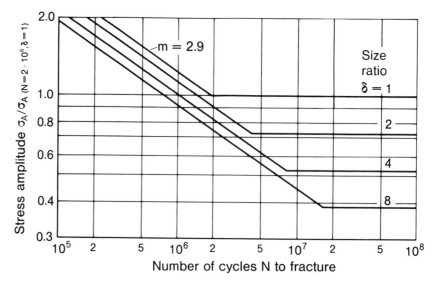

230 Effect of specimen size on fatigue strength on the basis of fracture mechanical calculation, after Pook in Ref.311.

The marked influence of residual stresses, particularly of welding residual stresses, on the high cycle fatigue strength of welded components (see section 2.3.2) can, in principle, also be explained[283] by means of fracture mechanics. As only the tensile amplitude range causes crack propagation, and as ΔK_{th} increases with compressive mean stress, stress-relief annealing or specifically generated compressive residual stresses have a strength-increasing effect.

The scatter of fatigue strength values for finite and infinite life resulting from varying fabrication quality can be explained in terms of fracture mechanics by scattering initial crack sizes in connection with different seam profiles (e.g. resulting from different welding positions); a quantified verification for this interpretation, however, is not available.

The conversion of the S-N curve for constant amplitude loading to service life curves for variable amplitude load spectra of various shapes on the basis of the simple fracture mechanical equations remains on the safe side, provided that crack-closure and amplitude interaction effects can be neglected. The applicability of the conversion for the service life curves of welded joints in the German crane design standard DIN 15018[152] has been demonstrated in Ref.293, 294. The above fracture mechanical approach has been further developed[296] into a non-linear damage accumulation assessment which includes the (linear) Miner's rule as a special case. However, it is precisely the interaction effect which produces the extremely varied operational durability of the materials in the presence of fatigue crack propagation, so that the above-mentioned more complex crack propagation calculations[284-286] including the interaction effect are recommended in individual cases.

Surface hardening by rolling or shot peening the weld toe is covered by taking residual compressive stresses, $\sigma = 0.5\text{-}1.0\sigma_Y$, into account over a depth $t^* = 0.1\text{-}0.5$mm. The crack propagation of sufficiently short cracks is retarded considerably.[313] In corrosive environments ΔK_{th} decreases and da/dN increases, which causes a corresponding decrease of fatigue strength for finite and infinite life.

The approach suggested by Heckel and Ziebart[316] for the notch and size effect which starts from a random distribution of crack-like surface defects and transfers the fatigue strength values obtained from smooth specimens to notched specimens via the fracture mechanical equations is less suitable for welded joints because the surface defects have a different size and distribution in the weld notches than they have in the parent metal plate, from which the smooth comparative specimens are taken.

9.5 Application for assessment of safety, residual life, fitness-for-purpose and back-tracing of failures

9.5.1 *Assessment of defects*

The use of fracture mechanical calculations of fatigue crack propagation for assessment of cracks and defects with respect to failure probability of the component turns out to be much more realistic than the described determination of fatigue strength for finite and infinite life. This is because it is possible to work with the actual size, shape and position of the crack or defect. Moreover, materials' characteristic parameters which are determined in crack propagation tests are taken as a basis. Fracture mechanics makes it possible to assess planar cracks and defects. Three dimensional defects such as pores and shrinkage cavities cannot be directly assessed by this method.

It is the objective of the above assessment to determine whether or not a crack or defect, detected in terms of size, shape and position, propagates when subjected to fatigue loading with a given amplitude or an amplitude spectrum. If it propagates it has to be proved that — within the inspection interval or within the total operating life — the increasing crack size remains sufficiently below the critical crack size which leads to structural failure by brittle fracture, global yield, instability or leakage (fail-safe or safe-life concept in aircraft design, leak-before-break concept in pressure vessel design). The fatigue crack propagation is assessed in accordance with the fracture mechanical equations stated above. Within this procedure either the detected initial crack is taken as a basis, or, if no crack has been detected, an initial crack size is introduced which is well above the resolution capacity of the non-destructive testing method (including a safety margin). When carrying out these calculations, it has to be taken into account that in corrosive environments (*e.g.* humid air or sea water) the crack growth may be considerably accelerated (stress corrosion or fatigue corrosion cracking). Crack propagation calculations of this kind are usually carried out for expensive high quality vessels and tanks (*e.g.* nuclear reactor pressure vessels and rocket fuel tanks).

9.5.2   *Residual life assessment*

Residual life calculations have been carried out, *e.g.* for butt welded joints of different materials (structural steels and aluminium alloys) and combined with crack initiation analysis, Fig.210 and 211. The results have been verified by testing.

More complex residual life calculations have been carried out for welded components in shipbuilding, *e.g.* liquid natural gas tankers. Structural stress analysis for crack initiation was continued on the basis of fracture mechanics up to final instantaneous fracture. These investigations referred to lap jointed L and T members, stiffener ends and frame corners. The calculations started with an initial surface crack, approximately 1mm deep and 10mm wide, which later expands into a through-crack, Fig.231. The crack fronts of the individual cross section members can be connected to form a curve similar to an ellipse, which simplifies the calculation of the stress intensity factors. Fracture mechanical investigations on residual life have also been conducted for plates with longitudinal ribs and transverse crack propagation.[319]

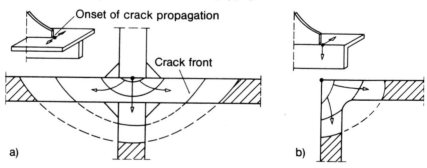

231 Crack propagation in hull girders used in shipbuilding, after Matoba and Inoue.[317]

9.5.3 Fitness-for-purpose assessment

The fitness-for-purpose concept for welded joints[177-179] is of paramount importance for practical application. In view of the tightening of quality standards in manufacturing it seems reasonable to accept those defects which are revealed only by tightening quality standards, if the harmless nature of these defects can be ascertained under relevant operating conditions by a fracture mechanical fitness-for-purpose assessment. It is not expedient to repair such defects. There are several recommendations on this subject.[177-182] The British recommendation[177-179] deals with crack propagation as a result of fatigue and its conversion into quality classes in detail. A comprehensive presentation of the fatigue part of the recommendation would go beyond the limits of this book. The procedure is only briefly outlined on the basis of Ref.179.

The welded joint under consideration is assigned to one of ten 'quality classes', each of which has its characteristic S-N curve, Fig.146. The assignment depends on manufacturing quality, inspection possibility and accepted failure probability (no details in Ref.179). Slightly different families of S-N curves apply to the states with and without welding residual stresses. The endurable stress amplitudes only are presented. In components with welding residual stresses the influence of mean stress is completely neglected. In components without welding residual stresses the stress amplitudes of the components — which have to be compared with the S-N curves — are modified so that the influence of mean stress on fatigue strength is taken into account. The S-N curves of aluminium alloys are lower by the factor $1/3$ than those of steels. The upper curves Q1 to Q6 are identical with the notch classes C to G of the British standard BS 5400 for fatigue resistant welded joints of bridges, failure probability $P_f = 0.025$. The lower curves represent the lower failure probability $P_f = 0.005$. With variable load amplitudes the Miner's rule is considered valid.

When a crack or defect has been detected, it is necessary to prove that the S-N curve or quality class which — on the basis of the calculated residual life — is to be assigned to the crack or defect concerned, is above the S-N curve or quality class guaranteed for the considered welded joint by design and manufacture. This proof is carried out as follows:

- the detected (elliptical or semi-elliptical) internal or surface crack is converted into a crack of infinite width with equal stress intensity factor. This results in a modified crack depth (obtainable from a graph). The final crack size which is considered as largest possible is handled in the same manner.
- depending on plate thickness, different stress amplitudes, σ_{ai} and σ_{ac} (for N $= 10^5$), are assigned to the above-mentioned initial and final crack sizes in the case of butt welded or fillet welded joints, assuming a definite failure probability ($P_f = 0.025$ or 0.005) (obtainable from a graph, the transition from the internal or surface crack to the through-crack is considered as stable).

— based on the S-N curve stresses, σ_{ai} and σ_{ac}, and the material's parameter, m = 4.0, the permissible stress amplitude is:

$$\sigma_{aper} = (\sigma_{ai}^4 - \sigma_{ac}^4)^{1/4} \qquad (142)$$

This stress in turn, defines the highest possible quality class or S-N curve to be used.

A well designed recommendation for the assessment of weld imperfections with respect to fatigue prepared by Hobbacher[182] has recently been passed by the IIW. The guideline of the German Welding Society[181] on the fracture mechanical evaluation of defects in welded joints does not include a code-relevant assessment procedure. It is more a well contrived commentary on a procedure not yet devised. The available approaches for defect evaluation on the basis of fracture mechanics are reviewed and explained for common use. A detailed assessment procedure is under preparation by the Austrian Steel Engineering Society. The fatigue reliable design is discussed by Pellini[183] mainly on the basis of a fracture mechanical assessment of defects.

9.5.4   *Back-tracing of failures*

If fatigue fracture has occurred in spite of the strength and safety assessments carried out, both the instantaneous final fracture and the preceding fatigue crack propagation can be traced back by fracture mechanical calculation (failure analysis). The appertaining material's parameters and geometrical data can be determined at the area of the structure where the fracture occurred, and in general, information is also available on the service load history. For this reason the failure analysis by fracture mechanics is highly informative, as has been demonstrated for fatigue and fracture in bridges by Fisher[318] and in ship structures by Fricke.[319]

10

Structural stress, notch stress and stress intensity approach for assessment of fatigue strength of spot welded joints

10.1 Development status of local approaches

Spot welded joints can be assessed with respect to fatigue strength in the same way as seam welded joints, proceeding from the local stress parameters, (*i.e.* structural stress, notch stress and stress intensity) at the edge of the weld spot. Certain differences arise, though, as a result of the different geometry and force flow, which justify treating these joints separately. A close general relationship can also be established between structural stresses and stress intensity factors. The sections which follow are based on the research results of the author into local stress parameters at the weld spot,[320-326] supplemented by the contribution of other authors, notably with regard to endurable local stress parameters. Although particular aspects still need to be clarified, a well-rounded general picture can nevertheless be obtained. The local concept is suited to establishing a general theory of the fatigue strength of spot welded joints, which may serve as a basis for practical action in more complex cases also.

When a spot welded joint is subjected to external load the maximum local stresses generally occur on the inside of the plate at the edge of the weld spot nugget. This applies in the first place to structural stresses at the weld spot edge, which can be estimated on the basis of approximation formulae, more exactly calculated using finite element methods, and also measured at the individual spot. This also applies, however, to the notch stresses and to the stress intensity factors at the crack-like end of the slit, which can be calculated on the basis of further approximation formulae or using the boundary element method. The stress increase at the weld spot is caused primarily by the concentration of the force flow on to the jointing face of the weld spot, flow concentration both in the plane of the plate and in planes perpendicular to it, the latter particularly with tensile shear load. The notch effect of the slit faces with gap or the crack effect of the slit faces without gap are superimposed. All three approaches are directly justifiable[320] in terms of crack initiation:

—   the structural stresses as the basis of notch stresses and stress intensity factors or even directly initiating cracking, if the crack occurs outside the slit notch;
—   the notch stresses as crack-initiating quantity on slit faces with gap;
—   the stress intensity factors as crack-initiating quantity on slit faces without gap.

The three approaches are also used independently from the above restrictions in a more technical manner.

The crack initiating stress parameters mentioned above are adequate for predicting fatigue strength for infinite life (*i.e.* the fatigue endurance limit). In the finite life regime, local strain quantities also need to be included to describe crack initiation, and crack propagation also requires coverage. In all cases, not only the local stress parameters (including residual stresses) should be known but also the local strength characteristic values of the material (for instance by measuring the local hardness).

10.2   Basic loading modes at weld spot

Proceeding from the three crack opening modes, I, II and III, of fracture mechanics, a distinction can first be made in the case of the weld spot edge subjected to homogeneous, *i.e.* axisymmetrical loading between the basic loading modes of cross tension (I), all-sided compression-tension (II) and torsion (III), Fig.232. In the case of the weld spot edge subjected to inhomogeneous, *i.e.* single sided, loading, the basic loading modes of peel tension (I. 1) single sided compression-tension (II. 1) and single sided

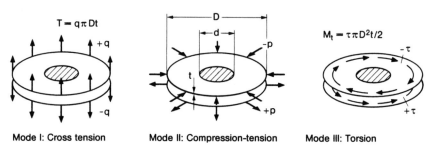

Mode I: Cross tension   Mode II: Compression-tension   Mode III: Torsion

232 Basic loading modes at weld spot based on fracture mechanics, homogeneous case.

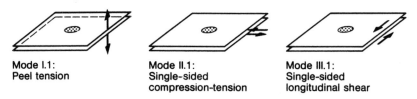

Mode I.1:   Mode II.1:   Mode III.1:
Peel tension   Single-sided   Single-sided
                        compression-tension   longitudinal shear

233 Basic loading modes at weld spot based on fracture mechanics, inhomogeneous case.

longitudinal shear (III. 1) occur, Fig.233. In view of the fact that single row lap and flange joints predominate in spot welding, the latter classification is performed on a square lap or flange element. An extended systematics is shown in Fig.234, *i.e.* the single, two and four sided edge loading (the arrows signify edge displacements) as well as additional types of basic loading modes of a higher order, which are of practical significance.

A further question of the systematic description of local stress at the weld spot is posed when proceeding from the resultant force state in the weld spot jointing face, where there are six joint forces, as shown in Fig.235. This system of joint forces is not a means of completely identifying the stress state at the weld spot. The six forces in the jointing face need to be supplemented by six 'eigenforces' in the plate centre plane, running-through tension and running-through bending in the two orthogonal directions, as well as two shear loading cases in the plate plane.[325] Only by adding the eigenforces is it possible

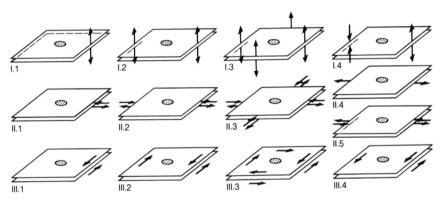

234 Basic loading modes at weld spot, extended systematics.

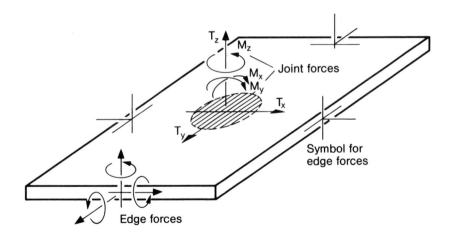

235 Resultant joint forces in weld spot jointing face supported by edge forces.

to establish the reversibly single-valued relation between the resultant forces at the weld spot (global quantities) and the stress distribution in the adjacent plates (local quantities). If only the joint forces are stated, the stress state is infinitely multi-valued, dependent on the support conditions at the outer edges of the plate. The difference between one-sided and all-sided support of the joint forces can be cited as an example, as can the possibility of bracing forces in the plate without influencing the joint forces.

10.3   Elementary mechanics of tensile shear and cross tension loading

In preparing for what is presented in section 10.4 regarding structural stresses, the elementary mechanical peculiarities of the spot welded joint are made clear by taking the two principal basic loading modes, tensile shear and cross tension loading.

Tensile shear loading is considered at the lap joint. The mechanical peculiarity of this joint is the transversely offset position of the plates — the difference in position being equal to the thickness of the plate — which produces bending moments in the plates when the joint is subjected to external tension. This causes considerable bending stresses to be superimposed on the tensile stresses, resulting in deflections, which are large relative to the thickness of the plate, or which lead to contact of the two plates. This non-linear secondary effect of large deformations is generally ignored in the numerical solutions stated below.

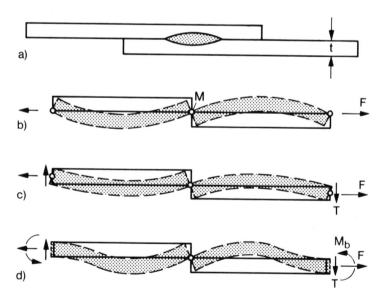

236 Plate bending by offset centre planes in tensile shear specimen (a), simplified beam models (b, c, d).

The coupling of the tension and bending effect can be presented on the transverse contour model of the lap joint, which in turn can be conceived as a beam model, Fig.236. In this case, the weld spot, which is actually extensive, is contracted to the point M. The point M can also be conceived as being localised in the weld spot edge. The beam model is symmetrical relative to the point M, if the dimensions and loads on the left and right are identical. The point symmetry relative to the point M excludes a bending moment around this point, for an internal force of this kind would violate the point symmetry. It is therefore possible to conceive a hinge in point M. The resulting bending deformation is shown greatly enlarged in Fig.235. The equilibrium of the internal forces is considered on the undeformed beam in the sections which follow.

Three different load cases are analysed: shear force in the centreplane of the weld spot (achievable by packing plates in the grip jaw); shear force in the centreplane of the plate (causing secondary transverse force); and shear force with bending-rigid clamping (producing increased secondary transverse force and secondary bending moment). As a consequence of the offset position of the plates, the forces acting in all three cases in the plate at point M include not only the tensile shear force, F (and possibly cross tension force, T) but also the offset bending moment, M_b:

$$M_b = \frac{Ft}{2} \tag{143}$$

In the plate strip (beam) of depth 1 the following tensile and bending stresses thus occur:

$$\sigma = \frac{F}{t} \tag{144}$$

$$\sigma_b = \pm \frac{Ft}{2} \frac{6}{t^2} \tag{145}$$

This leads to the following stresses on the inside and outside of the plate respectively:

$$\sigma_i = \frac{4F}{t} = 4\sigma \tag{146}$$

$$\sigma_o = -\frac{2F}{t} = -2\sigma \tag{147}$$

The tensile stress on the inside is thus four times as large as the mean tensile stress while the compressive stress on the outside is twice as large.

The bending stresses at the point M are identical in the three load cases; they only vary towards the clamping position. In the first case they remain constant, in the second case they drop to zero, in the third case they change their sign.

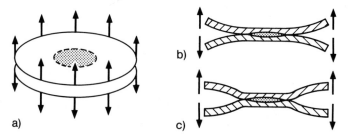

237 Plate bending in cross tension specimen, simplified model of circular
plate (a), outer edge support bending-free (b) or bending-rigid (c).

The transverse force which increases from case to case, has already been
included in the bending stress approach presented above.

The cross tension stress is approximated by a circular plate which is clamped
bending-rigid at the inner edge, subjected to transverse force at the outer edge
and at the same time supported bending-rigid or bending-free, Fig.237. The
obvious modelling as a beam is inadequate because the beam does not correctly
reproduce the internal forces per unit of circumferential length, which decrease
with increasing circular plate circumference. The plate bending radial stress
$\sigma_{r\,max}$ at the inner edge (diameter d) of the circular plate — plate membrane
stresses do not occur assuming small deformations — develops under
transverse load, T, and bending-rigid or bending-free supporting of the outer
edge (diameter, D) to yield:

$$\sigma_{rmax} = \frac{3T}{\pi t^2} \left[\frac{\ln\dfrac{D}{d}}{1 - \left(\dfrac{d}{D}\right)^2} - \frac{1}{2} \right] \tag{148}$$

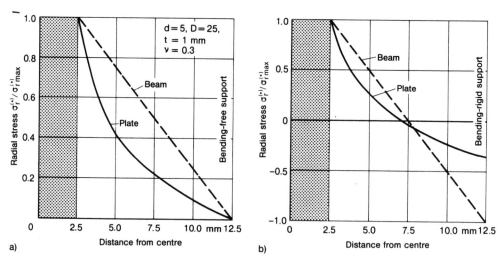

238 Radial stress decrease in circular plate model of cross tension specimen
in comparison with beam model, outer edge support bending-free (a) and
bending-rigid (b).

$$\sigma_{r\,max} = \frac{3T}{\pi t^2} \left[\frac{1-(1+\nu)\ln\dfrac{D}{d}}{1+\nu+(1-\nu)\left(\dfrac{d}{D}\right)^2} - \frac{1}{2} \right] \tag{149}$$

The stress, $\sigma_{r\,max}$, increases with D/d, as is more clearly shown from the simplified equations (151) and (152). The decrease of σ_r over the radius is significantly steeper in the plate than in the beam, Fig.238. On the other hand it is not quite as steep as the decrease of notch stress at the circular hole. The stresses on the inside and outside of the plate are opposite and equally large.

10.4   Structural stresses at weld spot

Simple formulae reflecting the influence of the principal dimensional parameters (nominal structural stresses, σ_{ns} or τ_{ns}) are used to estimate the maximum structural stress values at the weld spot edge. They are derived from simplified models relating to mechanical weld spot behaviour in accordance with technical plate, shell or beam theory.

The structural stresses are proportional to the joint forces and eigenforces acting at the weld spot. These forces are the first to be determined. Equilibrium conditions are sufficient for this only in the simpler cases of single and multiple spot specimens. In more complex cases of specimens and components, a finite element calculation is required. In practice, Fig.239, the relevant stiffness and joint force model works satisfactorily with a relatively coarse mesh. At least one nodal point should be provided at each weld spot. The transversely offset position of the plates must not be suppressed in the model. The nodal points

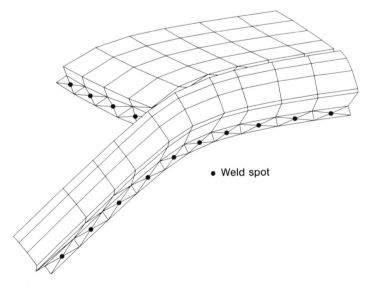

• Weld spot

239 Stiffness and joint force model of roof pillar of passenger car with weld spots marked.

superimposed one above the other at the weld spot are connected by a short rigid bar. The internal forces of the weld spot jointing face (tensile shear force, F; cross tension force, T; bending moment, M_b; torsional moment, M_t) are part of the output at the middle section of the bar. The eigenforces (running-through forces: tensile force, F'; bending moment, M_b'; shear force, S') result from the element forces in the weld spot nodal point of the considered plate after subtraction of the rigid bar end forces.

For tensile shear force, F, at the weld spot, the formula derived from the rigid circular core subjected to load in the plane of the plate[324,327] with factor 4 for the stress increase by plate bending is:

$$\sigma_{ns} = 1.27\frac{F}{td} \tag{150}$$

For cross tension force, T, at the weld spot, the formula derived from the rigid core subjected to transverse load in the circular plate with the outer edge supported bending-rigid or bending-free (diameter D), is:

$$\sigma_{ns} = 0.69\frac{T}{t^2}\ln\frac{D}{d} \tag{151}$$

$$\sigma_{ns} = 1.03\frac{T}{t^2}\ln\frac{D}{d} \tag{152}$$

These stresses are approximately doubled if the outer edge is supported on one side only provided the slanting-free transverse displacement of the weld spot is maintained.

For cross bending moment, M_b, at the weld spot, the formula derived from the rigid circular core subjected to bending load in the circular plate with the outer edge supported bending-rigid or bending-free, is:

$$\sigma_{ns} = 25.4\frac{M_b\,d}{t^2 d\,D}e^{-4.8d/D} \tag{153}$$

$$\sigma_{ns} = 22.7\frac{M_b\,d}{t^2 d\,D}e^{-4.0d/D} \tag{154}$$

These stresses, too, are approximately doubled if the outer edge is supported on only one side.

For torsional moment, M_t, at the weld spot, the formula derived from the condition of rotational equilibrium, is:

$$\tau_{ns} = 0.64\frac{M_t}{td^2} \tag{155}$$

For running-through tensile force, F', or running-through bending moment, M_b', in the plate connected by a weld spot, the formula derived from the rigid circular core in the tension or bending plate (width, b), is:

$$\sigma_{ns} = 1.54\frac{F'}{tb} \tag{156}$$

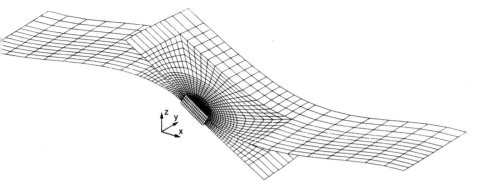

240 Structural stress model of tensile shear specimen, symmetry half, deformed state, slanting of weld spot.

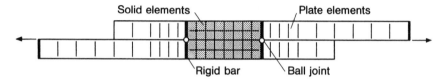

241 Structural stress model of tensile shear specimen, cross section.

$$\sigma_{ns} = 9.24 \frac{M'_b}{t^2 b} \tag{157}$$

Finite element models with a finer mesh than with the joint force models are used (within research projects) for more accurately determining structural stresses. A proven method is to model the interior of the weld spot with solid elements and the adjoining outside areas with plate elements. The translational degrees of freedom of the solid elements are coupled at the edge of the weld spot to the rotational and translational degrees of freedom of the plate elements by rigid elements which are mounted bending-free in the plane of the jointing face. By modelling plate elements up to the edge of the weld spot, the notch or crack stress which occurs here in reality is completely suppressed, which corresponds to the structural stress definition. The finite element mesh of the one half of the symmetrical model of the tensile shear specimen is shown in Fig.240 and 241. The elastic solid elements and rigid edge elements of this model may be replaced in the considered special load case by a rigid cylinder without a major loss in accuracy, but which greatly simplifies the calculation.

The structural stress concentration factor, K_{ns}, follows from the maximum value, $\sigma_{s\,max}$, of the equivalent structural stress (after von Mises) related to the nominal structural stress, σ_{ns}. It characterises the ratio of the accurately calculated structural stresses to those determined by the approximation formula. It represents a correction factor to the simplified parameter dependence of the nominal structural stresses in the formula. If, in contrast, $\sigma_{s\,max}$ is related to the conventional nominal stress, σ_n, the averaged basic stress, this produces the real structural stress concentration factor, K_s. The structural

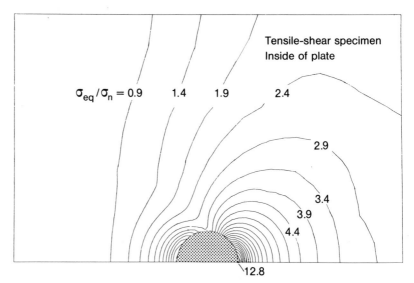

242 Equivalent structural stresses (according to von Mises) in tensile shear specimen, inside of plate.

stress result of the finite element calculation for the tensile shear specimen presented in the example, is shown in Fig.242 (nominal stress in clamping cross section $\sigma_n = 1.0$, $K_s = \sigma_{s\,max}/\sigma_n = 12.8$, $K_{ns} = \sigma_{s\,max}/\sigma_{ns} = 1.26$).

10.5   Notch stresses at weld spot

Although consideration of notch stresses at the weld spot is particularly suited to joints having gaps between the two plates, it is applied within the framework of this fundamental analysis, freed from specific applications, to joints without gap, in other words to ideal slit faces. The investigation is kept more compact by eliminating the gap width as an influencing parameter. The slit face end at the weld spot edge is fictitiously rounded in accordance with the Neuber approach to the microstructural support effect. In the case here of low-alloy deep drawing sheet steel the value of the radius is $\rho_f \approx 0.25\text{mm}$.[320] As this radius is no longer small relative to the plate thickness, the notch stress, σ_k, preferably calculated by the boundary element method, must be corrected by the stress increase caused by the reduction of the plate thickness to give the effective notch stress, σ_k^*. The correction formula with the notch stress concentration factors $K_k = \sigma_k/\sigma_n$ and $K_k^* = \sigma_k^*/\sigma_n$, and also $\rho = \rho_f/t$ and $\sigma^* = \sigma_o/\sigma_i$ is:

$$K_k^* = K_k \frac{(1-\rho^*)^2}{1 + \rho^*(1 + \sigma^*)} \tag{158}$$

For instance, in the case of the beam model of the tensile shear specimen with $\rho_f = 0.2\text{mm}$ (slightly smaller than recommended above), $t = 1\text{mm}$, $\sigma_i = -2\sigma_o$, Fig.243, the stress concentration factor $K_{ks} = 3.15$ (referred to σ_{ns}) and hence $K_{ks}^* = 0.58K_{ks} = 1.83$. What was obtained with $\rho_f = 0.25\text{mm}$ for the real tensile shear specimen, on the other hand, was $K_k^* = 23.2$, which with $K_s = 12.8$

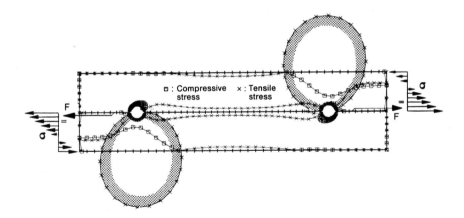

243 Beam model of tensile shear specimen with fictitious radius, $\rho_f = 0.2t$, at weld spot edge.

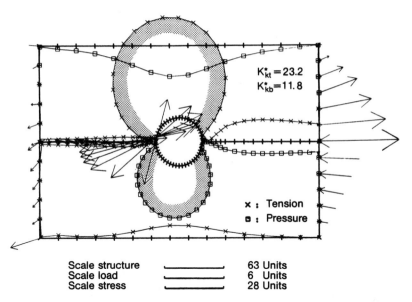

244 Notch stresses in tensile shear specimen, vertex of weld spot edge, $\rho_f = 0.25t$.

leads to the factor $K_{ks}^* = 1.81$. With the plate thickness $t = 1\,mm$, the fatigue effective notch stress, σ_k^*, should generally be little less than twice the value of the structural stress, $\sigma_{s\,max}$. With a smaller plate thickness, the increase is likely to occur in only a weakened form, in an intensified form, by contrast, with a greater plate thickness. These data also apply to spot welded joints with a gap.

The notch stress pattern in the vertex of the weld spot edge in the case of the tensile shear specimen and cross tension specimen (conforming to DIN 50124 and DIN 50164) is shown in Fig.244 and 245. These are two particularly important basic types of notch stress pattern at spot welded joints. In the case of

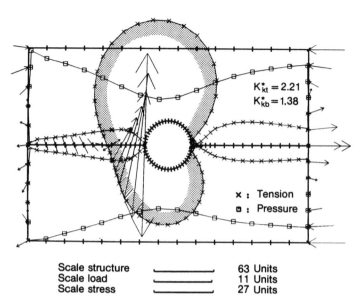

Scale structure ├───────────┤ 63 Units
Scale load ├───────────┤ 11 Units
Scale stress ├───────────┤ 27 Units

245 Notch stresses in cross tension specimen, vertex of weld spot edge,
$\rho_f = 0.25t$.

the tensile shear specimen, a tension maximum appears at the top compared
with a smaller compression maximum at the bottom while the stresses are close
to zero in the continuation of the slit. In the case of the cross tension specimen,
tension maxima appear at the top and bottom, which are linked by stresses
which are also high in the continuation of the slit.

The structural stress occurring at the weld spot edge, which can be measured
or calculated, is taken as the basis for calculating the notch stress at the
fictitiously rounded notch at the weld spot edge. The approximation for the
maximum notch stress after von Mises is obtained edge point by edge point
proceeding from the absolutely larger principal stresses on the surfaces of the
plates transferred on to the contour model perpendicular to the weld spot edge,
with the boundary element method being applied in accordance with the
procedures presented with respect to the seam welded joints in section 8.2.9.
This notch stress or notch stress concentration factor then requires correction
considering the cross section weakening caused by the notch depth, in
conformity with equation (158). Such a corrected notch stress distribution is
included in Fig.246 for the tensile shear specimen.

A basically different method for determining the notch stress concentration
consists of using the formulae derived by Creager for blunt cracks proceeding
from the stress intensity factors, K_I and K_{II}. It has been applied in Ref.360, 361
but this approach seems to neglect the notch effect of the non-singular term of
equation (159).

10.6   Stress intensity factors at weld spot

At the geometrically ideal weld spot edge with gap-free slit faces, the stresses are singular by analogy to the behaviour at crack fronts. The singularity generally varies from edge point to edge point, but may also occur uniformly in special cases. It can be described in tensor form at each edge point by the stress intensity factors, K_I, K_{II}, K_{III} (transverse tension, transverse shear, longitudinal shear (r, ϕ polar coordinates in axially-radially aligned plane through the edge point):

$$\sigma_{k1} = \frac{1}{\sqrt{2\pi r}}[K_I f_{k1}^I(\varphi) + K_{II} f_{k1}^{II}(\varphi) + K_{III} f_{k1}^{III}(\varphi)] + \sigma_0 + O(\sqrt{r}) \qquad (159)$$

In addition to the singular stresses at $r = 0$, non-singular components, σ_0, also occur, which signify the locally identical membrane stress state in both plates, consisting of two normal stress components and a shear stress component. Singular terms of a higher order are designated with $O(\sqrt{r})$. They can be disregarded as small quantities.

The stress intensity factors, K_I, K_{II}, K_{III}, can be determined using an approach developed by the author based on the local structural stresses at the weld spot edge:[320,321]

$$K_I = (0.58\sigma_b^{++} + 2.23\tau_q^{+-})\sqrt{t} \qquad (160)$$

$$K_{II} = (0.50\sigma^{+-} + 0.50\sigma_b^{+-} + 0.5\tau_q^{++})\sqrt{t} \qquad (161)$$

$$K_{III} = 1.41\tau_l^{+-}\sqrt{t} \qquad (162)$$

σ_b^{++} and σ_b^{+-} designate the symmetrical and antisymmetrical radial bending stress portions in plate points of the weld spot edge lying one above the other, τ_q^{+-} and τ_q^{++} the corresponding shear stress portions (normal to plate plane), σ^{+-} the antisymmetrical radial membrane stress portion and τ_l^{+-} the antisymmetrical longitudinal shear stress portion (in plate plane) (σ^{++} and τ_l^{++} form part of σ_0). The above equations are derived for load stress cases, but they can also be applied to residual stress cases provided these can be described by a plate model.

In the equations (160) to (162) the square root from the plate thickness, t, occurs repeatedly and plays the same part in spot welded joints as the crack length in traditional fracture mechanics. The stress intensity factors increase with the plate thickness if the basic stress in the plate remains unchanged. This approach is correct as long as the weld spot diameter is markedly greater than the plate thickness, which is the case in practice. It differs essentially from the older approach of Pook[328,329] on the basis of Chang and Muki[330] for the tensile shear specimen, according to which the mean shear stress in the weld spot jointing face should be multiplied by the square root of the weld spot diameter, d, supplemented by a correction factor dependent on d/t. Whereas the author's approach proceeds from the local structural stresses, the approach of Pook with its averaged shear stress is based on a global quantity, with the result that the local conditions are not precisely reflected. Solutions with finite solid elements, on the other hand, are not very accurate in consequence of coarse

Table 23. Stress intensity factors at the front face (K_I, K_{II}) and the side face (K_{III}) vertex of the weld spot edge in the tensile shear specimen ($F = 1kN$, $t = 1mm$, $d = 5mm$, other dimensions to some extent differing)

Authors	K_I, N/mm$^{3/2}$	K_{II}, N/mm$^{3/2}$	K_{III}, N/mm$^{3/2}$
Radaj	54.0	137.5	64.3
Pook[328,329]	163.6	199.8	199.8
Yuuki[331]	48.6	106.8	
Smith and Cooper[332,333]	63.2	132.4	99.0

meshing and, without recourse to formulae of the type stated, do not supply any parameter dependence from which a generalisation could be made. The author's results for the tensile shear specimen are compared in Table 23 with those of Pook,[328] as well as those obtained by Yuuki[331] and Smith and Cooper[332,333] using the finite element method.

In addition, a remarkable aspect of equations (160) to (162) is that the stress intensity factor is not only dependent on the bending and membrane stresses at the weld spot edge but additionally on the transverse shear stress, τ_q, normal to plate plane. For this reason, the distribution of the stress intensity factor may also deviate more strongly from the distribution of the structural stress on the inside of the plate. On the other hand, the relation is simplified in anticipation of equations (166) to (171) with a relatively small transverse shear stress:

$$K_I = 0.15(\sigma_{ti} - \sigma_{to} + \sigma_{bi} - \sigma_{bo})\sqrt{t} \tag{163}$$
$$K_{II} = 0.25(\sigma_{ti} - \sigma_{bi})\sqrt{t} \tag{164}$$
$$K_{III} = 0.71(\tau_{ti} - \tau_{bi})\sqrt{t} \tag{165}$$

The symmetrical and asymmetrical stress portions contained in equations (160) to (162) are determined from the radial (σ), axial (τ_q) and tangential (τ_t) structural stress components at the weld spot edge.[320] The indices t and b signify the top and bottom plate, the indices o and i their outer and inner surface. The same thickness and material are assumed for the connected plates:

$$\sigma^{+-} = 0.25(\sigma_{ti} + \sigma_{to} - \sigma_{bi} - \sigma_{bo}) \tag{166}$$
$$\sigma_b^{++} = 0.25(\sigma_{ti} - \sigma_{to} + \sigma_{bi} - \sigma_{bo}) \tag{167}$$
$$\sigma_b^{+-} = 0.25(\sigma_{ti} - \sigma_{to} - \sigma_{bi} + \sigma_{bo}) \tag{168}$$
$$\tau_q^{++} = 0.5(\tau_{qt} + \tau_{qb}) \tag{169}$$
$$\tau_q^{+-} = 0.5(\tau_{qt} - \tau_{qb}) \tag{170}$$
$$\tau_t^{+-} = 0.5(\tau_{ti} - \tau_{bi}) \tag{171}$$

If numerical solutions are used for the structural stress problem, the stresses on the righthand sides of the equations are completely known. If a procedure based on measurement is employed, eight measuring points are generally required for each edge point: four measuring points directly in front of the weld spot edge (inside and outside on both plates) to distinguish tension from bending and four other measuring points at a slightly greater distance to determine the transverse shear force from the gradient of the bending moment. With negligible transverse shear, the number of measuring points is reduced.

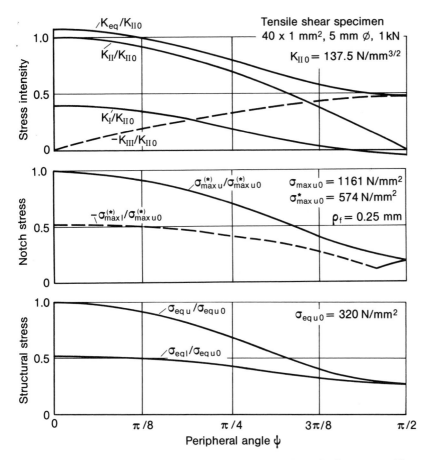

246 Structural stresses, notch stresses and stress intensity factors at weld spot of tensile shear specimen.

Only in exceptional cases, though, is the single measuring point on the outside of the plate, as frequently used in practice, sufficient.

The standardised tensile shear specimen was taken as a typical application for analysis using the methods stated, Fig.246. The resulting equivalent structural stress, σ_s, is determined according to von Mises, the resulting equivalent stress intensity factor K_{eq} in the simplified form of equation (185). All three local stress parameters, structural stress, notch stress and stress intensity factor, decrease from the maximum value on the front face to a markedly lower value on the side face of the weld spot. The three quantities behave in a similar manner.

A remarkable aspect of the stress intensities is that quite a severe slit opening, K_I, occurs at the front face in addition to transverse shearing, K_{II}, this slit opening changing into weak slit closing on the side face. In addition, a relatively large longitudinal shearing, K_{III}, is noticeable on the side face. The ratio K_I/K_{II} = 0.39 determined is significantly smaller than K_I/K_{II} = 0.88 for the centre section model of the specimen in which the beam effect takes the place of the plate effect.

10.7   Local stress parameters for common weld spot specimens

The local stress parameters at the weld spot edge (structural stress, notch stress, stress intensity factor) were determined numerically for common weld spot specimens, specimen dimensions and specimen loads,[326] Fig.247.

The group of simple specimens comprises:

— single face tensile shear specimen (SS) and variant with edge stiffened by folding (ES);
— two spot tensile shear specimen (TS);
— double face tensile shear specimen (DS);
— through-tension specimen (TT);
— cross tension specimen in cross shape (CT) and U shape (UT);
— peel tension specimen as double angle specimen (DA) and as angle to plate variant (AP).

The group of structural component specimens comprises:

— double hat profile specimen (DH) subjected to torsional moment and internal pressure;

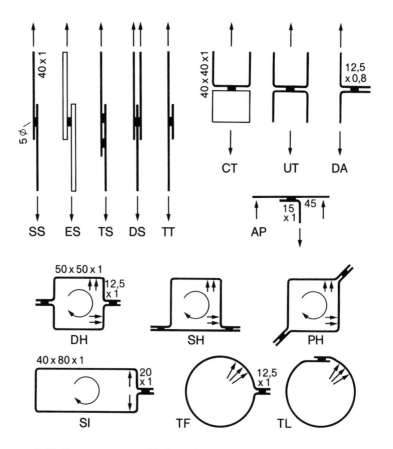

247 Common spot welded specimens, scope of numerical investigation.

— single hat profile specimen (SH) subjected to torsional moment and internal pressure;
— peaked hat profile specimen (PH) subjected to torsional moment and internal pressure;
— single sided hat profile specimen (SI) subjected to torsional moment and straddling force;
— tubular flange joint specimen or tubular peel tension specimen (TF) and tubular lap joint specimen or tubular tensile shear specimen (TS) subjected to internal pressure.

It is only possible to summarise the results of the calculations here. This summary is restricted to the stress intensity factors, the resulting equivalent value of which, K_{eq}, according to equation (185) is used for assessing the relative load carrying capacity of the particular weld spot. The equivalent structural stress and notch stress not shown here, behave in a similar way to the equivalent stress intensity factor. The larger deviations in some cases are caused particularly by the fact that the shear stress normal to the plate plane does not have any effect on the equivalent stress in the plate surface, but does affect the stress intensity. On the other hand, the eigenforces manifest themselves primarily in the equivalent stress and scarcely in the stress intensity. As regards K_{eq}, it can be stated that the strength hypothesis for multi-axial stressing on which it is based is questionable and a modification on the basis of future experimental results appears probable.

The distribution of the stress intensity factors, K_I, K_{II}, K_{III}, at the weld spot edge with the same tensile shear force is presented in Fig.248 for the tensile shear specimens including the tubular tensile shear specimen. The K_{III} stress intensity is dominant. $K_{II\,max}$ appears at the front face, in many cases combined with $K_{I\,max}$ (markedly smaller), while $K_{III\,max}$ can be found at the side face. Stiffening the edge of the specimen by folding reduces $K_{I\,max}$ to approximately two thirds. The two spot specimen displays a reduced K_{II} as a consequence of the distribution of the load on to two spots. In the case of the double shear specimen, K_{II} is reduced to one third (with the same spot resultant force as in the single shear specimen) as a consequence of the completely suppressed weld spot slanting whereas K_{III} at the side face remains unchanged and now determines the load carrying capacity. K_{II} on the through-tension specimen is, as is to be expected, very small. The stress intensity factors in the tubular tensile shear specimen are partly enlarged, because of the particularly short lap length and the slightly asymmetrical stressing of the joint. A striking element in the comparative evaluation of weld spot load carrying capacity according to K_{eq}, Table 24, is the high load carrying capacity of the double shear specimen ($2 \times 2.09 = 4.18$). From the greatly differing load carrying capacity of the weld spots in the various specimens it follows that, purely for this reason, there cannot be a uniform tensile shear load carrying capacity transferable from the specimen to the structural component.

The stress intensity factors for the cross tension specimens with the same cross tension force, are presented in Fig.249. The K_I stress intensity is dominant. K_I is constant over the circumference of the weld spot in the case of the standardised cross tension specimen, and is slightly increased or decreased

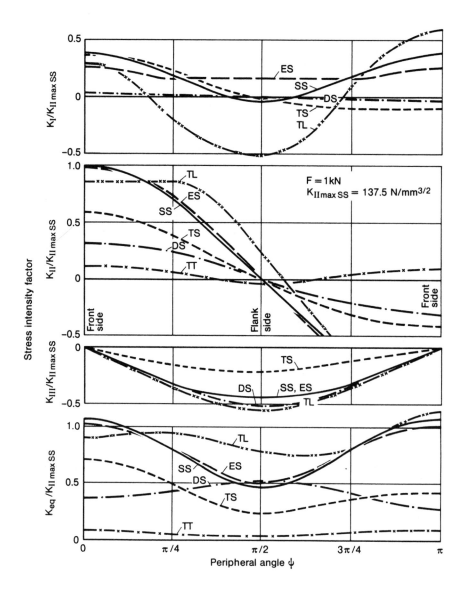

248 Stress intensity factors at weld spot of different tensile shear specimens subjected to same tensile shear force.

Table 24. Load carrying capacity of weld spot, relation of various tensile shear specimens according to K_{eq}, tubular lap joint specimen included

Specimen type	SS	ES	TS	DS	TT	TL
Load carrying capacity	1.0	1.06	1.54	2.09	10.4	0.93

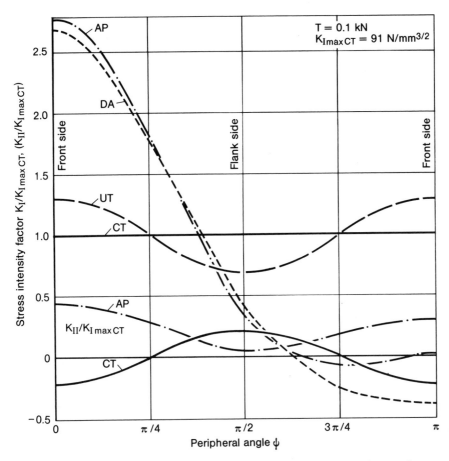

249 Stress intensity factors at weld spot of different cross tension specimens subjected to same cross tension force.

Table 25. Load carrying capacity of weld spot, relation of various cross tension specimens according to K_{eq}, relation of tensile shear specimen

Specimen type	CT	UT	DA	AP	CT/SS
Load carrying capacity	1.0	0.79	0.38	0.37	0.16

in the vertices in the case of the U shaped cross tension specimen. In the case of the two peel tension specimens, a sharp increase occurs, as is to be expected, in the vertex facing the peel force. The weld spot load carrying capacities are compared in Table 25. The peel tension specimens resist to less than half the pure cross tension specimens. The cross tension load carrying capacity is significantly less than the tensile shear load carrying capacity. The relations stated vary greatly with plate thickness and support spacing. The difference between cross and peel tension load carrying capacity can be sharply reduced but not totally eliminated by including the resulting bending moment of the weld spot joint face into the characteristic parameter of the load carrying

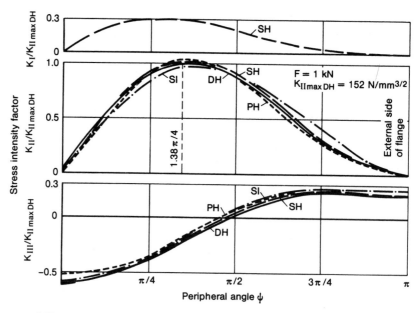

250 Stress intensity factors at weld spot of different hat profile specimens subjected to torsion loading for same shear force at weld spot.

Table 26. Load carrying capacity of weld spot, relation of various hat profile specimens subjected to torsional moment according to K_{eq}, relation to tensile shear specimen

Specimen type	DH	SH	PH	SI	DH/SS
Load carrying capacity	1.0	0.95	0.99	1.04	0.95

capacity. It is, therefore, not possible to state a uniform peel tension load carrying capacity transferable from the specimen to the structural component.

The stress intensity factors for hat profile specimens subjected to torsional moment with the same shear force at the weld spot (according to Bredt's formula) are shown in Fig.250. The K_{II} stress intensity dominates. $K_{II\ max}$ appears approximately 30 degrees in the direction of the web near to the vertex of the weld spot at the front face. $K_{III\ max}$ appears at the side face on the inside of the flange. A slight K_I stress intensity appears only with the single hat profile specimen. The relations of the weld spot load carrying capacities (weld spot shear force according to Bredt's formula) indicate only slight differences, both to each other and also compared with the tensile shear specimen, Table 26.

For the hat profile specimens including the tubular flange joint specimen subjected to internal pressure, the stress intensity factors with the same cross tension force at the weld spot show the single sided increase known from the peel tension specimens, Fig.251. The relations of the weld spot load carrying capacities reveal particularly favourable conditions for the double hat profile specimen, Table 27, as a result of the favourable location of the flanges relative to the bending moment distribution in the profile contour. The double hat profile specimen is for this reason (and as a result of a shorter clamping distance) also practically equivalent to the cross tension specimen.

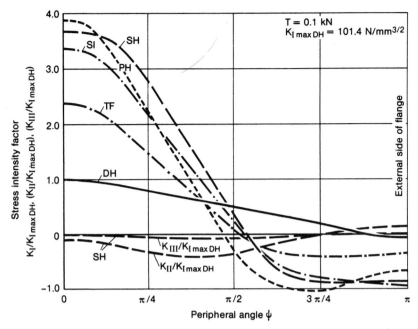

251 Stress intensity factors at weld spot of different hat profile specimens subjected to internal pressure for same cross tension force at weld spot.

Table 27. Load carrying capacity of weld spot, relation of various hat profile specimens subjected to internal pressure according to K_{eq}, tubular flange joint specimen included, relation to cross tension specimen

Specimen type	DH	SH	PH	SI	TF	DH/CT
Load carrying capacity	1.0	0.27	0.26	0.30	0.42	0.92

Within the framework of the analysis[326] summarised above, an answer is also given to the question of what effect a variation of plate thickness, t, or of weld spot diameter, d, in the case of the tensile shear and cross tension specimen has on the local stress parameters at the weld spot edge. This is done, proceeding from t = 1mm, and d = 5mm, by halving and doubling t on the one hand (with constant d), on the other hand by halving and doubling d (with constant t). In addition, consideration is given to the parameter coupling d = $5\sqrt{t}$ often used in practice.

A first approximation for the parameter dependence is obtained on the basis of the nominal structural stress formulae. Hence for the tensile shear specimen, it follows from equation (150):

$$\sigma_{ns} \propto 1/t \text{ with d constant,}$$

$$\sigma_{ns} \propto 1/d \text{ with t constant,}$$

$$\sigma_{ns} \propto 1/t^{2/3} \text{ with d } = 5\sqrt{t}.$$

Table 28. Load carrying capacity of weld spot in tensile shear specimen on the basis of approximation formulae (subscript ns) and more precisely on the basis of finite element analysis (subscript fe), relation for variable plate thickness, t, and weld spot diameter, d, strength criterion, K_{eq}, (the subscript 1/5 means t = 1mm, d = 5mm)

	Plate thickness, t, mm			Spot diameter, d, mm		
	0.5	1.0	2.0	2.5	5.0	10.0
$(F/F_{1/5})_{ns}$	0.71	1.0	1.41	0.5	1.0	2.0
$(F/F_{1/5})_{fe}$	0.74	1.0	1.29	0.55	1.0	1.87

Table 29. Load carrying capacity of weld spot in cross tension specimen on the basis of approximation formulae (subscript ns) and more precisely on the basis of finite element analysis (subscript fe), relation for variable plate thickness t and weld spot diameter d, strength criterion, K_{eq} (the subscript 1/5 means t = 1mm, d = 5mm)

	Plate thickness, t, mm			Spot diameter, d, mm		
	0.5	1.0	2.0	2.5	5.0	10.0
$(T/T_{1/5})_{ns}$	0.35	1.0	2.83	0.75	1.0	1.50
$(T/T_{1/5})_{fe}$	0.35	1.0	2.79	0.80	1.0	1.52

Hence for the cross tension specimen, it follows from equation (151) or (152) with D/d constant:

$$\sigma_{ns} \propto 1/t^2 \text{ with d constant,}$$

$$\sigma_{ns} \text{ independent of d with t constant,}$$

$$\sigma_{ns} \propto 1/t^2 \text{ with d } = 5\sqrt{t}.$$

The reciprocal values of the proportionalities presented above directly state the load carrying capacity dependence of the weld spot if σ_{ns} is regarded as being decisive for local strength. If, on the other hand, the stress intensity factor is regarded as being decisive for the strength, the proportionalities stated have to be multiplied by $\sqrt{t}$. This weakens the influence of the plate thickness compared with the above expressions. The conclusion which can additionally be drawn from the different exponents for the influence of thickness in the case of the tensile shear and cross tension specimen, is that the cross tension strength increases relative to the tensile shear strength as plate thickness increases.

The parameter dependencies presented are approximations, which, however, are confirmed with a deviation of only a few per cent by the exact numerical solutions, Tables 28 and 29.

Smith and Cooper[332,333] have calculated the stress intensity factor, K_{eq}, for the tensile shear specimen (K_{eq} in accordance with the tangential stress criterion, equations (180), (181)), using finite solid element models and thus achieved more accurate results relative to Pook[328,329] and Yuuki,[331] Fig. 252. The parameter range of d and t covers the cases of particular importance in practice. The broken curve identifies the commonly used dimensioning based on d = $5\sqrt{t}$. The curve patterns in this diagram agree approximately with the

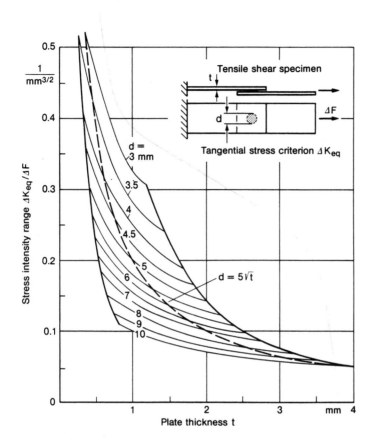

252 Equivalent stress intensity range at weld spot of tensile shear specimen dependent on plate thickness and weld spot diameter, after Smith and Cooper.[332]

proportionalities substantiated in the preceding sections, $K_{eq} \propto \sigma_s\sqrt{t}$ and $\sigma_s \propto F/td$, from which follows $K_{eq}/F \propto 1/d\sqrt{t}$.

The finite element results of Yuuki for the stress intensity factors, K_I and K_{II}, of the single face tensile shear, double face tensile shear and cross tension specimen are compiled in a handbook[290] including variation of plate thickness and plate width. A new efficient boundary element procedure for determining the stress intensity factors at spot welds has recently been published by Yuuki and Ohira.[359]

The contributions of Fujimoto, Mori and Sakuma[334] and of Hahn and Wender[335] are worth emphasising as historically early finite element solutions regarding structural stress distribution in the tensile shear specimen. As regards weld spot stress in the hat profile specimen subjected to torsional moment, reference is also made to the solutions achieved by Niisawa and Tomioka[336,337] on the basis of a shear-lag scheme.

10.8   Assessment of load carrying capacity of weld spots in structural components

The methods and knowledge in respect of assessing the load carrying capacity of weld spots in more complex structural components on the basis of local stress parameters are still in the stage of development and tentative applications testing.

The assessment by calculation can principally be performed in the same way as for the comparison of weld spot specimens presented in the preceding section, especially as structural component specimens and load cases were included in this investigation, for instance the box beam consisting of a hat profile and a cover plate which is often used in automotive engineering. It is good practice to proceed in the case of the component from the resultant joint forces and eigenforces acting at the weld spot, which can either be approximated using simplified engineering formulae or also more accurately calculated using finite element models. As already mentioned in section 3, a relatively coarsely discretised model is adequate in the latter case, a model in which only one double nodal point is provided for each weld spot.

In simpler cases, a static equilibrium consideration (without reference to deformation behaviour) is sufficient to determine the resultant joint forces at the weld spots from the external load and from the position of the weld spots in the welded joint. Mori *et al*[338] have demonstrated that the fatigue strength of a spot welded joint between hat profile box beams can be derived from the fatigue strength of the tensile shear specimen and angle-to-plate specimen on the basis of these weld spot forces, Fig.253. Characteristically, transfer problems were already encountered, however, with the local strains measured

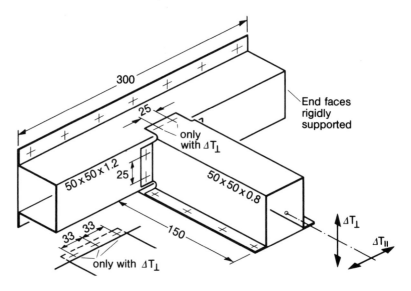

253 Spot welded joint between hat profile beams, loading in fatigue test, after Mori *et al.*[338]

at a certain small distance from the weld spot, mainly on the outside of the plate. Harmonisation was only achievable with special tensile shear and cross tension specimens, which reflected the geometrical details of the beam joint.

As was already shown in the comparison of specimens in the preceding section, it is not sufficient here simply to record the tensile shear and cross tension force. Equally important are the secondary resultant forces in the jointing face, in particular the cross bending moment, but also the eigenforces at the weld spot in the plate centre plane. Equivalent forces in the plate centre plane can also be used in place of the forces in the weld spot jointing face.[338,339] They differ by the offset bending moment which is produced in transferring the shear forces from the weld spot jointing face to the plate centre plane.

The functional-analytic solution for a simple, statically indeterminate joint force distribution problem is mentioned for the sake of completeness.[340,341] The solution was achieved for the model of two infinitely extended elastic tension plates which are connected to each other by a row of weld spots or by a weld spot field with no transverse offset of the plates. The weld spots at the ends of the rows running longitudinally or transversely to the direction of tension are subjected to especially high forces, the same being true of weld spots at the corners of the rectangular fields. For instance, the concentration factor for the shear force in the square field of 3 × 3 weld spots in the infinite tension plate is 1.32.

More widely used in practice than numerical solutions is the examination of local strains at the weld spot edge on the outside or inside of the plates based on measurement, Fig.254. On the outside of the plate, the strain gauges are attached close to the edge of the electrode indentation area and aligned radially, on the inside of the plate as close as possible to the weld spot edge notch after bending up the unloaded lapped plate[334] (this slightly alters the structural behaviour) or by means of holes drilled in the mating plate.[342] The inside

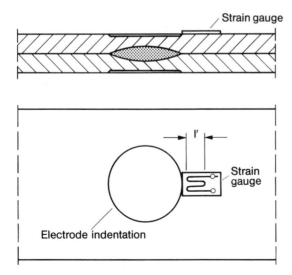

254 Strain gauge on outer surface of plate at edge of electrode indentation.

measurement is closest to the actual crack initiation point, and can thus be correlated with the fatigue strength of the weld spot better than other quantities. In the case of outside measurement, the inside strain has to be estimated on the basis of the outside strain. For single face tensile shear the ratio is $\sigma_i/\sigma_o \approx -2.0$, for cross tension $\sigma_i/\sigma_o \approx -1.0$, for double face tensile shear and through tension $\sigma_i/\sigma_o \approx 1.0$ assuming identical plate thickness and identical plate material in the joint. It is therefore necessary to know which type of loading dominates at the considered weld spot. The dominant type of loading can be determined, for example, from the ratio of radial strain at the front face and tangential strain at the side face, provided it is only necessary to distinguish between tensile shear and cross tension loading in the single face joint.[343] Another proposal is based on distinguishing between tensile shear and cross tension loading by means of the photo-elastic surface coating method.[344] In addition, for single measurements on the outside or inside, it is necessary to know the principal direction of the loading of the weld spot to be able to record the maximum strain value. Otherwise, several radially aligned strain gauges should be distributed at the circumference of the weld spot or suitable ring circular rosettes are used. As the strain gauges are not small compared with the plate thickness and weld spot diameter (usual strain gauge length 0.6 to 2.0mm), the result of the strain measurement is an averaging over strain distributions with steep gradients. The strain measured depends on the size of the strain gauge and on how far away it is positioned from the edge of the electrode indentation or weld spot. It is in any case less than the actual maximum value. Relative statements in respect of load carrying capacity are at least possible on the basis of the strains measured.

10.9   Effect of residual stresses and hardness distribution

In addition to the dominant influence of the local cyclic stress on weld spot load carrying capacity, expressed by structural stress, notch stress and stress intensity and compared with the relevant local material characteristic values, there are a variety of additional, often only secondarily effective, local influencing variables, which can be estimated only roughly initially. These include residual stresses, hardness distribution, large deflections and local yielding. These influences are presented in this and in the following two sections to the extent that relevant knowledge exists.

The local residual stresses, which can be expressed by structural stress, notch stress and stress intensity, are comparable with the effect of a static prestress. Compressive prestress has a strength-increasing effect, tensile prestress a strength-decreasing effect both on crack initiation and crack propagation. The strength increase by crack closure occurs in the stress relieved state for $R \lesssim 0$, in the as-welded state for $R \lesssim -1$.

Available knowledge on the residual stress state at the weld spot is small and, at this stage, not sufficient for generalisations. On the one hand, this is surprising because the stress state, to the extent that it is caused by the welding process, can be approximated as being axisymmetrical and is thus more easily

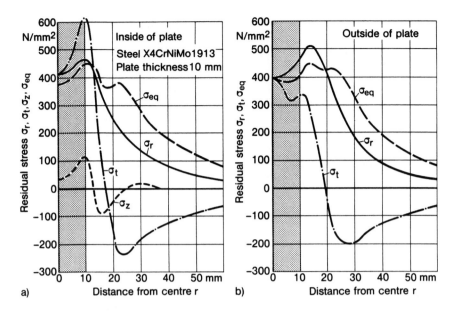

255 Residual stresses at weld spot, after Schröder and Macherauch.[345]

accessible for numerical analysis and to measurement. On the other hand, however, formation of residual stresses is a complex, non-linear, transient phenomenon influenced by a multitude of parameters. Attempts to model it for the purpose of obtaining generally valid statements encounter unusual difficulties. Moreover, the welding residual stresses are subject to change as a consequence of service loading.

One of the few welding residual stress analyses which has been conducted hitherto on the weld spot, is considered below by way of an example. It concerns the solution of Schröder and Macherauch,[345,346] verified numerically by measurement, for spot welding of plates made of X4CrNiMo1913 alloyed steel (plate thickness 10mm, weld spot diameter 20mm), Fig.255. The finite element model on which this is based, is characterised by the requirement that the different thermomechanical conditions on the surface and in the interior of the plates have to be taken into consideration. The temperature field specified close to reality is characterised by maximum temperatures in the area of the weld nugget and maximum temperature gradients in the electrode contact area and at the edge of the weld nugget. No microstructural transformations of relevance to the residual stresses occur. The weld spot cooled down without electrode transverse pressure. Biaxial tension exists in the weld spot. The radial tensile stresses, σ_r, decrease slowly to zero outside the weld spot. The tangential tensile stresses, σ_t, decrease at a faster rate, switch into the compressive range and only then approach zero level. Of particular interest is the increase in the tangential stress at the weld spot edge on the insides of the plates beyond the value of the uniaxial yield stress ($\sigma_Y = 440\,\text{N/mm}^2$, yield hypothesis after von Mises). The increase is caused by the triaxial tensile stress state (there are also axial tensile stresses) which occurs in this case.

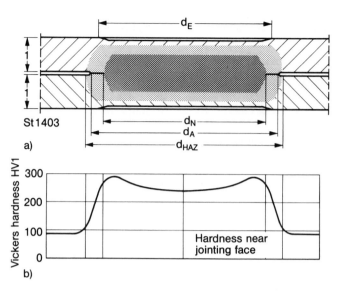

256 Microstructural zones (a) and hardness profile (b) at weld spot; diameters d_E, d_N, d_A and d_{HAZ} explained in text.

In respect of the unknown residual stress intensity factor at the weld spot edge, it can be stated that, assuming axisymmetry, only K_I occurs and that K_I can be scarcely or only partially determined from the bending portion of the radial residual stresses according to equation (160) (the bending portion does not correspond to the difference of the surface stresses, owing to the non-linear distribution of these stresses; by the way, according to Fig.255, slit closing can be expected), but more readily from the distribution of the axial stresses in the slit faces, which in the mechanical model are assumed to remain in contact. Similar statements can be made regarding the notch residual stresses at the weld spot edge, the radial residual stress being fully notch-effective (with K_I it was only the bending portion).

Locally increased hardness means locally increased fatigue strength. The hardness distribution is explained by considering the microstructural state in a spot welded joint of sheet steel, Fig.256a. The weld nugget (diameter d_N) with cast microstructure is surrounded by the heat affected zone (diameter d_{HAZ}). It may extend at the top and bottom up or down to the electrode indentation area (diameter d_E). The free slit faces extend into an annular adhesive face (diameter d_A) with reduced bonding strength surrounding the weld spot within the heat affected zone and only then reach the actual weld nugget. The weld spot edge notch is located at the outer edge of the adhesive face. A typical hardness profile is shown in Fig.256b, relatively high hardness in the weld nugget, steep increase of hardness in the heat affected zone, maximum hardness at the edge of the weld nugget. This rough assessment already reveals three possible types of failure: fracture outside the heat affected zone in the unnotched plate with a relatively low hardness and fatigue strength, fracture in the heat affected zone at the point of the edge notch with increased (widely scattering) hardness levels, fracture at the edge of the weld spot nugget under strongly increased hardness and fatigue strength after the bonding face has torn open. The above analysis reveals that a notch stress or stress intensity assessment with the fatigue strength values of the non-hardened base material is likely to be on the safe side. In the case of light metal alloys, the hardening is replaced in many cases by a softening with locally reduced fatigue strength.

10.10   Large deflections and buckling

The modelling concepts and numerical solutions described with respect to local stress at spot welded joints include the assumption of small displacements and rotations. In other words, the equilibrium and boundary conditions are formulated for the undeformed state of the structure, which is in harmony with the general linearisation of structural behaviour. The resulting linear solution requires checking for applicability and expanding as need be, by a step-by-step iteration into the non-linear range.

Non-linear effects in the form of large deflections (large relative to plate thickness) and large rotations occur particularly with thin sheet metal. The large deflections, mainly in the immediate vicinity of the weld spots, may, after overcoming the gap width, cause the plates to touch and support each other. On the other hand, they may cause buckling, primarily at some distance from the weld spots. Assuming ideal geometry without predeformation, the buckling occurs as an instability phenomenon. A characteristic feature for corresponding numerical models on the other hand is that predeformation, notably gaps and bending distortion, has to be simulated in the initial state, and that the numerical results (stresses, strains, deflections, rotations, reaction forces) depend non-linearly on the load level. The large rotations mentioned, for example, have an influence on the mechanical behaviour of weld spots subjected to tensile shear loading and also require simulation by a non-linear model if necessary.

Geometrically non-linear models of the structure are suited to a numerical solution even in complex cases, more roughly using approximate formulations or more precisely using the finite element method. The local stress parameters can, as hitherto, be determined from the joint forces and eigenforces in the approximate formulation or in the finite element model. Allowance only requires to be given to the large deformations in this formulation or model.

The slanting angle occurring at the weld spot of the standardised tensile shear specimen ($t = 1mm$) in the linear model, is about 0.5 angular degrees when subjected to a load approximately equal to the fatigue strength ($F = 1kN$). This amount has to be multiplied in the fatigue strength for finite life range. Owing to the slanting angle, not only tensile shear $F_\parallel$ is produced at the weld spot but also cross tension $F_\perp$, which reduces the fatigue strength, Fig.257:

$$F_\parallel = F \cos \gamma \qquad\qquad\qquad (172)$$

$$F_\perp = F \sin \gamma \qquad\qquad\qquad (173)$$

A geometrically conditioned maximum value, γ_{max}, in the static tensile shear test is achieved after Koenigsberger[347] if the initially offset plates are in

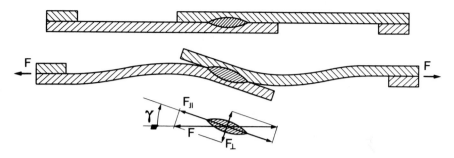

257 Slanting of weld spot subjected to tensile shear, superimposed cross tension by slanting.

alignment finally as a result of the fact that the weld spot has rotated accordingly, with plastic hinges being formed on its end faces, Fig.258:

$$\tan \gamma_{max} = \frac{t}{\sqrt{d^2 - t^2}} \tag{174}$$

The approximation d = 4t results in γ_{max} = 14.5 angular degrees. This limit value determined in accordance with equation (174) is in fact exceeded slightly when the weld spot fails by buttoning under static load, but it can be considered applicable for considerations in the fatigue strength for finite life range. In the case of weld spots with hardened heat affected zone, the diameter of the heat affected zone, and not the nugget diameter, should be used as the weld spot diameter, d, in equation (174), provided the fractures are shifted to the outside as a result of the hardening.

Systematic calculations and statements regarding the influence of mutual plate contact on the local stress of spot welded joints are not known. It is likely that this influence becomes important at higher loads (after overcoming the gap width), *i.e.* in the fatigue strength for finite life and in the low cycle fatigue strength range.

The important case in automotive engineering of large deflections of spot welded box section members manufactured from hat profiles has, in contrast, been successfully dealt with in a rough numerical solution.[348] If a bending load is applied to members of this type with spot welded flanges in the region of maximum bending stresses, the flanges are subjected to longitudinal compression and tension. Assuming small deformations, this does not cause

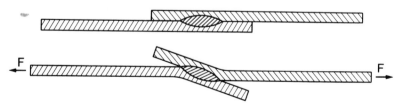

258 Aligned tension plates after formation of plastic hinges at weld spot, limit state.

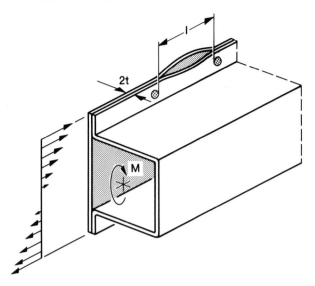

259 Flange gap formation caused by longitudinal bending-compression.

any local increase in stress at the weld spots. In reality though, large deflections with high bending stress do occur at the weld spot as well as in the middle of the flange section between two weld spots, especially when subjected to longitudinal compression. The gap formation at the flanges under longitudinal compression, Fig.259, can be interpreted as a rod or plate buckling process. In the case of a box section member manufactured from hat profile and cover plate, it is the cover plate particularly which is at risk of buckling. It deflects in the shape of a half wave between two weld spots even when subjected to relatively low longitudinal stress. Cracks are initiated at the weld spot and in the middle between two weld spots at the outer edge of the flange. In the case of the box section member manufactured from two hat profiles, the flanges restrained by the web deflect in a multiwave shape when subjected to relatively higher longitudinal stress. Cracks occur over the entire length of the flanges and at the weld spots.

Proceeding from the buckling load formula for rods clamped bending rigid at both ends, with rod length, l, equal to the flange section length between two weld spots, flange width, b_f, flange thickness, t, and elastic modulus, E, the critical longitudinal stress, σ_{cr}, is obtained:

$$\sigma_{cr} = \frac{\pi^2 E t^2}{3 l^2} \tag{175}$$

With continuous transverse elastic support of the rod to simulate the supporting effect of the residual plate or of the web on the flange, expressed by the spring constant, k (per unit of flange length), which in turn depends on the support spacing, l_f, of the flange centreline, Fig.260, this produces according to Ref.348 ($k = F_f/wl = Et^3/4l_f^3$ determined according to beam bending theory):

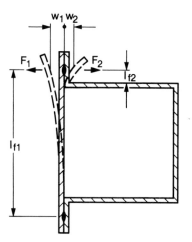

260 Elastic transverse support of compressive loaded flanges forming a gap, support spacing, l_{f1} and l_{f2}, after Oshima and Kitagawa.[348]

$$\sigma_{cr}^* = \frac{\pi^2 E t^2}{3l^2 \left(m^2 + \frac{3l^4}{m^2 \pi^2 b_f l_f^3} \right)} \tag{176}$$

The number of half waves, m, should be selected so that the minimum value of σ_{cr}^* appears. For the cover plate in Ref.348 m = 1 and $\sigma_{cr}^*/\sigma_{cr} = 1.68$. For the hat profile flange m = 5 and $\sigma_{cr}^*/\sigma_{cr} = 42.6$. This means that the cover plate buckles significantly more easily than the hat profile flange, yet is prevented from deflecting in combination with the hat profile with the result that the longitudinal stresses in the hat profile flange increase initially until buckling also occurs here. Immediate deflection of the cover plate is conceivable, however, in those cases where the longitudinal shortening in the flange direction necessary for the buckling deflection to occur, for instance, as a result of global flange plate bending. The (in some cases) different location of the crack in the cover plate and in the hat profile flange can be explained from the single and multiwave buckling behaviour.

On the basis of the buckling stress and buckling deformations, it is now possible to determine the stresses in the inner and outer faces of the flanges, as well as at the weld spot edge, as shown below for the half wave buckling of the cover plate.[348] The flange longitudinal force is designated with $f = F/F_{cr}$, the manufacture related initial deflection in the middle between two weld spots with $\delta = t_0/t$. Indexed m and e for middle and end of the flange section and i and o for inner and outer face of the flanges, this results in the following formulae if the additional structural stress increase caused by the weld spot is disregarded:

$$\frac{\sigma_{ei}}{\sigma_{cr}} = \frac{\sigma_{mo}}{\sigma_{cr}} = f\frac{1 - 3\delta}{1 - f} \tag{177}$$

$$\frac{\sigma_{eo}}{\sigma_{cr}} = \frac{\sigma_{mi}}{\sigma_{cr}} = f\frac{1 + 3\delta}{1 - f} \tag{178}$$

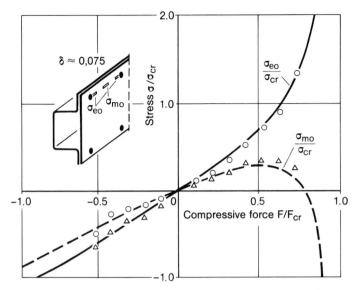

261 Stresses on outer surface of cover plate as a function of flange
compressive force, calculated curves compared with measured points,
after Oshima and Kitagawa.[348]

The result of the calculation for the cover plate of the box section member is
presented in Fig.261, compressive forces and compressive stresses appearing
positive as dimensionless variables. Obviously δ = 0.75 was selected.

The bending portion of the stress, σ_{ei}, on the inside of the plate at the weld
spot still has to be multiplied by the structural stress concentration factor, K_{sb},
for plate bending:

$$\frac{\sigma_{ei}^*}{\sigma_{cr}} = f\frac{1-3K_{sb}\delta}{1-f} \tag{179}$$

The (in some cases) strongly non-linear behaviour of the stresses, as shown,
for instance, in Fig.261, has to be considered when converting them in
accordance with equations (177) to (179) first of all into cyclic amplitudes and
mean stresses, and then into fatigue strength values. When subjecting the
member to alternating load globally, practically pulsating stresses are possible
locally.

With the given quantities f, δ and K_{sb}, it is thus possible to determine at least
the location of primary crack initiation. Crack initiation at the weld spot
dominates in the case of large initial deflection, in the example of the box section
member according to Ref.348 for δ > 0.25. No data are presented in this
reference in respect of a quantitative fatigue strength analysis on the basis of the
above derivations.

10.11   Crack initiation with local yielding and crack propagation

When loading in the fatigue strength for finite life range, plastic deformations
occur in addition to large deflections. The former are limited initially to surface

zones close to the weld spot, although they later also cover the entire thickness of the plate (plastic hinge) and extend over the cross section of the specimen (global yielding). It is not necessary to deal in detail here with the state of fully developed yielding because it is of relevance only for static load carrying capacity and, at the most, for extreme low cycle fatigue strength. As far as the fatigue strength for finite life question is concerned, the process of elastically restrained local yielding is of main interest.

The purely numerical and also the combined numerical-experimental procedure can be adopted in accordance with the methods for assessing the risk of crack initiation stated in connection with seam welded joints (refer to section 8.3) and is to be supplemented by a crack propagation calculation. In view of the fact, though, that these methods describe the crack as being initiated as soon as it has achieved a depth of about 0.25mm and the crack propagation down to this depth represents the greater part of the life, the crack propagation analysis may be dispensed with entirely in the case of high cycle fatigue of very thin sheets. The notch root approach with the fictitious rounding of the weld spot edge notch offers the best basis for incorporating local yielding. The structural stress approach in the modified form of a structural strain analysis can be applied advantageously if the fracture occurs in the unnotched region outside the heat affected zone. The fatigue fracture mechanics approach can, in turn, be applied at the weld spot edge if elastic-plastically defined stress parameters such as the J integral, can be determined here. The possibilities mentioned have hitherto only partly been realised, however. On the other hand, the crack propagation concept of Smith and Cooper[332,333] (refer to section 10.12), offers interesting aspects, also combined with local yielding.

The two contributions from Oh[349] and Kan[350] for determining the weld spot load carrying capacity of the tensile shear specimen on the basis of local elastic-plastic stress and strain (macrostructural support effect, cyclic stress-strain curve, four parameter formulation for strain S-N curve) have to be considered as being deficient in method. Both papers are based on a membrane model although, in the single shear tensile shear specimen, the plate bending causes an increase in membrane stresses by the factor 4 in the elastic range. It remains unclear in what way agreement has nevertheless been achieved with test results. The membrane stress concentration factor $K = 1 + 0.37b/d$ determined by functional analysis in Ref.349 before application of the Neuber macrostructural support effect formula, with specimen width, b, and weld spot diameter, d, can be regarded as correct (for $b/d = 8$ it follows $K = 3.96$, maximum equivalent stress after von Mises, reference stress in specimen cross section). The stress concentration factor $K = 6.13 \times 3/\pi = 5.82$ where $b/d = 3$, calculated in Ref.350 by means of finite elements, is, on the other hand, inexplicably large ($K \approx 1.5$ would have to be expected). The yield zones calculated dependent on load level are correlated in Ref.350 to the observed different fracture patterns (at first front face plate fracture, then front and flank face mixed fracture in the circumferential face and finally shearing off in the jointing face when the load range is increased).

A complete model for crack initiation and crack propagation at tensile shear loaded weld spots has been proposed by Lawrence, Wang and Corten.[360,361] The fracturing process is subdivided into the phases of crack initiation up to 0.25 mm crack depth (dominating for $N \geq 10^6$), of through-thickness crack propagation (dominating in the low cycle range) and of crack propagation across the specimen width (to be considered in cases of larger plate width). The notch stresses and notch strains are calculated using the Peterson formula for the microstructural support effect and the Neuber formula for the support effect of local yielding. According to the Peterson formula the weld spot edge is considered as fictitiously rounded with that small radius which results in the worst possible local fatigue notch factor. The notch effect is determined proceeding from the stress intensity factors according to Pook applying the formulae for blunt cracks according to Creager. The elastic part of the local strain S-N curve is evaluated (*i.e.* the Basquin formula).

The investigation[360] shows the influence of various parameters on the fatigue strength for infinite life ($N = 10^7$, endurance limit). The endurable load amplitude increases with weld spot diameter, d, and plate thickness, t. It is nearly independent of plate width, b, for larger b/d ratios. Residual stresses have a strong influence on the load carrying capacity, increasing strength with compressive residual stresses, decreasing strength with tensile residual stresses. A higher ultimate strength of the steel has no marked direct influence on the fatigue strength of the weld spot. But it has a strong indirect influence via the residual stresses which can be higher in the higher tensile steel. The investigation[361] presents finite life calculations for a preset load amplitude.

10.12   Local fatigue strength of weld spot

To be able to determine the fatigue strength of a spot welded joint on the basis of local stress or strain parameters without looking into the details of crack initiation and crack propagation, it is necessary to know the local fatigue strength characteristic values, expressed by the parameters stated. For notch or fracture mechanics analysis, the characteristic values of the base material are principally suitable, modified if necessary by the influence of the locally changed microstructure. By contrast, in analysing structural stresses or structural strains at the weld spot, endurable values are required, which have been determined at comparable spot welded joints, with the exception of fracturing outside the heat affected zone. Endurable values, which have been determined on spot welded specimens, are also closer to practical application in the case of notch and fracture mechanics analysis. In essence, such procedures then constitute evaluations relative to a certain specimen. If the comparison with the specimen is replaced finally by the comparison with the in-service proven structure, it may then be possible, with an identical type of material, to dispense entirely with precise knowledge of the local strength values.

Guide values for endurable local stress and strain parameters determined on spot welded specimens in the Wöhler fatigue test, have been obtained here by evaluating relevant literature, and are stated below. In Japan, the individual values constituting the basis have even been compiled in a data bank. Reference is made to the appropriate sections (e.g. 1.4 and 9.3) of this book with respect to the base material characteristic values. The statements are preceded in each case by a discussion of the applicable strength hypothesis for the (in the majority of cases) multiaxial local stress states.

The endurable strain range, $\Delta\varepsilon_{r\,en}$, (measured radially) at the weld spot edge on the inside of the plate, is usually stated without reference to multiaxiality. It is an obvious step to take the multiaxiality influence into account by using the von Mises distortion energy hypothesis. The hypothesis of the maximum tensile principal stress has also been proposed. The local strain state at the weld spot edge can be approximated as radially plane, in other words, the strain in circumferential direction is completely suppressed.

Endurable values, $\Delta\varepsilon_{r\,en}$, are available mainly for pulsating loads. The conversion to radial and circumferential structural stresses can be achieved purely elastically, $\Delta\sigma_r = E\Delta\varepsilon_r/(1-v^2)$ and $\Delta\sigma_t = v\Delta\sigma_r$, with respect to fatigue strength for infinite life (not with respect to fatigue strength for finite life). Hence, the von Mises equivalent stress from these stresses is $\Delta\sigma_{eq} = E\Delta\varepsilon_r(1 - v + v^2)^{1/2}/(1-v^2)$, or $\Delta\sigma_{eq} \approx E\Delta\varepsilon_r$ with $v = 0.3$. Hence, $\Delta\varepsilon_{r\,en}$, converted uniaxially and elastically to a stress, results in the endurable equivalent structural stress, $\sigma_{eq\,en}$, with the restriction, however, that $\Delta\varepsilon_r$, as an averaged value at a certain distance from the weld spot edge is smaller than the actual maximum value, in other words, $\Delta\sigma_{eq\,en}$ cannot also be directly related to the numerically determined maximum stress values. In addition, as the result for $\Delta\varepsilon_r$ also varies depending on the strain gauge length and distance of the strain gauge from the weld spot edge, differing literature data with respect to the value of $\Delta\varepsilon_{r\,en}$ are explainable by this fact alone.

The evaluation of the published endurable strain and stress ranges at the weld spot[119,338,339,343,351,354] does not result in a uniform picture for the reasons mentioned and in view of the need to calculate back from the strain or stress measured on the outside of the plate to the strain or stress on the inside of the plate. Some of the endurable values differ scarcely at all from the values of the unnotched base material. The data presented below are therefore restricted to two specific investigations, substantiated by relatively many test data.

The investigation conducted by Kitagawa, Satoh and Fujimoto[351] was the source for endurable radial strains, determined in tensile shear and cross tension specimens subjected to pulsating load, the strains being measured with 1 mm long strain gauges on the outside of the plate, immediately outside the edge of the electrode indentation. Conversion to the values for the inside of the plate stated in Table 30 was based on the factor 2.0 in the case of tensile shear, and on the factor 1.0 in the case of cross tension. The values $\Delta\varepsilon_{r\,en}$ for N = 10^7 characterise the fatigue strength for infinite life. These correspond to the stresses $\Delta\sigma_{eq\,en}$ = 168-336 N/mm² for unalloyed steel and $\Delta\sigma_{eq\,en}$ = 105-210 N/mm² for higher strength steel, which thus comes off worse. The elastic conversion of the strain values in the fatigue strength for finite life range results in unrealistically high stress values because of the yield phenomena not taken into account.

Table 30. Endurable strain range, $\Delta\varepsilon_{en}$, on plate inside at weld spot edge dependent on number of cycles, N, mean values from tensile shear and cross tension specimens made of steel, after Kitagawa *et al*[351]

Steel	Yield limit, σ_Y, N/mm²	Thickness, t, mm	$\Delta\varepsilon_{en} \times 10^4$		
			N = 10^5	N = 10^6	N = 10^7
Unalloyed	170-195	0.8-1.6	20-34	12-21	8-16
Low alloy	225-255	0.8-1.6	16-29	10-17	5-10

Table 31. Endurable structural stress range, $\Delta\sigma_{en} = 2\sigma_{a\,en}$, on plate inside at weld spot edge dependent on number of cycles, N, scatter band from hat profile specimens subjected to torsional moment and internal pressure, after Rupp and Grubisic[119,339] (the values for N = 10^7 are extrapolations)

Steel	Yield limit, σ_Y, N/mm²	Thickness, t, mm	$\Delta\sigma_{en}$, N/mm²		
			N = 10^5	N = 10^6	N = 10^7
Unalloyed	150-175	0.8-2.5	230-490	180-340	130-190

The recent investigation of Rupp and Grubisic[119,339] on hat profile specimens of unalloyed steel under torsion and internal pressure, subjected to pulsating load, in which the strain measured on the outside with 0.6mm long strain gauge located at a distance of 1mm (centre of strain gauge) from the edge of the electrode indentation, converted via a finite element solution to the strain on the inside of the plate, resulted in the endurable stresses $\Delta\sigma_{en} = E\Delta\varepsilon_{r\,en}$ summarised in Table 31 (ignoring individual shear fractures at lower stress in the thick walled torsion specimens). The higher values apply to the thinner plates, the lower values to the thicker plates. On the basis of the narrow scatter band of the values converted to $\sigma_a/\sqrt{t/0.8}$, which are independent of thickness, as shown in Ref.119, endurable values $\Delta K_{en} = 190\text{-}215\,\text{N/mm}^{3/2}$ for N = 10^5 and $\Delta K_{en} = 125\text{-}150\,\text{N/mm}^{3/2}$ for N = 10^6 are obtained with the roughly correct formula for the stress intensity factor $K_a \approx 0.5\,\sigma_a/\sqrt{t}$ and with $\Delta K = 2K_a$. These figures are lower than the values in Table 32, which is explainable to the extent that the strains and stresses used above remain significantly below the actual maximum values at the weld spot edge (estimated factor 1/1.3) and that they are related to a state of major cracks, not of complete fracture (factor according to Rupp 1/1.15). Applying the estimated factors results in $\Delta K_{en} = 290\text{-}320\,\text{N/mm}^{3/2}$ for N = 10^5 and $\Delta K_{en} = 185\text{-}230\,\text{N/mm}^{3/2}$ for N = 10^6, in other words values which coincide with the lower limit of the scattering range of the values in Table 32.

Fracture mechanics strength values from fatigue tests with spot welded specimens are reported in Ref.331, 343, 352 and 353. These are S-N curves with a scatter band for endurable ranges of the stress intensity, ΔK_{eq}, ΔK_I and ΔK_{II}, or of the J integral, ΔJ. These ranges characterise the fatigue strength of the particular specimens analysed more globally than locally, as the life values up to fracture, not the local processes, are evaluated. The result of the evaluation depends, of course, on the correctness and accuracy of the ΔK or ΔJ

value calculated for the particular specimen. Where the approach of Pook[328,329] is used, ΔK and thus also ΔK_{en} are determined too high. In the case of the finite element solutions, there may be objections regarding accuracy if too coarse a discretisation was selected. Finally, the strength hypothesis for simultaneous loading according to all three slit opening modes, on which the evaluations are based must be regarded as being insufficiently clarified, including the possible influence of σ_0 from equation (159) on the strength characteristic value. The test evaluation on the basis of ΔK_{en} or ΔJ_{en} must, therefore, for the time being be regarded rather as a semi-empirical method for normalising test results, by which the dominant influencing parameters are revealed, than consideration of the local processes.

The fatigue fracture mechanics strength hypotheses for load cases characterised by the stress intensities, K_I, K_{II} and K_{III} ('mixed mode loading') determine the onset and the rate of crack propagation including the initially unknown crack propagation direction from a critical value of the deformation energy, of the stress or of the energy release rate. The criteria are scarcely verified in the case of cyclic loading. More detailed knowledge exists, at the time being, only for static loading characterised by K_I and K_{II}. It is necessary, therefore, to check the applicability of a criterion in each individual case.

In the case of spot welded joints, a preferred criterion, notably in Japan, has been the tangential stress criterion after Erdogan and Sih.[355] It states that the crack subjected to K_I and K_{II} loading is propagated in the direction φ_0 in which the tangential stress reaches its maximum at a small distance from the crack tip (for all stresses are singular at the crack tip itself):

$$K_I \sin \varphi_0 + K_{II}(3\cos \varphi_0 - 1) = 0 \tag{180}$$

The strength-relevant equivalent stress intensity designated also 'resulting stress intensity' follows from:

$$K_{eq} = \left(K_I \cos^2 \frac{\varphi_0}{2} - 1.5 K_{II} \cos \varphi_0\right) \cos \frac{\varphi_0}{2} \tag{181}$$

The deformation energy density criterion after Sih[287] which comprises not only K_I and K_{II} but also K_{III}, is open to more varied physical interpretations and more generally adaptable to the results of tests performed for practical application. The criterion states that the crack propagates in the direction of the minimum total deformation energy density (approximately equivalent to the maximum volumetric deformation energy density[356]), and that a critical value of the stated energies is decisive for this. The strength-relevant equivalent stress intensity factor follows from:

$$K_{eq} = (A_{11}K_I^2 + A_{12}K_I K_{II} + A_{22}K_{II}^2 + A_{33}K_{III}^2)^{1/2} \tag{182}$$

The factors A_{11} to A_{33} are dependent on the elasticity constants of the material and (with the exception of A_{33}) on the crack propagation angle φ_0, which in turn is determined by K_{II}/K_I. Although the validity of the above hypothesis in the case of fatigue crack propagation can very well be questioned in view of the lack of verification in tests for the present, the form of equation (182) appears suitable to being adapted to global test results existing for single

Table 32. Endurable stress intensity range ΔK_{en} ($N/mm^{3/2}$) at weld spot dependent on number of cycles, N, mean values of tensile shear and cross tension specimens made of steel, diagram values from Ref.331, 343, 352 and 353, converted according to 1 ksi $in^{1/2}$ = 34.8 $N/mm^{3/2}$, 1 MPa $m^{1/2}$ = 31.6 $N/mm^{3/2}$, 1kJ = 1kNm, $\Delta K_J = [E\Delta J/(1 - v^2)]^{1/2}$

Authors	Steel	Yield limit, σ_Y, N/mm^2	Thickness, t, mm	Type, ΔK	ΔK_{en}, $N/mm^{3/2}$			
					N = 10^4	N = 10^5	N = 10^6	N = 10^7
Wang/Ewing[352]	Low alloy	157-357	0.8-3.2	ΔK_J	526	332	215	136
Pollard[353]	Low alloy	217-1085	0.9-2.6	$\Delta K_{I/II}$	852	487	278	160
Mizui et al[343]	Unalloyed	160-180	0.8-1.4	ΔK_{eq}		340	225	150
Yuuki et al[331]	Unalloyed	170-180	0.8-1.2	ΔK_{eq}	699	427	259	158

parameter cases without the introduction of crack angles. The factors A_{11} to A_{33} are then no longer dependent on K_{II}/K_I but are material constants.

A first estimation for the factors A_{11} to A_{33} in the simplified form follows from the application of the deformation energy density hypothesis to the 'pure' K_I, K_{II} and K_{III} load cases at the through-crack in the infinitely extended plate. The Poisson ratio $v = 0.28$ results in $K_{IIc}/K_{Ic} = 0.985$ and $K_{IIIc}/K_{Ic} = 0.663$ with the relations stated in Ref.287 as a result of which equation (182) is given the following simpler form:

$$K_{eq} = (K_I^2 + 1.03K_{II}^2 + 2.27K_{III}^2)^{1/2} \qquad (183)$$

A similar equation form is also obtained proceeding from the energy release rate hypothesis of Irwin,[288] which likewise ignores the influence of the crack propagation angle (plane strain for K_I and K_{II}, $v = 0.28$):

$$K_{eq} = (K_I^2 + K_{II}^2 + 1.39K_{III}^2)^{1/2} \qquad (184)$$

The form of equation (182) used by the author in Ref.326 and in section 10.7,

$$K_{eq} = (K_I^2 + K_{II}^2 + K_{III}^2)^{1/2} \qquad (185)$$

is, in formal terms, a particularly simple and possibly not unrealistic choice as a makeshift solution, which can be upheld only for as long as more precise experimental findings are unavailable.

The endurable stress intensity factors at weld spots dependent on numbers of stress cycles are summarised in Table 32 on the basis of published data.[331,343,352,353] In addition, the scatter band of the relatively comprehensive and accurate analysis of Yuuki in Ref.332 is presented in Fig.262. The endurable stress intensity factors stated for N = 10^7 agree quite well with the accepted threshold stress intensity factor for structural steels, K_{th0} = 180 $N/mm^{3/2}$ (for crack opening mode I and R = 0). It is once again emphasised that these are criteria relating to the total life of the specimens investigated, in other words local criteria for global phenomena, so that questions remain unanswered.

On the basis of the tangential stress criterion, the different crack paths dependent on the type of weld spot specimen, were explained in terms of tendencies, Fig.263. The crack directions calculated according to the stated criterion, are compared with the cracks actually occurring at the face end of the

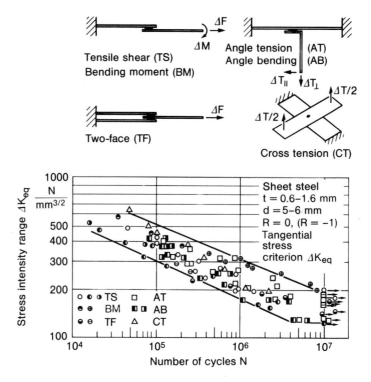

262 Endurable range of equivalent stress intensity factor at weld spot, scatter band resulting from different specimens and loadings, after Yuuki *et al* in Ref.332.

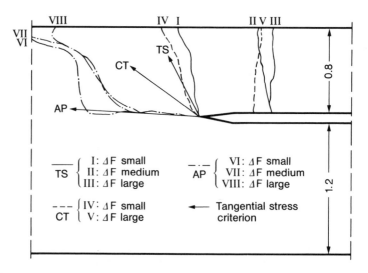

263 Fatigue crack paths in vertex of weld spot edge, different specimens (TS tensile shear, CT cross tension, AP angle-to-plate) and load levels, after Yuuki *et al*.[331]

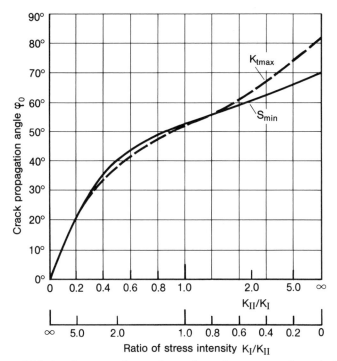

264 Crack propagation angle, φ_0, dependent on stress intensity factor K_{II}/K_I, tangential stress criterion (K_{tmax}) compared with deformation energy density criterion (S_{min}), after Yuuki et al.[331]

weld spot. The crack angle, φ_0, depends on the ratio K_{II}/K_I, Fig.264. Tangential stress criterion and deformation energy density criterion produce similar values, φ_0. In the direction of low cycle fatigue ($N \lesssim 10^5$) and thus higher load amplitudes, the crack no longer occurs at the weld spot edge notch but outside the heat affected zone. The occurrence of this shift can also be determined from the location of the crack emergence on the outside of the plate.

Under special loading conditions (e.g. pure peel or cross tension) the fatigue crack will propagate either into the jointing face or perpendicular to it into the plate. It is assumed that the latter process starts from small secondary cracks in the fractured jointing face. Comparison of the stress intensity factors of primary and secondary crack gives the minimum depth, a_{min}, for secondary crack propagation[322] (plate thickness, t, structural bending stress, σ_{sb}, structural total stress, σ_s, both stresses on the inner surface of the plate at the weld spot edge:

$$a_{min} = 0.145 \left(\frac{\sigma_{sb}}{\sigma_s}\right)^2 t \tag{186}$$

This formula has not yet been verified by testing. But it can be used to explain the tendency to plate fractures in tensile shear specimens ($(\sigma_{sb}/\sigma_s)^2 \approx 0.5$) and to jointing face fractures in peel and cross tension specimens ($(\sigma_{sb}/\sigma_s)^2 \approx 1.0$).

The surprising fact that the total life of the spot welded specimens, composed of a complex process of crack initiation at the weld spot edge notch and crack propagation through jointing face and specimen thickness can be determined on the basis of the geometrical and fracture mechanics conditions of the specimen at the stage of crack initiation, has been explained by Smith and Cooper.[332,333] The authors prove on the tensile shear specimen by means of a highly developed method of measuring crack propagation, that the crack propagation rate, da/dN, remains approximately constant from the onset of crack propagation on the inside of the plate under the initial geometry until the crack has reached the outside of the plate (da/dN measured in the vertex of the crack front). This means that the total life, N, is proportional to the plate thickness, t, and inversely proportional to the initial crack propagation rate, da_0/dN, $N \propto t/(da_0/dN)$. The initial rate, da_0/dN, is, according to the crack propagation law of Paris and Erdogan dependent on the initial stress intensity range, ΔK_0, at the weld spot front face, $da_0/dN = C(\Delta K_0)^m$. To determine the life, N, of the specimen at a given load range on the basis of the relations stated, it is necessary to know the material constants, C and m. According to Ref.332, the values of the base material can be used for this, provided the stress intensity, ΔK_0, has been sufficiently accurately determined, Fig.265 (the solutions of Pook[328,329] and Yuuki[331] are not sufficiently accurate).

A variant of the stress intensity factor approach has been used in fatigue tests with the tensile shear specimen.[357,358] It is based on the well known relation between stress intensity factor, K_I, and crack opening displacement, δ, at the crack tip (Poisson's ratio, ν; constraint factor, M; yield stress, σ_Y; elastic modulus, E):

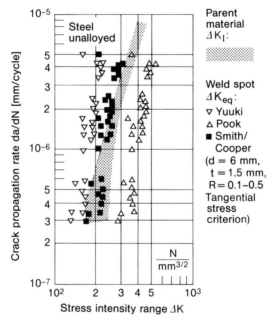

265 Crack propagation rate, da/dN, dependent on range of stress intensity factor, ΔK, for unalloyed steel, base materials and weld spots, ΔK according to different authors.[332]

$$\delta = \frac{K_I^2(1-v^2)}{M\sigma_Y E} \tag{187}$$

The slit opening displacement, δ, is proportional to the opening rotation, θ, against one another of the two plates forming the slit. It follows $K_I \propto \sqrt{\theta}$, so that $\sqrt{\Delta\theta}$ can be used instead of ΔK or ΔJ to characterise the fatigue strength of weld spots presupposing pure mode I loading. The tendency for jointing face shear failure instead of plate failure increases with the ratio $\Delta\tau/\Delta\theta$. The mean shear stress, $\Delta\tau$, in the joint face is an indication for the ΔK_{II} and ΔK_{III} stress intensities of the weld spot in this case. The above relations and statements are valid in the high cycle fatigue range.

10.13 Combination of spot welding and bonding

In the case of the weldbonded joint, the stress distribution at the weld spot is altered in a positive sense by the addition of the adhesive layer. The adhesive layer takes up part of the shear force previously borne by the weld spot alone, or conversely, the weld spot takes up part of the shear force previously borne by the adhesive layer alone, Fig.266. The exact data depend on the ratio of weld spot jointing face area to bonded face area, of joint width to lap length and of adhesive rigidity to thickness of adhesive layer. Bonding additionally prevents the occurrence of single plate bending and of secondary cross tension at the weld spot. Using high strength structural adhesives (epoxy adhesive, for instance) it is therefore possible for a weldbonded joint to achieve a multiple of the fatigue strength of a spot welded only joint if the conditions of the comparison are favourable.

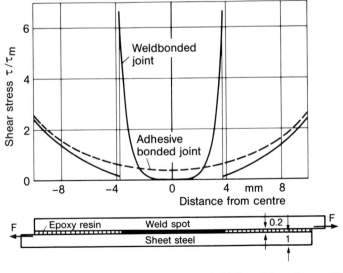

266 Shear stress distribution in weldbonded joint subjected to tensile shear, averaged shear stress, τ_m, cross sectional model, after Eichhorn *et al*.[362]

The shear stress in the adhesive layer represents the quantity responsible for failure of the adhesive layer. For the failure of the weld spot on the other hand, it is above all the radial stress in the plate at the weld spot edge which is the determining factor. The distribution of the radial stress at the weld spot (radius R_1) with surrounding annular bonded face (radius $R_2 = 4R_1$) between two infinitely extended membrane plates subjected to tension and compression (with no transverse offset, *i.e.* without plate bending effect) is presented in Fig.267 in the symmetry line section as to Ref.363, 364. The decisive element for the fatigue strength are the maximum values of the radial stress ahead of the weld spot and ahead of the annular bonded face.

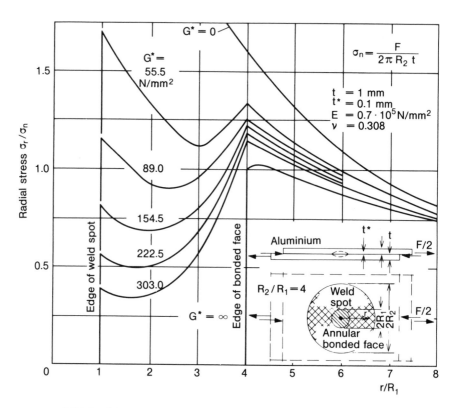

267 Radial stress, σ_r, related to nominal stress, σ_n, in the circumferential face of the weld spot (radius R_1) with annular bonded face (radius R_2), G^* shear modulus of adhesive, after Pfau.[364]

11

Corrosion and wear resistance of welded joints

11.1   General aspects

In addition to deformation resistance including structural stability and fracture, corrosion and wear resistance can play a decisive role in practice. The fatigue process, in particular, is usually superimposed by corrosion which justifies the loose connection of this section with fatigue. Fatigue strength and static strength are reduced by corrosion. The phenomena of corrosion and wear and the measures for reducing or avoiding them have been investigated on unwelded material for the most part. Therefore, only an elementary outline is provided below with suggestions on how to solve problems connected with welded joints. Quantitative data can only be found to a limited extent in the open literature. Additional tests are always necessary in individual situations because of the large number of possible variations and combinations of the numerous influencing parameters.

11.2   Corrosion resistance

Corrosion is the undesired chemical, or more precisely, electrolytic reaction of the material surface with the ambient medium, which leads to erosion or crack formation in the primary material, *e.g.* rust on steel (see also standardisation of terms in accordance with DIN 50900.[365] Corrosion is more prevalent with metals because here the high electrical conductivity provides an adequate speed of reaction of the electrolytic processes. The surface condition is of vital importance. The ambient corrosive medium can be gaseous, *e.g.* SO_2, liquid *e.g.* fresh or sea water, or solid, *e.g.* rust. If the corrosive erosion concentrates on the less electropositive of two contacting metals, this is referred to as 'contact corrosion'; if only individual material phases, grain boundaries (intercrystalline corrosion) or grain imperfections (pitting) are attacked, this is referred to as 'selective corrosion'. Generally, plastic deformation enhances corrosive processes. The crack propagation enhanced by high tensile stresses (mostly residual stresses) in the case of electrolytic corrosion is designated

Corrosion type	Designation	Scheme
Even		Me (Metal)
Uneven	Gap corrosion	Me / Me
	Contact corrosion	Me 1 / Me 2 Me 2 less electro-positive than Me 1
	Selective corrosion	Heterogeneous micro-structure
	Pitting	Me
	Intercrystalline corrosion	Crack Corrosion at grain-boundary
Uneven, combined with mechanical loading	Stress corrosion cracking	Crack σ Static loading
	Fatigue crack corrosion	Crack $\Delta\sigma$ Cyclic loading

268 Types of electrolytic corrosion, after Kloos *et al.*[374]

'stress corrosion cracking'. Corrosion frequently occurs combined with wear (fretting corrosion), fatigue (corrosion fatigue) or erosion and cavitation.

The accelerated propagation of fatigue cracks, which can take place even in an environment which is only slightly corrosive (*e.g.* moist air, hydrogen, saline solution) and the accelerated initiation of cracks in an extremely corrosive environment is designated 'corrosion fatigue' or 'fatigue crack corrosion' ('active material': the whole surface corrodes, 'passive material': only the fatigue cracks are accelerated by corrosion). Figure 268 gives an outline of the corrosion phenomena.

Hydrogen embrittlement is not corrosion in the sense described above. In this process atomic hydrogen diffuses from the surface (including internal cavities) into the material and initiates embrittlement and possibly microcracks ('flakes' and 'fish eyes') which, however, are then frequently enlarged by stress corrosion. Mill scale formation on steel (including its weaker initial form, the annealing colour), the reaction of the material surface with the oxygen of the air at elevated temperatures (without any water) and the similar process of oxide formation on aluminium, which unlike the mill scale acts as corrosion protection, are not corrosion in the above sense. Solder fracture, the crack forming reaction of surface-active melted metals with material surfaces which are subjected to tensile stress (without the formation of electrolyte) is also not corrosion in the above sense. Finally, the deliberate use of electro-chemical transformation in electrolytic polishing or in etching is not corrosion in the above sense.

Typical corrosion damage cases in welded joints are the following:

— contact corrosion in welded joints of dissimilar material;
— selective corrosion in welded joints made of similar materials with weld metal of the similar or dissimilar type as a consequence of inhomogeneous microstructures of the material (cast or rolled structure, local cold work, local microstructural changes in the heat affected zone) particularly after improper heat treatment;
— electrolytic element formation with slags and residual flux;
— stress corrosion cracking as a consequence of high tensile residual welding stresses in the weld metal and transition zone;
— pitting and gap corrosion in slits and gaps (including weld undercuts, coarse rippling, gaps under mill scale, residual paint and weld spatter) which retain the moisture and have a particularly high electrolytic potential;
— fatigue crack corrosion in moist air, in seawater and in other corrosive environments.

The articles by Class[366] and Wirtz[367] and Achten, Herbsleb and Wieling[368] give an insight into actual cases of corrosion in welded joints.

Even without definite corrosion damage the high cycle fatigue strength and endurance limit react particularly sensitively to the electrolytic surface conditions (mainly an influence of fatigue crack corrosion). Thus the fatigue strength of steel and aluminium alloys is higher in a vacuum than in dry air, noticeably reduced in moist air and very much reduced in a corrosive environment, *e.g.* in seawater. The reduction factor compared with air is 0.5-0.6 in smooth specimens without protective coating and about 0.8 in welded joints which corresponds to a life reduction factor of 0.5. Protective layers of plastics, *e.g.* epoxy, or high quality metals, *e.g.* nickel, and films of oil, which protect against moisture and oxygen, prevent a decrease of fatigue strength. To what extent this protective effect also applies to welded joints has only been investigated in isolated cases *e.g.* in fatigue tests on welds coated with epoxy in a salt spray solution. The fatigue behaviour of hot dip galvanized weld joints has been investigated by Olivier and Rückert.[369] The corrosion protection by the zinc coating is explained mainly by the separation effect and the cathodic potential of the coating.

Measures are taken to avoid or reduce corrosion by decreasing the corrosiveness of the attacking medium or by preventing its becoming locally established ('active corrosion protection') and by protecting the materials which are subjected to corrosive attack ('passive corrosion protection'), see also the compilation of information sheets, Ref.370 and 371. Materials protecting measures are:

— selection of materials with higher corrosion resistance (*e.g.* weather resistant copper-alloyed structural steels or stainless chrome-alloyed special steel) or of microstructural states with the same (*e.g.* austenitic or martensitic grain structure) including combinations of such materials and structural states;
— surface coating to protect against the corrosive medium: several layers of paint[373] or a layer of plastics on a surface which is free of mill scale, rust and impurities (sand-blasted or pickled), roll cladding, explosion cladding, surface layer welding, metal spraying,[372] anodic oxidation (anodising), electroplating (nickel plating, hard chrome plating) and hot galvanizing on the same type of surface;
— cathodic corrosion protection *e.g.* of ship's hulls, water tanks and underground pipelines: an electrode made from a less electropositive metal is connected to the structure (*e.g.* zinc or magnesium electrode to a steel structure) which dissolves instead of the structure to be protected (sacrificial electrode). This method is effective against pitting, gap and stress corrosion;
— filling of the slit or gap faces with adhesive when spot welding (weldbonding);
— avoiding slits or gaps open from one side by design measures;
— removal of weld spatter, welding scale, residual paint and defects at the open weld root;
— reduction of the high residual tensile stresses and of hydrogen diffusion if there is a risk of stress corrosion, generation of residual compressive stresses in the surface.

The corrosion protection layers applied by welding or metal spraying have a direct relation to welding engineering. The risk of cracking in the bonding layer requires special inspection with respect to residual stresses and material ductility.

Welding over 'primer coatings' applied in manufacture lasting several months has been investigated several times with regard to its effect on strength, because manufacture would be cheaper if welding over primer were acceptable (see section 2.1.9). As the linear porosity and elongated cavities which arise especially in fillet welds, as a consequence of the gasification of the primer coating when it is welded over, reduce the fatigue strength of the weld, especially when subjected to transverse loading, in an uncontrollable way (the conventional reference to the proportion of the area occupied by pores is not sufficient), welding over primer coatings is not permissible for high quality welds.

11.3 Wear resistance

Wear is the undesired mechanical erosion of the material surface characterised by small particles braking off or simply by permanent deformation. The wear process includes base body and counter body (the wear pairing), the intermediate layer (granular, liquid or gaseous) and finally frictional movement combined with external loading. It is affected by these factors in a complex manner. Therefore the correlation with the conventional mechanical characteristic values of the material can only be stated in exceptional cases.

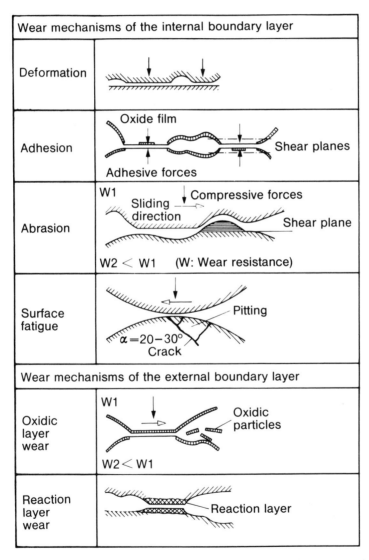

269 Wear mechanisms of the internal and external boundary layer, after Kloos *et al.*[374]

Typical examples of wear damage are:

— abrasion by breaking off first of all of exposed surface peaks continuing right up to 'seizure', the fusion welding and breaking loose again of the friction surfaces (*e.g.* in the case of sliding bearings and gear pairings);

— erosion wear as a consequence of impact from solid particles (*e.g.* sand) out of flowing liquids or gases or from liquid particles (*e.g.* water droplets) from flowing gases or vapours;

— cavitation wear as a consequence of the wake effect (with degasification or evaporation) in flowing liquids (*e.g.* at pump and turbine blades and also at ship propellers);

— fretting corrosion in screwed and riveted joints and also in pin joints (not in welded joints).

A summary of some wear mechanisms is provided by Fig.269.

Wear has a direct relation to welding engineering in so far as wear resistant protective coatings are frequently applied by welding or metal spraying. Particular attention must be paid to the risk of cracking in the bonding layer.

12

Example for the development of a fatigue resistant welded structure

12.1   Introduction, general outline, requirements

The fatigue strength of welded structures is presented in the previous chapters in a general and systematic form. The industrial use of the presented methods and data will now be sketched briefly using a practical example. The frame of a commercial vehicle (truck or trailer, site vehicle, low loader, agricultural vehicle) was selected as the example. Vehicle frames of this type are frequently welded. However, truck frames in particular are mainly riveted or screwed.

The example is deliberately taken from an engineering sector where high demands are made on fatigue strength with the smallest possible expenditure in material (to increase the economic efficiency of vehicle operation), where no relevant code exists and where the assessment of nominal stress is completely inadequate. The procedures are simpler and more obvious in other engineering sectors with lower demands, relevant codes and well proven assessments of nominal stress, so it is possible here to dispense with illustrating them.

The vehicle frame acts as the load-bearing structure between the wheel system, body and drive system. Bearing, spring and damping elements establish the mutual connections. It is mostly manufactured from higher tensile structural steel. The frame has a ladder-like shape in trucks and trailers, consisting of longitudinal and transverse bars, which are arranged in a single plane, apart from minor cranks and eccentricities (with the exception of low loading vehicles which have large cranks). Special purpose vehicles may use a frame type which deviates from the ladder shape (e.g. X-frame). Channel section bars are used for frames with moderate torsional stiffness, circular and rectangular section tubes are used for frames with high torsional stiffness. In general, frames are preferred which have a low torsional stiffness combined with a high bending stiffness because it makes this a design with a smaller spring travel possible for the axle suspension or allows smaller forces if twisting is enforced. Frames which have a high torsional stiffness are preferred for dumper trucks, box-type vans and tool carriers.

Apart from the characteristic dimensions for the bar cross sections, the torsional stiffness of the channel section ladder type frame is influenced decisively by the design of the joints between the longitudinal and transverse bars. The local cross section stiffeners also have a global stiffening effect.

The loading on the vehicle frame is caused by the weight of the body plus load, by the reaction forces from unevenness of the roadway (including potholes and kerb edges) and from driving manoeuvres (accelerating, braking, cornering, by the trailer coupling forces and, in exceptional cases, by the crashing forces in accidents. The loading of the frame is characterised by global bending and torsion moments (three components for each) in-plane shearing and longitudinal tension, the same as static and dynamic, cyclic and shocklike forces.

The task to be mastered by the designer consists of achieving a certain level of torsional stiffness and a high shearing and bending stiffness combined with sufficient fatigue strength, particularly with regard to the welded joints at the corners and intersection of the frame. Fatigue strength for infinite life is often required. Plastic deformations are only permissible in a localised manner and to a limited extent.

12.2   Development procedure, numerical and experimental simulation

The activities involved in the development of a fatigue resistant (welded) vehicle frame are embedded in an organisational development procedure which is illustrated in Table 33 (see also Ref.375). The development procedure between the order for development and the clearance for the developed product is entered horizontally and the development functions 'design', 'analysis', 'testing', and 'information' are entered vertically in a matrix chart. The models entered into the matrix for numerical and experimental simulation are explained briefly below (see also Ref.376 and 377).

The empirical knowledge gained from existing designs plays a major role in the preliminary design stage. The design is cross-checked by means of elementary nominal stress formulae and conventional quasi-static load assumptions. A detailed load specification is compiled as guidance for further design development work, which contains the information given below with regard to the strength and life of the frame to be guaranteed:

— vertical loading from pay load, weight of driver's cab, body, tank and driving gear, and from the reaction forces at the axles;
— vertical loading on single wheel from kerb step (200mm) or single impediment (400mm), causing frame torsion;
— transverse loading from cornering, road camber, side winds and trailer operation;
— longitudinal loading from braking, accelerating, trailer operation, producing pitch vibration with frame bending and longitudinal wheel impacts on one side, producing frame shearing;

Table 33. Organisational development procedure of a welded vehicle frame

Development functions	Development stages			
Design	Draft design	Main design	Detail design	Fault elimination, safety assessment
Analysis	Elementary formulae with nominal stresses, dimensioning and design based on experience	Additional: vehicle dynamics model; structural dynamics model; structural statics model; damage accumulation rule; collision model	Additional: notch stress model; stress intensity factor model	Additional: crack propagation model
Testing		Extrapolation from previous test results	Frame prototype testing: static stress and stiffness; fatigue strength; crash behaviour	Vehicle prototype testing: accelerated testing on rough road or test stand; operational tests; crash test
Information Basis	Experience and data on existing vehicles; load specifications; safety requirements	Design loads; load spectra; road roughness spectra; characteristics of connecting members and materials; endurable stresses	Shape and dimensional imperfections, surface and internal defects, residual stresses	Crack propagation data; loading histories; scatter ranges
Result	Design and main dimensions of frame	Shape and dimensions of frame and sections; design of corners and intersections; materials and manufacturing process	Design and manufacturing details: type, thickness, position of welds; welding plan; notch root improvements; corrosion protection	Strength and life data; acceleration spectra; failure probabilities; crash behaviour

— roadway unevenness spectra with correlation to speed of travel;
— support forces and excitation spectra from driving gear, support reactions from axle guides;
— extreme situations of vibrational loading; lifting, rolling and pitching vibrations;
— impact loading from 'active' collision of the vehicle itself (high frame loading, assessment obligatory for light vans) or 'passive' collision by other vehicles (relatively low frame loading via the lightweight protective structure preventing colliding vehicles from being crushed under the frame).

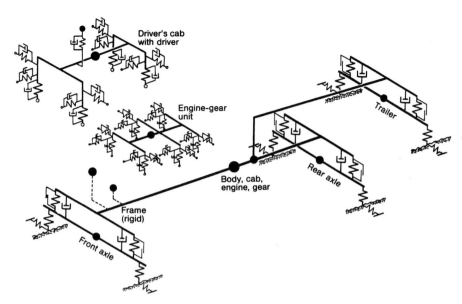

Driver's cab
with driver

Engine-gear
unit

Trailer

Body, cab,
engine, gear

Rear axle

Frame
(rigid)

Front axle

270 Three dimensional vibration model consisting of point masses,
springs and dampers for semi-trailer truck; driver's cab and engine
supported against the frame; after Geissler.[377]

The loading data are stated in a simplified form as quasi-static equivalent or maximum forces or, more differentiated, as load spectra.

Structural analysis by numerical methods provides particularly strong support for decisions in the main design stage. Within this, the differentiated models below are used in addition to the elementary formulae. The vehicle dynamics model is used for investigating the directional driving stability of the vehicle and the low frequency ($\lesssim 10$cps) vehicle vibrations. The basis is a three-dimensional point mass or a multi-body system with extremely non-linear kinematics, spring and damping elements and a relatively small number of degrees of freedom ($\lesssim 50$), Fig.270. Using this model it is also possible to determine the load-time functions or load spectra which act on the frame, for example, as a function of the spring characteristics and the roughness profile of the roadway. The structural dynamics model is used for complementing the higher frequency frame vibrations (up to ≈ 150cps). The basis is a linearised finite element model consisting of bar, plate and shell elements, Fig.271.

The structural statics model is similar to Fig.272 and serves to determine the structural stresses and the global stiffness of the frame. The structural stresses are to be limited to the extent that fatigue fractures are avoided within the preset vehicle life and that the local plastic deformations caused by the rare extreme loadings are restricted. Fatigue strength for infinite life is required for frames of trucks and trailers with regard to cyclic vertical loading ($N \approx 10^7$). The static stiffness of the frame should reach a preset value. As the design of the corners and intersections of the frame is decisive for the fatigue strength of these joints

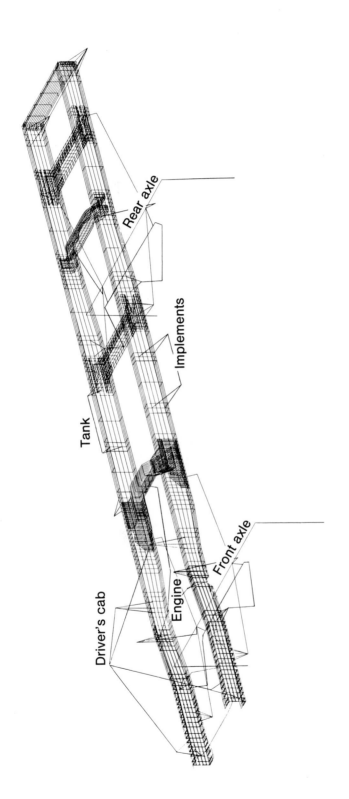

271 Finite element structural dynamics model for frame of semi-trailer truck (but with riveted joints).

272 Finite element structural statics model for truck frame (but with riveted joints).

and contributes to the stiffness of the frame, these structural parts are simulated in detail by the finite element model.[378,379,383] Bar element models, which include the torsion-bending moment and the derivation of twisting angle (the two additional parameters of torsion-bending theory) and which take the joint eccentricities into account,[383-385] are only suitable for a first estimate. The damage accumulation hypothesis (a modified Miner rule) combined with simplified S-N curves serves to estimate the life when the frame is subjected to the different load spectra with varying mean stress. The structural crash model, a bar element structure with plastic hinges, serves to estimate the impact force versus deformation behaviour of the protective structure (force-displacement curve and energy absorption).

When the loading processes are so diverse or complex that they cannot be simulated within the numerical model (a question of the feasible expenditure for numerical analysis and data processing) previous statistically evaluated measurements are used as a basis for extrapolation on to the new design, taking account of simple mechanical correlations. The reduced-scale frame models previously included in stress and stiffness investigations are out of date as a consequence of the more recent development of numerical analysis methods for structural analysis.

In the detail design stage, the frame prototype is available on which firstly analysis results can be checked (particularly in bending and torsion tests) and secondly a fatigue test can be carried out, as can a crash test if required. Further numerical analysis supports detail design by means of parallel detailing of the structural models, *e.g.* with regard to assessing the notch effect of the welds. At the stage of fault elimination and final safety assessment the vehicle prototype with the installed frame is available. To assess whether the service fatigue strength over ten operating years and a distance travelled of 300 000 miles is adequate, the vehicle is subjected to an accelerated process of testing — 6000 miles on a rough road track with excessive amplitudes. The load specifications for the frame can be checked by acceleration measurements in normal operation. The rough road track is being replaced increasingly nowadays by loading simulation on servo-controlled four wheel test stands. In the final testing stage described above, numerical crack propagation models are also occasionally useful for making predictions on safety and life. Crash tests round off the testing programme for the frame.

Bibliography

Chapter 1 — Introduction

1   DIN 50100: 'Fatigue test' (German). Publ Beuth Verlag, Berlin/Köln, 1978.
2   Gurney T R: 'Fatigue of welded structures'. Publ Cambridge University Press, Cambridge, 1979 (2nd ed).
3   Neumann A (Ed): 'Welding engineering handbook for designers' Vol I-IV (German). Publ VEB-Verlag Technik, Berlin, 1955 (1st ed) to 1976 (3rd ed), new edition Vol I-III, publ DVS-Verlag, Düsseldorf, 1984 (Vol II, 4th ed).
4   Neumann A (Ed): 'Principles of welding engineering' Vol II, Design (German). Publ VEB-Verlag Technik, Berlin, 1981 (8th ed).
5   Ruge J: 'Handbook of welding engineering' Vol III Construction — design of components, Vol IV Construction — calculation of joints (German). Publ Springer Verlag, Berlin, 1985 (Vol III) and 1988 (Vol IV).
6   Radaj D: 'Assessments of strength' Vol I and II (German). Publ DVS-Verlag, Düsseldorf, 1974.
7   Erker A, Hermsen H W and Stoll A: 'Design and calculation of welded structures' (German). Publ DVS-Verlag, Düsseldorf, 1971 (2nd ed).
8   Sahmel P and Veit H J: 'Principles of the design of welded steel structures' (German). DVS-Verlag, Düsseldorf, 1981 (6th ed).
9   Schwaigerer S: 'Strength calculations for boilers, tanks and pipelines' (German). Publ Springer Verlag, Berlin, 1978 (3rd ed).
10 Stüssi F and Dubas P: 'Principles of structural steel engineering' (German). Publ Springer Verlag, Berlin, 1971.
11 Hertel H: 'Fatigue strength of structures' (German). Publ Springer Verlag, Berlin, 1969.
12 Thum A and Erker A: 'Structural strength of welded joints' (German). Publ VDI-Verlag, Berlin, 1942.
13 Ros M and Eichinger A: 'The risk of fracture in solid bodies subjected to repeated loading'. Report 173, publ Eidgen Materialprüfungs u Versuchsanstalt, Zürich, 1950.
14 Rühl K H: 'Load-bearing capacity of metallic components' (German). Publ Verlag Ernst, Berlin, 1952.
15 Stüssi F: 'The theory of fatigue strength and August Wöhler's fatigue tests' (German). Publ VSB-Verlag, Zürich, 1955.
16 Neumann A: 'Problems of the fatigue strength of welded joints' (German). Publ VEB-Verlag, Berlin, 1960.

17   Harris W: 'Metallic fatigue'. Publ Pergamon Press, Oxford, 1961.

18   Forrest P G: 'Fatigue of metals'. Publ Pergamon Press, Oxford, 1962.

19   Frost N E, Marsh K J and Pook L P: 'Metal fatigue'. Publ Clarendon Press, Oxford, 1974.

20   Almen J O and Black P H: 'Residual stresses and fatigue in metals'. Publ McGraw Hill, New York, 1963.

21   Hempel M: 'Fatigue strength characteristics of materials'. *Z VDI* 1962 **104** (27) 1362-1377.

22   Jacoby G: 'Fundamentals of the fatigue behaviour of welded joints'. *Alumin* 1964 **40** (7) 427-433 and (12) 749-759.

23   Munz D, Schwalbe K and Mayr P: 'Fatigue strength behaviour of metallic materials' (German). Publ Vieweg Verlag, Braunschweig, 1971.

24   Tetelmann A S and McEvely A J: 'Fracture characteristics of materials used in engineering' (German). Publ Verlag Stahleisen, Düsseldorf, 1971.

25   VDI-Richtlinie 2226: 'Recommendation for strength calculations involving metallic components' (German). Publ VDI-Verlag, Düsseldorf, 1974.

26   'Fatigue strength for infinite and finite life' (German). VDI-Berichte 661, publ VDI-Verlag, Düsseldorf, 1988.

27   Wellinger K and Dietmann H: 'Strength calculation' (German). Publ Kröner Verlag, Stuttgart, 1976.

28   Dahl W (Ed): 'Behaviour of steel subjected to cyclic loading' (German). Publ Verlag Stahleisen, Düsseldorf, 1978.

29   Aurich D: 'Fracture processes in metallic materials' (German). Publ Werkstofftechnische Verlagsgesellschaft, Karlsruhe, 1978.

30   Buxbaum O: 'Service fatigue strength' (German). Publ Verlag Stahleisen, Düsseldorf, 1986.

31   Haibach E: 'Fatigue strength for infinite and finite life, service fatigue strength — methods and data for calculating components' (German). Publ VDI-Verlag, Düsseldorf, 1989.

32   'Guideline for a service fatigue strength calculation' (German). Publ Verlag Stahleisen, Düsseldorf, 1985.

33   Zammert W U: 'Service fatigue strength calculation' (German). Publ Vieweg Verlag, Braunschweig, 1985.

34   Proc int conf 'Fatigue of welded structures'. Publ The Welding Institute, Abington, Cambridge, 1971.

35   Maddox S J (Ed), Proc int conf 'Fatigue of welded constructions'. Publ The Welding Institute, Abington, Cambridge, 1988.

36   Proc int conf 'Aluminium weldments (III)' (German). Publ Aluminium-Verlag, Düsseldorf, 1985.

Chapter 2 — Fatigue strength for infinite life (see also Ref.2,3,6,15,16)

37   Olivier R and Ritter W: 'Atlas of S-N curves for welded joints of structural steels. Part I: Butt joints, Part 2: Transverse stiffeners, Part 3: Double T joints, Part 4: Longitudinal stiffeners, Part 5: Double fillet welds, flame cutting, stud welding, flange plates' (German). DVS-Berichte Vols 56/I-IV, DVS-Verlag, Düsseldorf, 1979-1985.

38   Haibach E, Olivier R and Ritter W: 'Influence of the weld and plate thicknesses on the fatigue strength of welded joints with fillet welds' (German). *Schw Schn* 1978 **30** (11) 442-446.

39   DIN 8524: 'Defects in welded joints made from metallic materials, Part 1: Fusion welded joints and Part 3: Cracks' (German). Publ Beuth Verlag, Berlin, 1984 and 1975.

40   Petershagen H: 'Evaluation of the fatigue strength of welded joints with imperfections, a review' (German and English). *Schw Schn* 1990 **42** (3) 134-137.

41   Petershagen H: 'Fatigue strength of welded joints with defects, a bibliography' (German). Document 2372, publ Institut für Schiffbau, Universität Hamburg, 1989.

42   IIW-Doc 340-69: 'Classification of defects in metallic fusion welds with explanations'. *Weld World* 1969 **7** (4) 200-210.

43   Pothoff F: 'Influence of pores on technological standards of quality' (German). *Der Prakt* 1975 **27** (8) 140-144.

44   Böhme D and Helwig R: 'Influence of pores on the fatigue strength of longitudinal fillet welds'. *Schw Schn* 1985 **37** (12) 670-674.

45   Böhme D, Helwig R, Olivier R and Ritter W: 'Influence of pores from manufacturing coatings on the fatigue strength of fillet welds subjected to longitudinal loading' (German and English). *Schw Schn* 1983 **35** (7) 308-312.

46   Petershagen H: 'The influence of undercuts on the fatigue strength of welded joints — a survey' (German and English). *Schw Schn* 1985 **37** (6) 270-274, English version IIW Doc XIII-1120-84.

47   Ros M: 'Strength and dimensioning of welded joints'. Report 135, publ Eidgen Materialprüfungs u Versuchsanstalt, Zürich, 1941.

48   Ros M: 'Strength and safety of welded joints'. Report 156, publ Eidgen Materialprüfungs u Versuchsanstalt, Zürich, 1946.

49   Stüssi F: 'On the endurance limit of welded joints' (German). *Schw Bauz* 1957 **75** (52) 806-808.

Chapter 3 — Fatigue strength for finite life (see also Ref.2,3,28,37)

50   ASTM Special Publication No 91-A: 'A guide for fatigue testing and the statistical analysis of fatigue data'. Publ ASTM, Philadelphia, 1964.

51   Weibull W: 'Fatigue testing and analysis of results'. Publ Pergamon Press, Oxford, 1961.

52   Haibach E: 'The endurance limit of welded joints in the case of limit load cycles greater than 2×10^6' (German). *Arch Eisenhütt* 1971 **42** (12) 901-908.

53   Haibach E: 'The fatigue strength of welded joints from the point of view of local strain measurement' (German). LBF Report FB-77, publ Laboratorium für Betriebsfestigkeit, Darmstadt, 1968.

54   Haibach E: 'Fatigue strength behaviour of welded joints' (German). VDI-Berichte 268, 1976, 179-192.

55   Bierett G: 'On the fatigue strength of welded and riveted steel joints' (German). *Stahl Eisen* 1967 **87** (24) 1465-1472.

56   Bierett G: 'Some important rules for the service fatigue strength of welded steel components' (German). *Schw Schn* 1972 **24** (11) 429-434.

57   Lachmann E: 'On the plasto-fatigue of parent materials and welded joints at a raised temperature — lifetime, application, transferability' (German). VDI-Forschungshefte 614, 1982, 1-64.

58   Rie K T and Schmidt R M: 'Fatigue of welded joints in the low cycle range' (German and English). *Schw Schn* 1986 **38** (10) 507-514.

59   Dengel D: 'The arcsin $\sqrt{P}$ transformation, a simple method for the graphic and numerical evaluation of planned fatigue tests' (German). *Z Werkstofft* 1975 **6** (8) 253-288.

60   Maennig W-W: 'Remarks on the evaluation of the endurance limit characteristics of steel and some investigations on determining the endurance limit range' (German). *Materialpr* 1970 **12** (4) 124-131.

61   Maennig W-W: 'The boundary method, a cost saving method for determining fatigue strength values — theory, practice and experiences' (German). *Materialpr* 1977 **19** (8) 280-289.

62   Westermann-Friedrich A and Zenner H: 'Counting methods for producing spectra from time functions, comparison of various methods and examples' (German). FVA Merkblatt 0/14, Forchungsvereinigung Antriebstechnik, Frankfurt M, 1988.

63   DIN 45667: 'Counting methods for evaluating random vibrations' (German). Publ Beuth Verlag, Berlin, 1985.

64   Gaßner E: 'Service fatigue strength, a basis for the dimensioning of construction components with statistically variable service loadings' (German). *Konstr* 1954 **6** (3) 97-104.

65   Gaßner E, Griese F W and Haibach E: 'Endurable stresses and life of a welded joint made from steel St 37 with different types of load spectra' (German). *Arch Eisenhütt* 1964 **35** (3) 255-267.

66   Haibach E, Ostermann H and Ruckert H: 'Service fatigue strength of welded joints made of structural steel subjected to a random sequence of loadings' (German). *Schw Schn* 1980 **32** (3) 93-98.

Chapter 4 — High tensile steels and aluminium alloys (see also Ref.2,3,11)

67   Haibach E: 'Fatigue strength of high strength fine grained structural steels in the as-welded state' (German). *Schw Schn* 1975 **27** (5) 179-181.

68   Minner H H: 'Increasing the fatigue strength of welded joints made from higher strength fine grained structural steel StE 460 and StE 690 by using the TIG dressing method' (German). Report 36, publ Institut für Statik und Stahlbau, TH Darmstadt, 1981.

69   Nowak B, Saal H and Seeger T: 'A proposal on dimensioning of components made of high tensile structural steels with respect to fatigue strength' (German). *Der Stahlb* 1975 **44** (9) 257-268 and (10) 306-313.

70   Schütz W and Winkler K: 'On the dimensioning of welded joints made from aluminium subjected to cyclic loading'. *Alumin* 1970 **46** (4) 311-321.

71   Haibach E and Atzori B: 'A statistical method for a renewed evaluation of results from fatigue strength tests and for deriving dimensioning rules, applied to welded joints made from AlMg5'. *Alumin* 1975 **51** (4) 267-272.

72   Buray E: 'Endurance limit of welded joints in aluminium with various filler materials'. *Alumin* 1981 **57** (5) 326-329.

73   Kosteas D and Graf E: 'Fatigue behaviour of welded aluminium components' (German and English). *Schw Schn* 1982 **34** (7) 331-335.

74   Poalas K, Graf U and Kosteas D: 'Fatigue behaviour of welded aluminium components' (German and English). *Schw Schn* 1988 **40** (3) 130-133.

75   Atzori B and Dattoma V: 'Fatigue strength of welded joints in aluminium alloys, a basis for Italian design rules'. IIW-Doc XIII-1088-83.

76   Atzori B and Dattoma V: 'A comparison of the fatigue behaviour of welded joints in steel and aluminium alloys'. IIW-Doc XIII-1089-83.

Chapter 5 — Design improvements (see also Ref.2,3,9,10,27)

77   Bramat M, Gerbeaux H and Vix J-M: 'Fatigue behaviour of girders built up by welding when subjected to in-plane bending'. *Soud et Techn Conn* 1973 (3/4) 129-140. Reviewed by Neumann D: 'Do stiffeners have to be cut out in corner joints?' (German). *Der Prakt* 1978 **30** (5) 90.

78   Haibach E and Plasil I: 'Investigations into the service fatigue strength of lightweight steel plates with hollow trapezoidal stiffeners in railway bridge design' (German). *Der Stahlb* 1983 **52** (9) 369-374.

79   Radaj D and Schilberth G: 'Notch stresses on cutouts in flanged girders subjected to bending' (German). *Schw Schn* 1978 **30** (8) 294-296.

80   Radaj D and Schilberth G: 'Comparative notch stress calculations for a heavy duty tie bar' (German and English). *Schw Schn* 1981 **33** (5) 230-233.

81   Kaufmann R: 'Investigations on the design of cruciform joints subjected to multi-axial loading using crosspieces' (German). Publ Institut für Stahlbau, TU Hannover, 1982.

82   Kloth W, Spangenberg D and Bergmann W: 'Guidelines for the design of welded vehicle frames made from lightweight sectional steel taking particular account of torsion loading' (German). Techn Blätter 4, publ Firma Wuppermann, Leverkusen, 1960.

83   Kloth W: 'Atlas of the stress fields in structural components' (German). Publ Verlag Stahleisen, Düsseldorf, 1961.

84   Mall G and Zirn R: 'Load bearing behaviour of tubular joints subjected to static loading' (German). *Schw Schn* 1969 **21** (12) 579-583.

85   Wellinger K, Mall G and Zirn R: 'Fatigue strength behaviour of welded tubular joints' (German). *Schw Schn* 1971 **23** (6) 215-217.

86   Wardenier J: 'Hollow section joints' (German). Publ Delft University Press, Delft, 1982.

87   Marshall P W: 'Connections for welded tubular structures'. *Weld Res Abr* 1985 **30** (10) 1-28.

88   Mang F, Bucak O and Klingler J: 'List of S-N curves for hollow section joints' (German). Publ Studiengesellschaft für Anwendungstechnik von Eisen und Stahl, Düsseldorf, 1987.

89   Wellinger K, Mall G and Zirn R: 'Dynamic load bearing behaviour of welded strapped tubular joints' (German). *Techn Mitt* 1971 **64** (7) 308-312.

90   Iida K and Matoba M: 'Evaluation of fatigue strength of hold frame ends in shiphulls'. IIW-Doc XIII-950-80.

91   Matoba M, Kawasaki T, Fuji T and Yamauchi T: 'Evaluation of fatigue strength of welded steel structures, hull's members, hollow section joints, piping and vessel joints'. IIW-Doc XIII-1081-83.

92   Petershagen H and Paetzold H: 'Investigations into the assessment of welded design shapes in shipbuilding' (German). Berichte 142, publ Institut für Schiffbau, Universität Hamburg, 1983.

93   Paetzold H: 'Fatigue strength behaviour of selected details from ship design' (German). Berichte 159, publ Institut für Schiffbau, Universität Hamburg, 1985.

94   Hapel K H and Just F: 'The stress concentration factor for plane hatch designs in container ships' (German). *Schiff Hafen* 1981 **33** (3) 99-106 and (8) 71-77.

95   Lehmann E: 'Cutouts in ship structural design'. Proc int sym 'Practical design in shipbuilding', Tokyo, 1983.

96   Fricke W: 'Determination of the local strain for shipbuilding design details' (German). *Jahrb Schiffb Gesell* 1985 **79** 295-314.

97 Paetzold H: 'Assessment of the service fatigue strength of cutouts for longitudinals on the basis of the local strains' (German). Berichte 455, Institut für Schiffbau, Universität Hamburg, 1985.

98 Schönfeldt H: 'Constructional design and guidelines for the calculation of highly stressed aluminium bracings on ships' (German). *Jahrb Schiffb Gesell* 1975 **69** 31-43.

99 'Welded joints in boiler, vessel and pipeline construction' (German). Publ DVS-Verlag, Düsseldorf, 1966.

100 Wellinger K and Sturm D: 'Seamless and welded steel tubes subjected to static and pulsating internal pressure loading' (German). *Techn Uberw* 1969 **10** (4) 114-118.

101 Wellinger K, Gaßmann H and Zenner H: 'Behaviour of welded tubes subjected to pulsating internal pressure dependent on loading state and the slope angle of the weld' (German). *Schw Schn* 1968 **20** (1) 8-17.

102 Wellinger K and Sturm D: 'Vessel with nozzle subjected to pulsating internal pressure loading' (German). *Mitt Verein Großkesselb* 1968 **48** (2) 99-105.

103 Wellinger K, Gaßmann H and Mall G: 'Pulsating internal pressure behaviour of pipelines with branches at right angles' (German). *Schw Schn* 1969 **21** (10) 444-454.

104 Wellinger K, Sturm D and Mall G: 'Behaviour of components in vessel, boiler and pipeline construction subjected to static and pulsating internal pressure loading' (German). *Schw Schn* 1970 **1** (14) 3-8.

105 Bartonicek J and Krägeloh E: 'Loading of hollow cylinders with eccentrically welded-in nozzle pipes subjected to internal pressure' (German). *Z Werkstofft* 1978 **9** 279-288.

106 Eßlinger M: 'Static calculation of boiler ends' (German). Publ Springer Verlag, Berlin, 1952.

107 Ringelstein K H and Bußhans L: 'Assessment by calculation of the stress state in dished ends with a nozzle in the transition to shell area' (German). Forschungsberichte TüV, Essen, 1984.

Chapter 6 — Spot, friction, flash butt and stud welded joints

108 Overbeeke J L and Draisma J: 'The F-N curve of heavy duty spot welded lap joints at R = 0' (German). Rep WH 72-3, Dep Mech Eng, TU Eindhoven. Reviewed in *Schw Schn* 1974 **26** (2), 64.

109 Overbeeke J L and Draisma J: 'Fatigue characteristics of heavy duty spot welded lap joints'. *Met Con* 1974 **6** (7) 213-219.

110 Overbeeke J L: 'Fatigue of spot welded lap joints'. *Met Con* 1976 **8** (7) 212-215.

111 Overbeeke J L and Draisma J: 'The influence of stress relieving on the fatigue of heavy duty spot welded lap joints'. *Weld Res Int* 1977 **7** (3) 241-253.

112 Overbeeke J L and Draisma J: 'Influence of stress relieving on fatigue of heavy duty spot welded lap joints'. *Met Con* 1978 **10** (9) 433-434.

113 Radaj D, Schlüter A and Baur A: 'Peel tension fatigue test for spot welded joints and fracture surface analysis' (German and English). *Schw Schn* 1986 **38** (3) 113-118.

114 DVS-Merkblatt 2709: 'Resistance spot welded specimens for determining the fatigue strength' (German). DVS-Verlag, Düsseldorf, 1983.

115 Buray E: 'Effect of stress ratio on the fatigue behaviour of AlMgSi 1 spot welded joints'. *Alumin* 1974 **50** (2) 154-158.

116 Eichhorn R, Hamm K and Schmitz G: 'Influence of the geometry of the specimen on the static and dynamic load carrying capacity of resistance spot welded joints on deep drawable plates of 1.2mm thickness made of aluminium materials' (German and English). *Schw Schn* 1988 **40** (10) 494-500.

117   Wellinger K, Eichhorn F and Gimmel P: 'Welding' (German). Publ Kröner Verlag, Stuttgart, 1964.

118   Kussmaul K and Mall G: 'The effect of defects on the strength behaviour of resistance welded multi-spot joints' (German). In 'Resistance welding and micro-joint processes' (German), publ DVS-Verlag, Düsseldorf, 1970, 39-51.

119   Rupp A, Grubisic V, Störzel K and Steinhilber H: 'Determination of endurable weld spot forces for the service fatigue strength analysis of spot welded joints in automobile engineering' (German). Report 78, Forschungsvereinigung Automobiltechnik, Frankfurt, 1989.

120   Amborn P: 'Investigations of the fatigue strength of spot welded joints' (German). Schweißtechnische Forschungsberichte, Vol 25, DVS-Verlag, Düsseldorf, 1988.

121   Sperle J O: 'Fatigue strength of non-load carrying spot welds'. Met Con 1984 16 (11) 678-679.

122   Krause H G, Preß H and Simon G: 'The dynamic behaviour of spot welds in higher tensile thin sheet' (German). Stahl Eisen 1987 107 (16) 31-37.

123   Aritotile R: 'Fatigue strength of automobile components made from spot welded thin sheet, influence on the selection of material'. Report EUR 10157 IT EEC Commission, Brussels, 1985.

124   Obuchow A W: 'Fatigue strength of multispot welded joints made from low carbon steel and the possibility of increasing it' (Russian). Westnik Masinastoenija 1953 11 81-84.

125   Satoh T, Katayama J and Morii Y: 'Fatigue strength of multispot welded joints'. IIW Doc III-638-80.

126   Davidson J A and Imhof E J: 'The effect of tensile strength on the fatigue life of spot welded sheet steels'. SAE Paper 840110, Warrendale Pa, 1984.

127   Sperle J O: 'Strength of spot welds in high strength steel sheet'. Met Con 1983 15 (4) 200-203.

128   Eichhorn F and Kunsmann A: 'The load bearing behaviour of resistance spot welds' (German). Werkstofft Z ind Fert 1970 60 (4) 179-185.

129   Pollard B: 'Spot welding characteristics of HSLA steel for automotive applications'. Weld J Res Supp 1974 343s-350s.

130   Shinozaki M, Kato T and Irie T: 'Fatigue of automotive high strength steel sheets and their welded joints'. SAE Paper 830032, Warrendale Pa, 1983.

131   Steffens H D, Dammer R and Kern H: 'Influence of the loading frequency on the life of spot welded joints' (German). Z Werkstofft 1983 14 19-24.

132   Orts D H: 'Fatigue strength of spot welded joints in a HSLA steel'. SAE Paper 810355, Warrendale Pa, 1981.

133   Barsom J M, Davidson J A and Imhof E J: 'Fatigue behaviour of spot welds under variable amplitude loading'. SAE Paper 850369, Warrendale Pa, 1985.

134   Davidson J A: 'A review of the fatigue properties of spot welded sheet steels'. SAE Paper 830033, Warrendale Pa, 1983.

135   Satoh T, Katayama J, Fujita M and Inoue S: 'Fatigue strength of spot welded joints in corrosive environments'. IIW-Doc III-773-84.

136   Eichhorn F and Stepanski H: 'Trends in the use of spot weld bonding, a state of the art review' (German). Therm Füg 1979/8013-17.

137   Eichhorn F and Schmitz B H: 'Comparative tests with standardised spot welded hollow sections made of steel plate with and without additional adhesive on the joint face' (German and English). Schw Schn 1984 36 (3) 113-116.

138   Jones T B and Williams N T: 'The fatigue properties of spot welded adhesive bonded and weld bonded joints in high strength steels'. SAE Paper 860583, Warrendale Pa, 1983.

139   Meisel D: 'Influence of friction welding on the torsional alternating fatigue strength of notched rotationally symmetrical components' (German). Mitt Inst Leichtb (Dresden) 1981 20 (2) 39-43.

140 Czuchryj J: 'Investigation of the fatigue strength of friction welded joints made of steel' (German and English). *Schw Schn* 1986 **38** (11) 561-564.

141 Welz W and Dietrich G: 'Strength investigations on stud welded joints' (German and English). *Schw Schn* 1971 **23** (8) 308-311.

142 Welz W and Dietrich G: 'Endurance limit of structures with welded on bolts' (German and English). *Schw Schn* 1981 **33** (2) 63-66.

Chapter 7 — Design codes, assessment of stress (see also Ref.3,15,16,49,75,86)

143 IIW-Doc 693-81: 'Design recommendations for cyclic loaded welded steel structures'. *Weld World* 1982 **20** (7/8) 153-165.

144 DIN 18800: 'Steel structures, dimensioning and design' (German). Part 1 and 6, publ Beuth Verlag, Berlin, 1981.

145 TGL 13500, issue dated 4.82: 'Steel engineering, steel structures' (German). GDR standard.

146 DIN 8563: 'Quality assurance of welding work' (German). Parts 1, 2, 3, 3A1, 10, 30, publ Beuth Verlag, Berlin, 1978/83.

147 DS 804 (draft): 'Specification for railway bridges and other engineering structures' (German). Publ Deutsche Bundesbahn, München, 1980.

148 DASt-Richtlinie 011: 'High tensile fine grained structural steel suitable for welding, StE 460 and StE 690, applications in steel structures' (German). Publ Stahlbau Verlag, Köln, 1979.

149 DIN 18808: 'Steel structures, load-bearing structures made from hollow steel sections mainly subjected to static loading' (German). Publ Beuth Verlag, Berlin, 1981.

150 Eurocode 3: 'Common uniform rules for steel structures' (German). Publ EEC Commission, Brussels, and Stahlbau-Verlagsgesellschaft, Köln, 1984.

151 ECCS-TC6-Fatigue: 'Recommendation for the fatigue design of steel structures'. Europ Convention, Constructional Steelwork, Brussels, 1985. Publ Schweizer Zentralstelle f Stahlbau, Zürich, 1987.

152 DIN 15018: 'Cranes, principles for load-bearing steel structures, dimensioning' (German). Publ Beuth Verlag, Berlin, 1984.

153 DIN 4132: 'Crane tracks, load bearing steel structures, principles for dimensioning, design and manufacture' (German). Publ Beuth Verlag, Berlin, 1981.

154 'Principles for dimensioning of large scale equipment in opencast mining' (German). Publ Minister für Wirtschaft und Verkehr, Nordrhein-Westfalen, Düsseldorf, 1960, revised by Landesoberbergamt Nordrhein-Westfalen, 1984.

155 German Lloyd: 'Specification for the classification and construction of seagoing vessels made of steel' (German). Volume 1, publ Hamburg, 1982.

156 API-RP2A: 'Recommended practice for planning, designing and constructing fixed offshore platforms'. Publ American Petroleum Institute, 1981.

157 AWS-D1-1: 'Structural welding code, Part 10, design of new tubular structures'. Publ American Welding Society, 1981.

158 DNV: 'Rules for the design, construction and inspection of offshore structures'. Publ Det Norske Veritas, Oslo, 1977.

159 UK-DoE-T (draft): 'Offshore installations — guidance on design and construction'. Publ United Kingdom Dept of Energy, London, 1981.

160 TRB, AD-Merkblatt S1: 'Technical rules for pressure vessels, working group for pressure vessels: delimitation of dimensioning for mainly static internal pressure loading from dimensioning for pulsating loading' (German). Publ Beuth Verlag, 1973.

161 TRB AD-Merkblatt S2: 'Technical rules for pressure vessels, working group for pressure vessels: dimensioning for cyclic loading' (German). Publ Beuth Verlag, Berlin, 1981.

162 TRD-Merkblatt 301: 'Technical rules for boilers: cylindrical shells subjected to internal pressure, dimensioning'. (German). Publ Beuth Verlag, Berlin, 1979.

163 DIN 2413 'Steel tubes, calculation of the wall thickness with regard to internal pressure' (German). Publ Beuth Verlag, Berlin, 1972.

164 ASME Boiler and Pressure Vessel Code: 'Section III Nuclear vessels; Section VIII, Div 2 Alternative rules for pressure vessels'. Publ Am Soc Mech Eng, 1968.

165 KTA 3201.2: 'Safety regulations by the Nuclear Engineering Committee: Components for the primary circuit of light water reactors, section layout, design and dimensioning' (German). Publ Heymanns Verlag, Köln, 1980.

166 TGL 22160: 'Steel pipelines, strength assessment' (German). GDR standard.

167 TGL 32903: 'Tanks and similar equipment, strength assessment' (German). GDR-standard.

168 DS 952: 'Specification for welding metallic materials in private works' (German). Publ Deutsche Bundesbahn, Hannover/Minden, 1977.

169 UIOC 897-13: 'Technical terms of delivery for the welding and testing of vehicle components' (German). Publ Foreign Standards Dept, DIN, Berlin.

170 DVS Merkblatt 1612: 'Design and assessment of butt and fillet welds in the construction of railway vehicles — Design shapes compilation' (German). Publ DVS-Verlag, Düsseldorf, 1984.

171 Merkblatt 415, Beratungsstelle für Stahlverwendung: 'Design loads and safety factors for railway vehicles' (German). Publ Düsseldorf, 1970.

172 Merkblatt 416, Beratungsstelle für Stahlverwendung: 'Design meeting the strength requirements in vehicle engineering' (German). Publ Düsseldorf, 1971.

173 TGL 14915/01, issue dated 12.74: 'Strength assessment for welded components' (German). Publ ZIS, Halle, 1974.

174 HSB: 'Aircraft engineering handbook for structural dimensioning' (German). Publ Industrieausschuß Strukturberechnungsunterlagen, Firma MBB, München, 1983.

175 Wirthgen G and Mogwitz H: 'Calculation of the service fatigue strength of machine components in accordance with the GDR standard TGL 19350' (German). *Konstr* 1989 **41** (3) 137-144.

176 'Service fatigue strength: design loads, assessment of life, experience in practice' (German). Publ DVM, Berlin, 1988.

177 IIW-Doc XIII-892-78: 'Proposed fatigue clause for British Standard, published document dealing with acceptance levels for weld defects'.

178 'IIW guidance on assessment of the fitness for purpose of welded structures'. Publ IIW, London, 1989.

179 BSI-PD 6493: 'Guidance on some methods for the derivation of acceptance levels for defects in fusion welded joints'. Publ BSI, London, 1980.

180 JWES 2805: 'Method of assessment for defects in fusion welded joints with respect to brittle fracture'. Publ JWES, Tokyo, 1980.

181 DVS Merkblatt 2401, Parts I and II: 'Fracture-mechanical assessment of defects in welded joints' (German). Publ DVS-Verlag, Düsseldorf, 1982 and 1989.

182 Hobbacher A: 'Recommendations for assessment of weld imperfections in respect to fatigue'. IIW-Doc XIII-1266-88/XV-659-88.

183 Pellini W S: 'Guidelines for fracture-safe and fatigue-reliable design of steel structures'. Publ The Welding Institute, Cambridge, 1983.

184   Siebke H: 'Description of a reference basis for dimensioning structures for service fatigue strength' (German). *Schw Schn* 1980 **32** (8) 304-314.

185   Herzog M: 'Single level simulation of the multilevel operational loading of bridges' (German). *Schw Schn* 1979 **31** (7) 289-291.

186   Heuler P and Schütz W: 'Fatigue life prediction in the crack initiation and crack propagation stages'. In 'Durability and damage tolerance in aircraft design', publ EMAS, Warley, 1985.

187   Heuler P, Vormwald M and Seeger T: 'Relative Miner rule and U_0 method, a comparative assessment' (German). *Materialprufung* 1986 **28** (3) 65-72.

188   Schönfeldt H: 'User problems with regard to service fatigue strength in ship design and ocean engineering' (German). In DVS-Berichte Vol 88 'Welded structures and service fatigue strength in practice', publ DVS-Verlag, Düsseldorf, 1984.

189   Petershagen H and Paetzold H: 'Service fatigue strength of ship structures on the basis of examples' (German). In DVS-Berichte Vol 88 'Welded structures and service fatigue strength in practice', publ DVS-Verlag, Düsseldorf, 1984.

190   Feder D: 'Experiences with a new guideline for ensuring the quality of welding work in large scale equipment in opencast mining' (German). In DVS-Berichte Vol 88 'Welded structures and service fatigue strength in practice', publ DVS-Verlag, Düsseldorf, 1984.

191   Neumann A: 'Quality stages of welded joints, thoughts on the assessment and selection of the quality of welded joints' (German and English). *Schw Schn* 1988 **40** (11) 544-550.

192   Schultze G: 'Influence of the defect size on the fatigue strength of welded joints made of high tensile steel' (German and English). *Schw Schn* 1986 **38** (1) 15-21.

193   Granzen H: 'On the quality assessment of welds in steel structures' (German). *Der Prakt* 1981 **33** (12) 333-334.

194   DIN 4100: 'Welded steel structures subjected to mainly static loading, dimensioning and design' (German). Publ Beuth Verlag, Berlin, 1968.

195   DIN-Bericht: 'Principles for establishing safety standards in building construction' (German). Publ Beuth Verlag, Berlin, 1981.

196   Schueller G I: 'Introduction to safety and reliability in load bearing structures' (German). Publ Verlag Ernst, Berlin, 1981.

197   Valtinat G: 'Safety in construction engineering', Schriftenreihe Institut für Fördertechnik, University of Karlsruhe, 2nd tech conf 'Integration of machinery and steel structures', 1983 (5) 21-30.

198   BS 5500: 'Unfired fusion welded pressure vessels'. Publ BSI, London.

199   Harrison J D and Maddox S J: 'Derivation of design rules for pressure vessels'. IIW-Doc XIII-941-80.

200   Iida K: 'Application of hot spot strain concept to fatigue life prediction'. IIW-Doc XIII-1103-83.

201   Radaj D and Möhrmann W: 'Notch effect on shoulder bars subjected to transverse shear' (German). *Konstruktion* 1984 **36** (10) 399-402.

202   Ritter W and Lipp W: 'Dimensioning welded chassis components in commercial vehicles for service fatigue strength' (German). In DVS-Berichte Vol 88 'Welded structures and service fatigue strength in practice', publ DVS-Verlag, Düsseldorf, 1984.

203   Grubisic V and Fischer G: 'Optimising the design of a steering knuckle of a commercial vehicle' (German). *VDI-Z* 1989 **131** (8) 111-115.

204   Fischer W: 'Investigations into railway vehicles' (German). In DVS-Berichte Vol 88 'Welded structures and service fatigue strength in practice', publ DVS-Verlag, Düsseldorf, 1984.

205 Haibach E: 'Assessment of the reliability of components subjected to cyclic loading' (German). *Luft Raumftechn* 1967 **13** (8) 188-193.

206 Haibach E: 'Required safety coefficients and quality assurance' (German). In the supplement to Ref.26.

Chapter 8 — Notch stress approach, seam welds (see also Ref.6, 25, 27, 38, 57)

207 Neuber H: 'Theory of notch stresses' (German). Publ Springer Verlag, Berlin, 1958 (2nd ed) and 1985 (3rd ed).

208 Peterson R E: 'Stress concentration factors'. Publ Wiley, New York, 1974 (2nd ed).

209 Radaj D and Schilberth G: 'Notch stresses for cutouts and inclusions' (German). Publ DVS-Verlag, Düsseldorf, 1977.

210 Zienkiewicz O C: 'The finite element method in structural and continuum mechanics'. Publ McGraw-Hill, London, 1967.

211 Gallagher R H: 'Finite-element analysis, fundamentals'. Publ Prentice-Hall, Englewood Cliffs, NJ, 1975.

212 Schwarz H R: 'The finite element method' (German). Publ Teubner Verlag, Stuttgart, 1980.

213 Kuhn G and Möhrmann W: 'Boundary element method in elastostatics'. *Appl Math Modell* 1983 **7** (4) 97-105.

214 Neuber H: 'Concerning the inclusion of the stress concentration in strength calculations' (German). *Konstr* 1968 **20** (7) 245-251.

215 Troost A and El-Magd E: 'General formulation of the fatigue strength amplitude in Haigh's diagram' (German). *Materialpr* 1975 **17** (2) 47-49.

216 Lang O: 'Dimensioning of complex steel components in the fatigue strength for finite life and infinite life ranges' (German). *Z Werkstofft* 1979 **10** (1) 24-29.

217 Hück M, Thrainer L and Schütz W: 'Calculation of S-N curves for components made of structural steel and casting steel, synthetic S-N curves' (German). Publ Report ABF11, Verein Deutscher Eisenhüttenleute, Düsseldorf, 1983.

218 Kloos K H and Velten E: 'Calculating the endurance limit of plasma-nitrided specimens similar to components taking into account the hardness and residual stress distribution' (German). *Konstr* 1983 **36** (5) 181-188.

219 Buxbaum O *et al*: 'Comparison of the life predicted by the notch root method and the nominal stress method' (German). Report FB-169, publ Fraunhofer Institut für Betriebsfestigkeit, Darmstadt, 1983.

220 DIN 8570: 'Tolerances for not-stated dimensions of welded structures' (German). Publ Beuth Verlag, Berlin, 1974/76.

221 Radaj D: 'Investigations into the geometrical shape of butt and fillet welds' (German). *Schw Schn* 1970 **22** (5) 206-209.

222 Wösle H: 'Undercutting in submerged-arc welding and its geometrical shape' (German). *Schw Schn* 1981 **33** (3) 127-129.

223 Radaj D: 'Stress concentration factors of cruciform joints with and without slits subjected to tension and bending' (German). *Konstr* 1965 **17** (7) 257-260.

224 Radaj D: 'On the issue of the stress concentration factors of cruciform joints' (German). *Konstr* 1967 **19** (8) 328-330.

225 Radaj D: 'Notch stress analysis for fillet welds using the finite element method' (German). *Schw Schn* 1975 **27** (3) 86-89.

226 Radaj D and Möhrmann G: 'Notch effect of welded joints subjected to transverse loading' (German and English). *Schw Schn* 1984 **36** (2) 57-63.

227 Radaj D: 'Notch effect of welded joints with regard to fatigue' (German). *Konstr* 1984 **36** (8) 285-292.

228 Radaj D: 'Notch stress proof for fatigue resistant welded structures'. IIW-Doc XIII-1157-85.

229 Petershagen H: 'A comparison of different approaches to the fatigue strength assessment of welded components'. IIW-Doc XIII-1208-86.

230 Rainer G: 'Calculation of stresses in welded joints using the finite element method' (German). Thesis, TH Darmstadt, 1978.

231 Olivier R, Köttgen V B and Seeger T: 'Assessment of fatigue strength for welded joints on the basis of local stresses' (German). Research Report 143, Forschungskuratorium Maschinenbau, Frankfurt M, 1989.

232 Bell R, Vosikovski O and Bain S A: 'The significance of weld toe undercuts in the fatigue of steel plate T joints'. *Int J Fat* 1989 **11** (1) 3-11.

233 Köttgen V B, Olivier R and Seeger T: 'The influence of plate thickness on fatigue strength of welded joints, a comparison of experiments with prediction by fatigue notch factors'. In 'Steel in marine structures', publ Elsevier Science, Amsterdam, 1987.

234 Mattos R J and Lawrence F V: 'Estimation of the fatigue crack initiation life in welds using low cycle fatigue concepts'. Publ Soc Aut Eng, SP-424, Warrendale, 1977.

235 Lawrence R V, Mattos R J, Higashida Y and Burk J D: 'Estimating the fatigue crack initiation life of welds'. ASTM STP 648, July 1978, 134-158.

236 Boller C and Seeger T: 'Materials data for cyclic loading. Part A: Unalloyed steels; Part B: Low alloy steels; Part C: High alloy steels; Part D: Aluminium and titanium alloys; Part E: Cast and welded metals'. Publ Elsevier Science, Amsterdam, 1987.

237 Boyer H E (Ed): 'Atlas of fatigue curves'. Publ Am Soc of Metals, Metals Park, Ohio, 1986.

238 Schnack E: 'Optimisation of tension strappings' (German). *Konstr* 1978 **30** (7) 277-281.

239 Schnack E and Spörl U: 'A mechanical dynamic programming algorithm for structure optimisation'. *Int J Num Met Eng* 1986 **23** S1985-2004.

240 Neuber H: 'The flat bar subjected to tensile loading with an optimum transition in section' (German). *Forsch Ing Wes* 1969 **35** (1) 29-30.

241 Baud R V: 'Fillet profiles for constant stress'. *Prod Eng* 1934 **5** (4) 133-134.

242 Ruge J and Drescher G: 'The cruciform joint of special quality with the endurance limit of the butt joint of special quality' (German and English). *Schw Schn* 1985 **37** (4) 149-157.

243 Radaj D: 'Notch effect of welded joints subjected to biaxial oblique loading' (German and English). *Schw Schn* 1984 **36** (6) 254-259.

244 Radaj D: 'Approximate calculation of the stress concentration factors of welded joints' (German). *Schw Schn* 1969 **21** (3) 97-103 and (4) 151-158.

245 Radaj D: 'Assessment of notch stress for welded vehicle structures' (German). In 'Calculation in automotive engineering', VDI-Berichte 537, publ VDI-Verlag, Düsseldorf, 1984.

246 Radaj D: 'Assessment of notch stress for fatigue resistant welded structures' (German). *Konstr* 1985 **37** (2) 53-59.

247 Wlassow W S: 'Thin walled elastic bars', Vol I and II (German). Publ VEB-Verlag Technik, Berlin, 1964/5.

248 Radaj D: 'Fatigue strength of girders with transverse stiffeners subjected to bending on the basis of the notch root concept' (German). *Der Stahlb* 1985 **54** (8) 243-249.

249 Radaj D: 'Effectiveness of the stress relief groove in welded vessel ends' (German). *Konstr* 1986 **38** (6) 237-242.

250 Radaj D, Gerlach H D and Gorsitzke B: 'Experimental-numerical assessment of notch stress for a welded boiler structure' (German). *Konstr* 1988 **40** (11) 447-452.

251 Gaßner E and Haibach E: 'The fatigue strength of welded joints from the point of view of a local measurement of strain' (German). In 'Determination of load-bearing capacity for welded joints', Vol I, publ DVS-Verlag, Düsseldorf, 1968.

252 Haibach E and Köbler H G: 'Evaluation of the fatigue strength of welded joints made from AlZnMg 1 by using a local strain measurement'. *Alumin* 1971 **47** (12) 725-730.

253 Klee S: 'The cyclic stress-strain and fracture behaviour of various steels' (German). Report 22, publ Institut für Statik und Stahlbau, TH Darmstadt, 1973.

254 Hanschmann D: 'A contribution on the computer-aided prediction of life up to crack initiation in aluminium vehicle components subjected to cyclic loading' (German). DFVLR Research Report, Köln, 1981.

255 Beste A: 'Elastic-plastic stress, strain and crack initiation behaviour in statically and cyclically loaded notched plates — a comparison between results from experiments and approximate calculations' (German). Report 34, publ Institut für Statik und Stahlbau, TH Darmstadt, 1981.

256 Bergmann J W: 'On the dimensioning of notched components with respect to service fatigue strength on the basis of the local stresses and strains' (German). Report 37, publ Institut für Statik und Stahlbau, TH Darmstadt, 1983.

257 Heuler P: 'Prediction of life up to crack initiation for random loading on the basis of local stresses and strains' (German). Report 40 publ Institut für Statik und Stahlbau, TH Darmstadt, 1983.

258 Seeger T: 'Material-mechanical approach to fatigue strength for finite and infinite life' (German). In the supplement to Ref.26.

259 Proceedings, int conf on biaxial/multi-axial fatigue. Materialprüfungsanstalt (MPA), Stuttgart, 1989.

260 Heuler P and Seeger T: 'Prediction of life on the basis of calculations and experiments using a welded component as an example' (German). *Konstr* 1983 **35** (1) 21-26.

Chapter 9 — Fracture mechanics approach, seam welds (see also Ref.2, 152, 177)

261 Heckel K: 'Introduction to the technical applications of fracture mechanics' (German). Publ Hanser Verlag, München, 1970.

262 Radaj D: 'Technical fracture mechanics' (German). In 'Assessments of strength' Vol II, publ DVS-Verlag, Düsseldorf, 1974.

263 Hahn H G: 'Fracture mechanics' (German). Publ Teubner Verlag, Stuttgart, 1976.

264 Schwalbe K-H: 'Fracture mechanics of metallic materials' (German). Publ Hanser Verlag, München, 1980.

265 Bäcklund J, Blau A F and Beevers L J (Ed): 'Fatigue thresholds, fundamentals and engineering applications'. Publ EMAS, Cradley Heath, 1982.

266 Lieurade H P: 'Application of fracture mechanics to the fatigue of welded structures'. *Weld World* 1983 **21** (11/12) 272-294.

267 Smith R A (Ed): 'Fatigue crack growth, 40 years of progress'. Publ Pergamon Press, Oxford, 1986.

268 Allen R J, Booth G S and Jutla T: 'A review of fatigue crack growth characterisation by linear elastic fracture mechanics. Part I: Principles and methods of data generation. Part II: Advisory documents and applications within national standards'. *Fat Fract Eng Mat Struct* 1988 **11** (1) 45-69 and (2) 71-108.

269 Kanazawa T and Kobayashi A S (Ed): 'Significance of defects in welded structures'. Publ University of Tokyo Press, 1974.

270 Maddox S J: 'Calculating the fatigue strength of a welded joint using fracture mechanics'. *Met Con* 1970 **17** (8) 327-331.

271 Maddox S J: 'Assessing the significance of flaws in welds subject to fatigue'. *Weld J* 1974 **53** (9) 401s-409s.

272 Maddox S J: 'An analysis of fatigue cracks in fillet welded joints'. *Int J Fract* 1975 **11** (2) 221-243.

273 Maddox S J: 'The effect of mean stress on fatigue crack propagation — a literature review'. *Int J Fract* 1975 **11** (3) 389-408.

274 Maddox S J and Webber D: 'Fatigue crack propagation in aluminium-zinc-magnesium alloy fillet welded joints'. Publ ASTM STP 648, 1978.

275 Heitmann H H, Vehoff H and Neumann P: 'Life prediction for random load fatigue based on the growth behaviour of microcracks'. Proc int conf on fracture, publ Pergamon Press, Oxford, 1985.

276 Neumann P, Heitmann H H and Vehoff H: 'Damage accumulation for random loading' (German). In 'Fatigue of metallic materials', publ Deutsche Gesellschaft für Metallkunde, Oberursel, 1985.

277 Christman T and Suresh S: 'Crack initiation under far field cyclic compression and the study of short fatigue cracks'. *Eng Fract Mech* 1986 **23** (6) 953-964.

278 Leis B N, Hopper A T, Ahmad J, Broek D and Kanninen M F: 'Critical review of the fatigue growth of short cracks'. *Eng Fract Mech* 1986 **23** (5) 883-898.

279 Hirt M A and Fisher J W: 'Fatigue crack growth in welded beams'. *Eng Fract Mech* 1973 **5** (5) 415-429.

280 Hirt M A and Kummer E: 'Influence of the stress concentration on the fatigue strength of welded structures' (German). In VDI-Bericht 313, 539-545, publ VDI-Verlag, Düsseldorf, 1978.

281 Hirt M A: 'Application of fracture mechanics to the determination of the fatigue behaviour of welded structures' (German). *Bauing* 1982 **57** 95-101.

282 Ohta A, Soya I, Nishijima S and Kosuge M: 'Statistical evaluation of fatigue crack propagation properties including threshold stress intensity factor'. *Eng Fract Mech* 1986 **24** (6) 789-802.

283 Harrison J D: 'The weld effect of residual stresses on fatigue behaviour'. *Proc Weld Inst* 1981 **47** 9-16.

284 Wheeler O E: 'Spectrum loading and crack growth' *Trans ASME J Basic Eng* 1972 **94** (3) 181-187.

285 Engle R E and Rudd J L: 'Spectrum crack growth analysis using the Willenborg model'. *J Aircr* 1976 **13** 462-466.

286 Schütz W: 'Life calculation for loadings with any desired load-time function' (German). VDI-Bericht 286, publ VDI-Verlag, Düsseldorf, 1976.

287 Sih G C: 'Handbook of stress intensity factors'. Publ Lehigh University, Bethlehem, Penns, 1973.

288 Tada H P, Paris C and Irwin G R: 'The stress analysis of cracks handbook'. Publ DEL Res Corp, Hellertown, Penns, 1973.

289 Rooke D P and Cartwright D J: 'Compendium of stress intensity factors'. Publ HMSO, London, 1976.

290 Murakami Y (Ed): 'Stress intensity factors handbook'. Publ Pergamon Press, Oxford, 1988.

291 Radaj D and Heib M: 'Stress intensity diagrams for practical purposes' (German). *Konstr* 1978 **30** (7) 268-270.

292 Radaj D: 'Adjustment for geometry of the stress intensity of elliptical cracks' (German). *Schw Schn* 1977 **29** (10) 398-402.

293 Hobbacher A: 'The service fatigue strength of welded joints considered on a fracture-mechanical basis' (German). *Arch Eisenhütt* 1977 **48** (2) 109-114.

294 Hobbacher A: 'Cumulative fatigue by fracture mechanics'. *Trans ASME J App Mech* 1977 **12** 769-771.

295 Hanel J J: 'Crack propagation in single and multistage cyclically loaded plates, taking special account of partial crack closure' (German). Report 27, publ Institut für Statik und Stahlbau, TH Darmstadt, 1975.

296 Franke L: 'Damage accumulation rule for dynamically loaded materials and components' (German). *Bauing* 1985 **60** 271-279.

297 Gurney T R: 'Finite element analyses of some joints with the welds transverse to the direction of stress'. *Weld Res Int* 1976 **6** (4) 40-72.

298 Sprung I and Zilberstein V A: 'Analysis of a backing bar notch effect'. *Eng Fract Mech* 1986 **24** (1) 145-152.

299 Niu X and Glinka G: 'The weld profile effect on stress intensity factors in weldments'. *Int J Fract* 1987 **35** 3-20.

300 Zwerneman F J and Frank K H: 'Stress intensity factor for a T shaped weldment'. *Eng Fract Mech* 1989 **32** (4) 561-572.

301 Zettlemoyer N and Fisher J W: 'Stress gradient correction factor for stress intensity at welded stiffeners and coverplates'. *Weld J* 1977 **56** (12) 393s-398s.

302 Albrecht P and Yamada K: 'Rapid calculation of stress intensity factors'. *Proc Am Soc Civil Engrs* 1977 (ST2) 377-389.

303 Terada H: 'An analysis of the stress intensity factor of a crack perpendicular to the welding bead'. *Eng Fract Mech* 1976 **8** 441-444.

304 Tada H and Paris P C: 'The stress intensity factor for a crack perpendicular to the welding bead'. *Int J Fract* 1983 **21** 279-284.

305 Itoh Y Z, Suruga S and Kashiwaya H: 'Prediction of fatigue crack growth rate in welding residual stress field'. *Eng Fract Mech* 1989 **33** (3) 397-407.

306 Wu X R: 'Welding residual stress intensity factors for half-elliptical surface cracks in thin and thick plates'. *Eng Fract Mech* 1984 **19** (3) 407-426.

307 Greasley A and Naylor S G: 'Influence of residual welding stresses on fatigue crack growth under compressive loading'. *Eng Fract Mech* 1986 **24** (5) 717-726.

308 Petershagen H: 'Investigations into the influence of defects caused by welding on the strength of welded joints in shipbuilding' (German). Report 430, publ Institut für Schiffbau, Universität Hamburg, 1983.

309 Harrison J D: 'An analysis of the fatigue behaviour of cruciform joints'. *Weld Res Int* 1971 **1**(1) 1-7.

310 Petershagen H: 'Cruciform joints and their optimisation for fatigue strength — a literature survey'. *Weld World* 1975 **13** (516) 143-154.

311 Haibach E: 'Questions concerning the fatigue strength of welded joints considered from a conventional and a fracture-mechanical point of view' (German). *Schw Schn* 1977 **29** (4) 140-142.

312 Radaj D: 'On the fracture mechanical assignment of welded joints to notch cases' (German). *Der Stahlb* 1986 **55** (8) 247-252.

313 Smith I F and Smith R A: 'Fatigue crack growth in a fillet welded joint'. *Eng Fract Mech* 1983 **18** (4) 861-869.

314 Smith I F and Smith R A: 'Defects and crack shape development in fillet weld joints'. *Fat Eng Mat Struct* 1982 **5** (2) 151-165.

315 Smith I F and Smith R A: 'Measuring fatigue cracks in fillet welded joints'. *Int J Fat* 1982 **4** (1) 41-45.

316 Heckel K and Ziebart W: 'A new method for the calculation of notch and size effect in fatigue'. Proc int conf on fracture, Waterloo, Canada, 1977.

317 Matoba M and Inoue K: 'Some stress intensity factors for hull members in relation to fatigue crack propagation'. IIW-Doc XIII-1081-83.

318 Fisher J W: 'Fatigue and fracture in steel bridges'. Publ J Wiley, New York, 1984.

319 Fricke W: 'Crack propagation investigations on reinforced plate areas' (German). Report 469, publ Institut für Schiffbau, Universität Hamburg, 1986.

Chapter 10 — Local approaches, spot welds (see also Ref.119, 287, 288, 290)

320 Radaj D: 'Strength assessment of spot welded joints on the basis of local stresses' (German and English). *Schw Schn* 1989 **41** (1) 26-31.

321 Radaj D: 'Stress singularity and notch stress on the contour model of the spot welded joint'. *Schw Schn* 1989 **41** (3) 117-119 (see also *Eng Fract Mech* 1989 **34** (2) 495-506).

322 Radaj D: 'The increase of structural stress in spot welded joints' (German and English). *Konstr* 1986 **38** (2) 41-47, (10) 397-404, 1987 **39** (2) 51-63, 1988 **40** (4) 159-164.

323 Radaj D: 'Hot spot structural stress calculation for the spot welded tensile-shear and hat profile specimen' (German and English). *Schw Schn* 1987 **39** (1) 7-12.

324 Radaj D: 'Nominal structural stresses for common spot welded specimens' (German). *Konstr* 1989 **41** (8) 255-259.

325 Radaj D: 'Complete determination of forces at the weld spot' (German). *Ing Arch* 1990 **59**, accepted for publication.

326 Radaj D, Zheng Zh and Möhrmann W: 'Local stress parameters at the weld spot of various specimens' (German). *Konstr* 1990 **42** (5/6), English version IIW-Doc III-929-89/XIII-1311-89 and *Eng Fract Mech*, accepted for publication.

327 Radaj D and Schilberth G: 'Notch stresses at cutouts and inclusions' (German). Publ DVS-Verlag, Düsseldorf, 1977.

328 Pook L P: 'Approximate stress intensity factors for spot and similar welds'. NEL Report 588, publ Nat Eng Lab, Glasgow, 1975.

329 Pook L P: 'Fracture mechanics analysis of the fatigue behaviour of spot welds'. *Int J Fract* 1975 **11** (2) 173-176.

330 Chang D J and Muki R: 'Stress distribution in a lap joint under tension shear'. *Int J Sol Struct* 1974 **10** 503-517.

331 Yuuki R, Nakatsukasa H and Yi W: 'Fracture mechanics analysis and evaluation of fatigue strength of spot welded joints' (Japanese).

332 Smith R A and Cooper J F: 'Theoretical predictions of the fatigue life of shear spot welds'. In 'Fatigue of welded constructions', publ The Welding Institute, Abington, Cambridge, 1988.

333 Cooper J F and Smith R A: 'Initial fatigue crack growth at spot welds'. In 'Fatigue of engineering materials and structures', publ Institution of Mechanical Engineers, London, 1986.

334 Fujimoto M, Mori N and Sakuma S: 'Stress distribution analysis of spot welded joints under tension-shear load'. IIW-Doc III-721-82, 1982.

335 Hahn O and Wender B: 'Finite element analysis of a resistance spot welded joint subjected to tensile shear loading using various structural models' (German and English). *Schw Schn* 1983 **35** (4) 174-178.

336 Niisawa J, Tomioka N and Yi W: 'Analytical method of rigidities of thin walled beams with spot welding joints and its application to torsion problems'. JSAE Review, 1984, No 13, 76-83.

337 Tomioka N, Niisawa J and Mabuchi A: 'On theoretical analysis of stress at welding flange of spot welded box section member under torsion'. SAE-Paper 860602, Warrendale Pa, 1986.

338 Mori N, Amago T, Ono M, Sasanabe M and Hiraide T: 'Fatigue life prediction methods for spot welds in T-shaped members under bending'. SAE Paper 860604, Warrendale Pa, 1986.

339 Rupp A, Grubisic V and Radaj D: 'Service fatigue strength of spot welded joints' (German). In 'Kerben und Betriebsfestigkeit', publ DVM, Berlin, 1989.

340 Pfau P: 'Force transmission between two plates, each of constant thickness, which are joined by spot shaped adhesion spots (force transmission between two spot welded plates).' (German). *Wiss Zeit* (TH Karl-Marx-Stadt) 1965 **7** (2) 1-15.

341 Pfau P: 'Force transmission between two spot welded plates (numerical examples)' (German). *Wiss Zeit* (TH Karl-Marx-Stadt) 1966 **8** (3) 161-164.

342 Satoh T, Katayama J and Fujino K: 'Fatigue strength of multi-spot welds in mild steel'. IIW-Doc III-584-78, 1978.

343 Mizui M, Sekine T, Tsujimura A and Shimazaki Y: 'An evaluation of fatigue strength for various kinds of spot welded test specimens'. SAE-Paper 880375, Warrendale Pa, 1988.

344 Radaj D, Glatzel G and Kitzler W: 'Photoelastic load type fringe patterns of spot welded joints' (German and English). *Schw Schn* 1988 **40** (1) 7-13.

345 Schröder R and Macherauch E: 'Calculation of the thermal and residual stresses in resistance spot welded joints using differing mechanical-thermal materials data as a basis' (German and English). *Schw Schn* 1983 **35** (6) 270-276.

346 Radaj D: 'Thermal effects of welding — temperature field, residual stresses, distortion' (German). Publ Springer-Verlag, Berlin, 1988.

347 Koenigsberger F: 'Design for welding in mechanical engineering'. Publ Longmans Green, London, 1948.

348 Oshima M and Kitagawa H: 'Buckling assisted fatigue of spot welded box beam under bending'. SAE-Paper 860605, Warrendale Pa, 1986.

349 Oh H L: 'Fatigue life prediction for spot welds using Neuber's rule'. In 'Design of fatigue and fracture resistant structures', ASTM STP 761, 1982.

350 Kan Y-R: 'Fatigue resistance of spot welds — an analytical study'. *Met Eng Quart* 1976 (11) 26-36.

351 Kitagawa H, Satoh T and Fujimoto M: 'Fatigue strength of single spot welded joints of rephosphorised high strength and low carbon steel sheets'. SAE-Paper 850371, Warrendale Pa, 1985.

352 Wang P-C and Ewing K-M: 'A J integral approach to fatigue resistance of a tensile-shear spot weld'. SAE-Paper 880373, Warrendale Pa, 1988.

353 Pollard B: 'Fatigue strength of spot welds in titanium-bearing HSLA steels'. SAE-Paper 820284, Warrendale Pa, 1982.

354 Mabuchi A, Niisawa J and Tomioka N: 'Fatigue life prediction of spot welded box section beams under repeated torsion'. SAE-Paper 860603, Warrendale Pa, 1986.

355 Erdogan F and Sih G C: 'On the crack extension in plates under plane loading and transverse shear'. Trans ASME 85 (1963), Ser D J Basic Eng, 519-527.

356 Radaj D and Heib M: 'Energy density fracture criteria for cracks under mixed mode loading' (German). *Materialpr* 1978 **20** (7) 256-262.

357 Davidson J A and Imhof E J: 'A fracture mechanics and system stiffness approach to fatigue performance of spot welded steel sheets'. SAE-Paper 830034, Warrendale Pa, 1983.

358 Davidson J A: 'Design related methodology to determine the fatigue life and related failure mode of spot welded sheet steels'. In 'HSLA steels, technology and application', publ American Society of Metals, Ohio, 1984.

359 Yuuki R and Ohira T: 'Development of the method to evaluate the fatigue life of spot welded structures by fracture mechanics'. IIW-Doc XIII-1358-89.

360 Lawrence F V, Wang P C, Corten H T: 'An empirical method for estimating the fatigue resistance of tensile-shear spot welds'. SAE-Paper 830035, Warrendale Pa, 1983.

361 Wang P C, Corten H T and Lawrence F V: 'A fatigue life prediction method for tensile-shear spot welds'. SAE-Paper 850370, Warrendale Pa, 1985.

362 Eichhorn F, Hahn O and Stepanski H: 'Influence of the geometry of specimens taking account of different types of adhesives' (German). *Schw Schn* 1979 **31** (1) 23-26.

363 Pfau P: 'Stress distribution in weldbonded joints, calculation on the basis of a one dimensional model' (German). *Wiss Zeit* (TH Karl-Marx-Stadt) 1975 **17** (1) 71-84.

364 Pfau P: 'Stress distribution in weldbonded joints, calculation on the basis of a two dimensional model' (German). *Wiss Zeit* (TH Karl-Marx-Stadt) 1977 **19** (3) 341-364.

Chapter 11 — Corrosion and wear resistance

365 DIN 50900 Part 1: 'Corrosion of metals' (German). Publ Beuth Verlag, Berlin, 1975.

366 Class I: 'Effects of corrosion on welded joints' (German). *Werk Korr* 1970 **21** (7) 559-568.

367 Wirtz H: 'Corrosion phenomena on welded joints made of steel' (German). *Bänd Blech Rohr* 1968 **9** (9) 526-532.

368 Achten A, Herbsleb G and Wieling N: 'The importance of gaps between pipes and pipe plates for the operational behaviour of heat exchangers'. *Chem Ing Tech* 1986 **58** (9) 732-738.

369 Olivier R and Rückert H: 'Fatigue strength of hot galvanised welded joints with and without corrosion' (German and English). *Schw Schn* 1985 **37** (10) 519-523.

370 'Design with respect to corrosion' (German). Merkblätter Arbeitsgemeinschaft Korrosion (AGK), publ DECHEMA, Frankfurt, 1981.

371 Bühler H-E, Litzkendorf M, Schubert B and Zerweck K: 'Design and manufacturing measures for preventing corrosion damage in equipment, machinery and pipeline construction' (German). *Konstr* 1982 **34** (10) 381-386.

372 Wirtz H and Hess H: 'Protective surfaces by means of welding and spraying metal' (German). Publ DVS-Verlag, Düsseldorf, 1969.

373 McKelvie A N: 'Corrosion protection of welds by painting'. *Met Con* 1981 **13** (11) 693-696 and (12) 744-748.

374 Kloos K H, Diehl H, Nieth F, Thomala W and Düßler W: 'Materials engineering' (German). In Dubbel 'Taschenbuch für den Maschinenbau' 15th ed, publ Springer Verlag, Berlin, 1983.

Chapter 12 — Development of a fatigue resistant structure

375 Müller-Berner A H, Mitschke A, Steiner A and Strifler P: 'The course of the development of modern commercial vehicles' (German). *ATZ* 1971 **73** (11) 406-414.

376 Radaj D: 'Status of calculation in the automotive industry' (German). *ATZ* 1982 **84** (11) 535-539.

377 Geißler H: 'Calculation of loads dependent on operation and component-specific design in motor vehicle construction' (German). *Autom Indust* 1983 (1) 31-36.

378 Oehlschläger H: 'Calculation of torsion flexible frames in vehicles taking account of the design of the intersections' (German). Thesis, TU Braunschweig, 1981.

379 Oehlschläger H: 'Flexible intersections in torsion calculations for commercial vehicle frames made of open sections' (German). *ATZ* 1984 **86** (3) 105-108.

380 Argyris H J and Radaj D: 'Stiffness matrices of thin walled bars and bar systems' (German). *Ing Arch* 1971 **40** (3) 198-210.

381 Radaj D: 'Thin walled bars and bar systems with general system line and warping elastic supports' (German). *Der Stahlb* 1971 **40** (3) 27-31.

382 Radaj D and Zafirakopoulos N: 'Forces and deformations at the transverse stiffener of thin walled U section bars' (German). *Forsch Ing Wes* 1973 **39** (3) 69-80.

383 Beermann H J: 'Numerical analysis of load bearing structures of commercial vehicles' (German). Publ TUV Rheinland, Köln, 1986.

Index